The MOLECULAR AND CELLULAR NEUROBIOLOGY series

Series Advisers

G.L. Collingridge, *Department of Anatomy, University of Bristol, Bristol, UK*

R.W. Davies, *IBLS, Division of Molecular Genetics, University of Glasgow, Glasgow, UK*

S.P. Hunt, *Molecular Neurobiology Unit, MRC Centre, Cambridge, UK*

Neurobiology of Alzheimer's Disease
Immune Responses in the Nervous System
Glial Cell Development
Cortical Plasticity: LTP and LTD

Forthcoming title
Molecular Biology of the Neuron

Cortical Plasticity
LTP and LTD

Sam Fazeli
Department of Experimental Pathology, United Medical and Dental Schools, Guy's Hospital, London, UK

Graham L. Collingridge
Department of Anatomy, University of Bristol, UK

© BIOS Scientific Publishers Limited, 1996

First published, 1996

All rights reserved. No part of this book may be reproduced or transmitted, in any form or by any means, without permission.

A CIP catalogue record for this book is available from the British Library.

ISBN 1 872748 74 0

BIOS Scientific Publishers Ltd
9 Newtec Place, Magdalen Road, Oxford OX4 1RE, UK.
Tel. +44 (0) 1865 726286. Fax. +44 (0) 1865 246823

DISTRIBUTORS

Australia and New Zealand
  DA Information Services
  648 Whitehorse Road, Mitcham
  Victoria 3132

Singapore and South East Asia
  Toppan Company (S) PTE Ltd
  38 Liu Fang Road, Jurong
  Singapore 2262

India
  Viva Books Private Limited
  4325/3 Ansari Road
  New Delhi 110002

USA and Canada
  BIOS Scientific Publishers
  PO Box 605, Herndon,
  VA 20172 – 0605

Typeset by Saxon Graphics Ltd, Derby, UK.
Printed by Information Press Ltd, Oxford, UK.

Contents

	Contributors	ix
	Abbreviations	xi
	Preface	xiii
1	**Introduction and overview.** *R.G.M. Morris*	1
	References	7
2	**The role of amino acid receptors in synaptic plasticity.** *R. Anwyl*	9
	Introduction	9
	LTP of AMPA receptor-mediated transmission	9
	LTP of NMDA receptor-mediated synaptic transmission	21
	LTD of AMPA receptor-mediated transmission	23
	LTD of NMDA receptor-mediated synaptic transmission	27
	LTD of $GABA_A$-mediated synaptic transmission	28
	References	28
3	**Second messengers in LTP and LTD.** *E.D. Roberson, J.D. English and J.D. Sweatt*	35
	Introduction	35
	Kinases	36
	Proteases	48
	Retrograde messengers	48
	Phosphatases	51
	Application of biochemical techniques to the study of LTP and LTD	52
	References	54
	Further reading	59
4	**The locus of expression of NMDA receptor-dependent LTP in the hippocampus.** *T.V.P. Bliss and M.S. Fazeli*	61
	Introduction	61
	Changes in transmitter release	62
	Changes in postsynaptic responsiveness	64
	Changes in synthesis or phosphorylation of pre- and post-synaptic proteins	66
	Evidence from quantal analysis	67
	Are NMDA and AMPA receptor-mediated components of the evoked response potentiated equally?	71
	Paired-pulse facilitation and depression	72
	Retrograde messengers	76
	Conclusions	77
	References	77

CONTENTS

5 **Post-translational mechanisms which could underlie the postsynaptic expression of LTP and LTD.** *T.A. Benke, I. Bresinck, V.J. Collett, A.J. Doherty, J.M. Henley and G.L. Collingridge* 83
   Introduction 83
   Glutamate receptors 84
   LTP in the CA1 region of the hippocampus 95
   Concluding remarks 98
   References 98

6 **Postsynaptic gene expression and long-term potentiation in the hippocampus.** *K.L. Thomas and S.P. Hunt* 103
   Introduction 103
   Immediate-early genes 105
   Secondary messenger pathways: serine/threonine kinases 110
   Signal transduction pathways: protein phosphatases 120
   Glutamate receptors 120
   LTP and growth-related changes in gene expression 126
   Summary 128
   References 129

7 **Neuromodulators of synaptic strength.** *M. Segal and J.M. Auerbach* 137
   Introduction 137
   Acetylcholine 138
   Noradrenaline 140
   Serotonin 144
   Conclusions 145
   References 145

8 **Epileptogenesis.** *W.W. Anderson* 149
   Introduction 149
   Blockade of kindling-like epileptogenesis by NMDA antagonists 150
   Anticonvulsant action of NMDA antagonists 164
   Mechanisms of epileptogenesis: the initial induction process 166
   Mechanisms of epileptogenesis: changes resulting in seizure expression 167
   Mechanisms of epileptogenesis: intracellular messenger mechanisms 172
   Mechanisms of epileptogenesis: gene expression 175
   Mechanisms of epileptogenesis: post-translational modifications of channels 176
   Blockade of kindling by anti-epileptics that are not NMDA antagonists 177
   Are there non-NMDA receptor mechanisms that can induce kindling epileptogenesis? 178
   Anticonvulsant and anti-epileptogenic therapy 179
   Reversal of permanent epileptogenesis 180
   Conclusions 181
   References 183

9 Bidirectional plasticity of cortical synapses. *M.F. Bear and A. Kirkwood* 191

- A problem and its theoretical solution — 191
- Evidence for bidirectional synaptic plasticity in hippocampus — 193
- Evidence for bidirectional synaptic plasticity in visual cortex — 194
- Evidence for bidirectional synaptic plasticity in neocortical areas other than visual cortex — 195
- Evidence for a sliding modification threshold — 197
- Developmental regulation of LTP and LTD in hippocampus and neocortex — 199
- Conclusions — 202
- References — 202

10 Synaptic plasticity in the cerebellum. *H. Daniel, O. Blond, D. Jaillard and F. Crepel* 207

- Introduction — 207
- Postsynaptic calcium release is required for LTD induction — 208
- Glutamate receptors are required for LTD induction — 210
- Second messenger cascades are required for LTD induction — 213
- Does LTD require desensitization of AMPA receptors of Purkinje cells? — 217
- Conclusions — 218
- References — 219

11 Roles of LTP and LTD in neuronal network operations in the brain. *E.T. Rolls* 223

- Introduction — 223
- Pattern associators — 223
- Autoassociative memory — 231
- Competitive networks — 236
- The hippocampus — 240
- Synaptic modification rules useful in learning invariant representations in the cerebral cortex — 245
- Conclusions — 248
- References — 248

Index — 251

Contributors

Anderson, W.W. Department of Anatomy, School of Medical Sciences, University of Bristol, University Walk, Bristol BS8 1TD, UK

Anwyl, R. Department of Physiology, Faculty of Health Sciences, Trinity College, Dublin 2, Ireland

Auerbach, J.M. Department of Neurobiology, The Weizmann Institute, Rehovot 76100, Israel

Bear, M.F. Department of Neuroscience, Howard Hughes Medical Institute and Institute for Brain and Neural Systems, Brown University, Box 1953, Providence, RI 02912, USA

Benke, T.A. Department of Anatomy, School of Medical Sciences, University of Bristol, University Walk, Bristol BS8 1TD, UK

Bliss, T.V.P. Division of Neurophysiology, National Institute for Medical Research, Mill Hill, London NW7 1AA, UK

Blond, O. Laboratoire de Neurobiologie et Neuropharmacologie du Développement, Université de Paris-Sud, Centre d'Orsay, Bâtiment 440, 91405 Orsay Cedex, France

Bresinck, I. Department of Anatomy, School of Medical Sciences, University of Bristol, University Walk, Bristol BS8 1TD, UK

Collett, V.J. Department of Anatomy, School of Medical Sciences, University of Bristol, University Walk, Bristol BS8 1TD, UK

Collingridge, G.L. Department of Anatomy, School of Medical Sciences, University of Bristol, University Walk, Bristol BS8 1TD, UK

Crepel, F. Laboratoire de Neurobiologie et Neuropharmacologie du Développement, Université de Paris-Sud, Centre d'Orsay, Bâtiment 440, 91405 Orsay Cedex, France

Daniel, H. Laboratoire de Neurobiologie et Neuropharmacologie du Développement, Université de Paris-Sud, Centre d'Orsay, Bâtiment 440, 91405 Orsay Cedex, France

Doherty, A.J. Department of Anatomy, School of Medical Sciences, University of Bristol, University Walk, Bristol BS8 1TD, UK

English, J.D. Baylor College of Medicine, Division of Neuroscience, One Baylor Plaza, Texas Medical Center, Houston, TX 77030–3498, USA

Contributors

Fazeli, M.S. Department of Experimental Pathology, 4th Floor Medical School, Guy's Hospital, London Bridge, London SE1 9RT, UK

Henley, J.M. Department of Anatomy, School of Medical Sciences, University of Bristol, University Walk, Bristol BS8 1TD, UK

Hunt, S.P. Division of Neurobiology, MRC Laboratory of Molecular Biology, Hills Road, Cambridge CB2 2QH, UK

Jaillard, D. Laboratoire de Neurobiologie et Neuropharmacologie du Développement, Université de Paris-Sud, Centre d'Orsay, Bâtiment 440, 91405 Orsay Cedex, France

Kirkwood, A. Department of Neuroscience, Howard Hughes Medical Institute and Institute for Brain and Neural Systems, Brown University, Box 1953, Providence, RI 02912, USA

Morris, R.G.M. Department of Pharmacology and Centre for Neuroscience, University of Edinburgh, Crichton Street, Edinburgh EH8 9LE, UK

Roberson, E.D. Baylor College of Medicine, Division of Neuroscience, One Baylor Plaza, Texas Medical Center, Houston, TX 77030–3498, USA

Rolls, E.T. Department of Experimental Psychology, University of Oxford, South Parks Road, Oxford OX1 3UD, UK

Segal, M. Department of Neurobiology, The Weizmann Institute, Rehovot 76100, Israel

Sweatt, J.D. Baylor College of Medicine, Division of Neuroscience, One Baylor Plaza, Texas Medical Center, Houston, TX 77030–3498, USA

Thomas, K.L. Division of Neurophysiology, National Institute for Medical Research, Mill Hill, London NW7 1AA, UK, and Division of Neurobiology, MRC Laboratory of Molecular Biology, Hills Road, Cambridge CB2 2QH, UK

Abbreviations

AA	arachidonic acid
ACh	acetylcholine
ACPD	amino-cyclopentane-1S,3R dicarboxylate
ACSF	artificial cerebrospinal fluid
ADPRT	ADP-ribosyltransferase
AHP	after-hyperpolarization
AMPA	α-amino-3-hydroxy-5-methyl-4-isoxazole proprionic acid
AP4	2-amino-4-phosphonobutanoate
AP5	2-amino-5-phosphonopentanoate
aPKC	atypical protein kinase C
BDNF	brain-derived neurotrophic factor
bZIP	basic leucine zipper
CaM	calmodulin
CAM	cell adhesion molecule
CaMKII	calcium/calmodulin-dependent protein kinase II
CaN	calcineurin
CCh	carbachol
C/EBP	CCAAT enhancer-binding protein
CF	climbing fiber
CNQX	6-cyano-7-nitroquinoxaline-2,3-dione
cPKC	'classical' protein kinase C
CRE	cAMP response element
CREB	CRE-binding protein
CV	coefficient of variation
DAG	diacylglycerol
E-LTP	early phase of long-term potentiation
EGS	electrographic seizure
EPSC	excitatory postsynaptic current
EPSP	excitatory postsynaptic potential
ERK	extracellular signal-regulated kinase
GABA	γ-aminobutyric acid
HEK	human embryo kidney
5-HT	5-hydroxytryptamine
I1	inhibitor protein 1
IEG	immediate-early gene
IGF	insulin-like growth factor
IP$_3$	inositol 1, 4, 5-trisphosphate
IPSP	inhibitory postsynaptic potential
L-LTP	late phase of long-term potentiation
L-NMMA	L-N-monomethylarginine
LTD	long-term depression
LTP	long-term potentiation
MAP2	microtubule-associated protein 2
MAPK	mitogen-activated protein kinase
MCPG	(RS)-α-methyl-4-carboxyphenylglycine

Abbreviations

mGlu	metabotropic glutamate
NA	noradrenaline
NCAM	neural cell adhesion molecule
NDGA	nordihydroguaiaretic acid
NGF	nerve growth factor
NMDA	N-methyl-D-aspartate
NOS	nitric oxide synthase
NT3	neurotrophin-3
nPKC	'new' protein kinase C
PAF	platelet-activating factor
PCP	phencyclidine
PDGF	platelet-derived growth factor
PF	parallel fiber
PI	phosphatidylinositol
PKA	protein kinase A
PKC	protein kinase C
PKG	cGMP-dependent protein kinase
PLA_2	phospholipase A_2
PP	protein phosphatase
PPD	paired-pulse depression
PPF	paired-pulse facilitation
PS	phosphatidylserine
PSD	postsynaptic density protein
PTK	protein tyrosine kinase
PTP	protein tyrosine phosphatase
PTZ	pentylenetetrazol
SNP	sodium nitroprusside
STP	short-term potentiation
tPA	tissue plasminogen activator

Preface

Many higher organisms have the ability to form new patterns of behavior through experience. The process of acquisition of new behavioral responses is referred to as learning. It became apparent by the end of the last century that the nervous system is made up of individual neurons, and its functional characteristics are a result of the integrated activity of these cells through the plethora of specialized neuronal contacts, namely synapses. The hypothesis that learning involves changes in the strength of central synapses was first put forward by Ramon y Cajal in his Croonian lecture to the Royal Society of London in 1894. Over 60 years later, Donald Hebb (1949) provided a working model of the mechanism by which changes in synaptic efficiency may encode information in the nervous system. He argued that the strength of the connections between a pair of neurons may be altered as a result of persistent activity. In 1973 Bliss and Lomo provided the first experimental evidence in support of the above hypothesis. Thus a brief high frequency stimulation of the perforant path resulted in a long-term potentiation (LTP) of the synaptic response of granule cells.

In the present book we have aimed to bring together diverse experimental approaches to the study of activity-dependent changes in neuronal connections, as exemplified by LTP and long-term depression (LTD). This book provides an insight into the rapidly growing field of cortical plasticity, by covering areas of research as diverse as molecular biology and neural networks.

Sam Fazeli (*London*)
Graham L. Collingridge (*Bristol*)

Introduction and overview

R.G.M. Morris

To outward appearances, one human brain is much like another, one monkey brain like another monkey brain, and one rat brain similar to another. The sizes and positions of the various component structures never change, their patterns of interconnectivity appear to be the same, their constituent cell types and the neurotransmitters they use likewise. One could be forgiven for supposing that the brain is very much a hard-wired device laid down according to a highly deterministic genetic program that unfolds through embryogenesis and development. But appearances are deceptive, for the brain, and particularly the overlying cortex, is highly 'plastic' – by which is meant that both its structure and its moment-to-moment operation can be changed in response to experience. Structural development of the nervous system does not proceed according to a single grand master plan, but is influenced fundamentally by the experience of the developing organism, and experience, in turn, affects the expression of genes at succeeding stages of development. Classical experiments have, for example, established that a visual cortex which grows up in a world having a particular visual structure will become responsive to distinctive features of that environment, but is less responsive, or even unresponsive, to features of different visual environments. Rewiring experiments indicate that even the sensory modality to which a cortical area responds can be influenced by the sensory properties of thalamo-cortical afferents. Such experience-dependent effects cannot be solely prewired but depend critically upon continuous gene–environment interactions (Bateson, 1996). Moreover, once development is complete, the mature central nervous system is hardly a system that responds to an individual event in the same stereotyped way every time that event occurs (though it does have a number of reflexes up its sleeve to do just that when necessary). Rather, it is a complex biological device whose patterns of activity can be altered by experience in a myriad of different ways so as to optimize its future response to events. In short, it can learn.

Importantly, the brain has a number of different ways by which it can learn and these are mediated by circuits in distinct anatomical locations. Some circuits are attuned to remembering small amounts of information accurately for short periods of time; in the human brain these seem to be located in areas of the frontal

---

Cortical Plasticity, edited by M.S. Fazeli and G.L. Collingridge.
© 1996 BIOS Scientific Publishers Ltd, Oxford.

lobe. Other circuits are devoted to remembering large amounts of information for long periods of time, typically at the cost of absolute accuracy. Long-term memory traces are probably widely distributed in the neocortex, but their creation depends on an interaction with an important group of structures in the medial temporal lobe responsible for remembering facts and events. Other neocortical circuits are involved in perceptual learning. Further circuits, including a structure called the amygdala, seem to have evolved to learn the biological significance of stimuli, particularly if danger lurks, while others are adept at the fine and rapid sequencing of motor movements. The cerebellum, at the back of the brain, is particularly implicated in this latter form of learning.

While anatomically discrete, there is a widespread belief amongst neuroscientists that the underlying neural mechanisms through which these circuits alter in response to experience share common principles. As Mark Bear and Alfredo Kirkwood note in their conclusion to Chapter 9, the expression of neuronal learning rules may differ in subtle if important ways, but the basic principles by which patterns of activity are registered, reacted to and give rise to learning are likely to have been conserved from circuit to circuit. Evolution, it is often said, is a tinkerer. Once it has come up with one trick, it adapts it for use elsewhere. Thus, intensive study of certain forms of neuronal plasticity in selected brain areas stands a chance of giving us general insight into an important aspect of brain function. But where to start?

The discovery of the phenomenon of long-term potentiation in Norway in 1966, and the subsequent detailed report by Tim Bliss and Terje Lømo in 1973, have had a deserved impact on research concerned with the neural mechanisms of learning and memory. Working with anesthetized rabbits, Bliss and Lømo (1973) found that brief periods of high-frequency tetanic stimulation could cause stable electrical potentials in the hippocampus to increase in amplitude for hours on end. They stimulated a major pathway from the neocortex into the dentate region of the hippocampal formation. Low frequency 'test' pulses induced a response that was recorded at the synaptic terminals of the pathway, and in the region of the target cell bodies, that repeatedly and reliably stayed constant in amplitude. But, following a short high-frequency train of impulses, further stimulation with test pulses at the same original low frequency yielded a larger or 'potentiated' response. As this potentiation lasted for hours, well beyond the duration of short-term memory, this phenomenon subsequently became known as long-term potentiation (LTP). LTP is an intriguing physiological phenomenon, but one that is typically induced under somewhat artificial circumstances. Whether its underlying mechanisms actually have anything to do with long-term memory is, in fact, now a field of research in its own right.

Over the past 25 years, research on LTP and other experience-dependent forms of neuronal plasticity has taken a number of directions. One body of work is concerned with its potential functional significance and typically explores its expression in behaving animals during various types of learning. A larger group of neuroscientists have focused on understanding its underlying mechanisms. A critical step in doing the latter was the development of *in vitro* brain slices which permit much better visual control of electrode placement, full experimental control of the ionic environment of the cells and synapses of interest, the ability to wash drugs in and out of the recording chamber, intracellular and patch recording, and a number of other advantages over experimentation *in vivo*. Electrophysiological and neuropharmacological studies have also been complemented by work at the biochemical and, more recently, the molecular level. Fundamental discoveries about the mechanisms of neuronal plasticity have come from this effort, with several of the key players being contributors to this book.

The first important discovery about LTP (though not in chronological order), was that synaptic efficacy can not only be increased, but also decreased. As the saying goes: what goes up, must come down. But whether what comes down does so passively over time as the initially potentiated response decays to a baseline, or does so actively in response to specific patterns of stimulation, has been debated over recent years. The discovery that specific patterns of low-frequency activity can induce a long-term depression (LTD) of synaptic efficacy seems to settle the argument in favor of the latter, but the greater expression of LTD in immature brain tissue and the considerable difficulty many have inducing LTD in freely moving animals *in vivo* gives grounds for caution. A second important discovery was the relative roles of distinct amino acid receptors in mediating fast synaptic transmission and in synaptic plasticity. Fast transmission is generally mediated by α-amino-3-hydroxy-5-methyl-4-isoxazole proprionic acid (AMPA) receptors that, upon binding the neurotransmitter glutamate, deliver sodium ions into the target cell and cause its depolarization. While changes in AMPA receptors (their number or efficacy) are implicated in the expression of LTP, a critical step in the induction of LTP is the activation of a different glutamatergic receptor. Graham Collingridge and his colleagues were the first to identify the critical role of the so-called *N*-methyl-D-aspartate (NMDA) receptor in hippocampal LTP (Collingridge *et al.*, 1983) using the then new selective agonists and antagonists developed by Jeff Watkins and his colleagues in Bristol. More recently, Collingridge's research group and several others have turned their attention to a further class of glutamate receptor – the metabotropic receptors that are coupled by G-proteins to a number of different second-messenger systems. Appropriately, this book begins with a comprehensive review by Roger Anwyl of both LTP and LTD and of the role of ionotropic and metabotropic glutamatergic receptors in their induction (Chapter 2).

Downstream of the membrane-associated receptors for glutamate are a host of second-messenger cascades that translate receptor activation into signals for kinase or phosphatase action. David Sweatt and his colleagues (Chapter 3) provide a comprehensive review of these messengers in LTP and LTD, focusing initially on a third key discovery – the role of kinases with respect to LTP and on phosphatases with respect to LTD. An intriguing feature of the effects of even broad-spectrum kinase inhibitors is that they do not prevent LTP from being induced, they merely limit the duration of potentiation to about an hour at most (generally referred to as short-term potentiation or STP). This indicates that the earliest phase of LTP expression does not involve phosphorylation although later stages clearly do. The chapter begins by focusing on the role of the two kinases CaMKII and PKC, the most specific evidence coming from experiments using pseudosubstrate inhibitor peptides such as $CaMKII_{273-302}$ and PKC_{19-31}. These strictly pharmacological studies have been complemented by studies in transgenic animals in which there is a targeted deletion of specific subunits of CaMKII and PKC. But David Sweatt has also long been interested in a possible role of cAMP and its substrate PKA. This takes us into later stages of LTP generally thought to occur hours after its induction. PKA induces the transcription of genes by activating a transcription factor, called CREB, which apparently has a single site for phosphorylation by PKA and which, when phosphorylated, can stimulate the transcription of genes involved in the late stages of LTP. Sweatt and his colleagues discuss a number of other kinases and phosphatases, and make a strong case for recognizing the complexity of the various biochemical cascades involved in the expression of synaptic plasticity.

But where do these cascades act? Is their action on the receiving, or postsynaptic, side of the synapse or on the sending, or presynaptic, side? Or both? Sam Fazeli and Tim Bliss (Chapter 4) tackle this old chestnut and land firmly on the fence. It is, they argue, probably a bit of both. In the process of coming to this classic compromise of an answer, they unravel the puzzle of the locus of control of LTP being on the postsynaptic side (where NMDA receptors are thought to be exclusively located) despite there being a body of evidence pointing to a component of presynaptic expression to the enhanced synaptic efficacy. This paradoxical state of affairs appears to call for the existence of a retrograde or intercellular messenger, a substance which travels in the 'wrong' direction at the synapse – from the postsynaptic side to the presynaptic side. Several elegant experiments have been conducted to establish the necessity for such a concept – and its identity – but so far the jury is out. In fact, the issue of the locus of expression of LTP has proved, over the years, to be something of an enigma wrapped in a conundrum. Certain techniques, such as quantal analysis, have appeared over the horizon like knights to the rescue; only for them to find, as in the possible existence of silent synapses, that cherished assumptions of this type of analysis (based on work at the neuromuscular junction) may not be so directly applicable to central synapses after all. In his memoirs, Francis Crick wrote that "in research, the frontline is almost always in a fog" (Crick, 1988, p. 35). Despite the imagination, ingenious experimentation and sheer hard work of many skilled laboratories, we are still to penetrate the haze engulfing this issue.

Jeremy Henley and Graham Collingridge and their colleagues (Chapter 5) take us on to the possible post-translation modifications of glutamatergic receptors on the postsynaptic side of the synapse. They summarize the current state of knowledge of receptor subunits for both AMPA and NMDA receptors, and discuss how each may be phosphorylated by a variety of kinases. They then discuss the mechanism by which post-translational modifications of these ionotropic glutamate receptors may underlie a postsynaptic component of expression of hippocampal LTP. Kerrie Thomas and Steve Hunt (Chapter 6) take on both transcriptional and translation modifications associated with the induction and expression of synaptic plasticity, stressing that certain gene-induced changes do not appear to occur in conscious animals until 48–72 h after the inducing stimulus. This raises an enormous puzzle about targeting, namely where to draw the boundary between gene-associated aspects of neuronal function that should be thought of as specific to plasticity and those which are more 'housekeeping' in character. The solution to the targeting problem may be via some kind of local but transient synaptic 'tag' that can catch somatically synthesized proteins if, and only if, a synapse has recently been potentiated, but there is still no published evidence for such a mechanism.

Hippocampal neurons have many afferents other than their extrinsic and intrinsic glutamatergic inputs. Menahem Segal and J. Auerbach (Chapter 7) summarize evidence that acetylcholine (ACh), noradrenaline (NA) and serotonin (5-HT) can each modulate plasticity. The action of ACh includes a potentiation of NMDA responses and is probably mediated via activation of M2 receptors and the inositol trisphosphate cascade. However, ACh clearly has a variety of cellular actions, and they also describe a form of LTP mediated by direct low-intensity activation of muscarinic receptors which appears to occlude tetanus-induced LTP. NA has a similar effect, while the actions of 5-HT depend upon a delicate interplay between the hyperpolarization of pyramidal cells and reduction of GABAergic inhibition.

William Anderson (Chapter 8) takes the book into the domain of pathology by considering the question of whether similar mechanisms underlie the development and expression of epilepsy as are responsible for the induction and expression of LTP. Making a sharp distinction between drugs that are anti-epileptogenic (slowing the development of epilepsy, at least in animal models) and anticonvulsant (reducing expressed seizure incidence), Anderson summarizes a wide body of evidence indicating that NMDA antagonists can be anti-epileptogenic. There are exceptions to this claim, and there are other drugs that can impede the development of seizures, but striking parallels, both at the receptor level and with respect to second-messenger and immediate-early gene action, point to the underlying mechanisms of epileptogenesis being some pathology of neuronal plasticity.

While the hippocampus has been the brain area of choice for many studies of activity-dependent synaptic plasticity, the cortex and the cerebellum have also been favored by a few. Mark Bear and Alfredo Kirkwood (Chapter 9) emphasize a learning rule for synaptic change developed by the physicist and Nobel laureate, Leon Cooper. This rule is characterized by the necessity for bidirectional change (i.e. both LTP and LTD) and a sliding threshold determining when and whether particular patterns of activity induce LTP or LTD. They summarize evidence for this model in hippocampus, visual cortex, and other cortical areas, and they explain how a slowly changing sliding threshold, adjusting in response to average patterns of activity, helps to maintain neuronal stability. They also discuss developmental aspects, particularly the potency of LTD in young animals at the time of greatest synapse elimination and the existence of numerous synapses in very young animals in which transmission appears to be mediated exclusively by NMDA receptors. This characteristic of immature synapses is interesting because NMDA receptors are only functionally effective in passing calcium into the cell when the membrane is partially depolarized. It suggests that, in wiring up the brain, only cells that actively participate in firing (or at least depolarizing) other cells have their interconnections stabilized by potentiation and, specifically, by the insertion of AMPA receptors into the postsynaptic membrane.

Francis Crepel and his colleagues (Chapter 10) turn to discussing the molecular mechanisms of LTD in the cerebellum, a depression of synaptic efficacy implicated in motor learning by the Marr-Albus theory. Their studies show a role for glutamate receptor activation, postsynaptic calcium and a number of second messengers in its induction, and support for Masao Ito's original idea that the expression of LTD in the cerebellum involves a 'desensitization' of AMPA receptors.

The focus of this book, and the series of which it is a part, is on cellular and molecular mechanisms of plasticity. But understanding these alone will not yield a full appreciation of the mechanisms of learning and memory. These mechanisms have to be embedded into local circuits, and these local circuits stacked up in parallel in areas such as the hippocampus, cortex and cerebellum. A full understanding also requires much more information about how stimuli are represented as spatio-temporal patterns of activity. Edmund Rolls (Chapter 11) ends the book with a survey of various circuits which have the capacity to store and recall information, given various arrangements of the afferents and Hebb-like learning rules that utilize LTP- and LTD-like changes. These distributed associative memory systems have the important property of 'pattern completion' by which is meant the ability to reconstruct a memory from partial information – a pretty typical state of affairs in real life. Edmund Rolls also discusses, in quantitative detail, the extent to which selected areas of the hippocampus (such as area CA3) might be thought of as autoassociative memory systems.

Are the mechanisms of LTP or LTD involved in learning and memory? We still don't know for sure. Studies looking for correlations between the expression and time-course of such changes and corresponding parameters of learning have yielded some fascinating parallels. Occlusion experiments reveal that prior saturation of hippocampal LTP can, under some circumstances, inhibit subsequent learning of a type thought (on the basis of ablation studies) to be mediated by the hippocampus, though not all studies are in agreement that such a saturation effect is either easy to see or reliable. Experiments examining the effects of drugs that block the induction of LTP reveal corresponding effects on learning and memory, though the aspects of learning that are actually affected by, for example, NMDA receptor blockade remain unclear. One new line of thought, following Gaffan (1991), is that NMDA receptor-dependent plasticity in hippocampus (though not necessarily elsewhere) is part of a mechanism for the automatic recording of attended experience (Morris, 1996). Work with transgenic animals, particularly mice in which there has been engineered a targeted deletion of genes relevant to synaptic plasticity offers an alternative approach that has caught the imagination of many, but this approach is now recognized as being somewhat difficult to interpret. Gene compensation potentially yields false negatives, and pleiotropic effects during development result in adult nervous systems that are altered in secondary ways. This powerful technology will soon be enhanced by inducible gene deletions, regionally specific changes, and the use of site-directed mutagenesis (an approach beyond the confines of conventional pharmacological approaches to altering receptor function), collectively rendering the transgenic approach too important to ignore. Finally, experiments examining the role of synaptic depression in learning, including depotentiation, remain on the drawing board, though their practical realization cannot be far away.

Each of these approaches to the fundamental issue of how neuronal and synaptic plasticity participates in information storage, coupled with theoretical ideas about network organization of the kind discussed by Edmund Rolls, will be illuminated and enhanced by the cellular and molecular research discussed in this book. It's an exciting time for the field.

References

Bateson P. (1996) Design for a life. In: *The Lifespan Development of Individuals: Behavioral, Neurobiological, and Psychosocial Perspectives* (ed. D. Magnusson). Cambridge University Press, Cambridge, pp. 1–20.

Bliss TVP, Lømo T. (1973) Long-lasting potentiation of synaptic transmission in the dentate area of the anaesthetized rabbit following stimulation of the perforant path. *J. Physiol.* **232**: 331–356.

Collingridge GL, Kehl SJ, McLennan H. (1983) Excitatory amino acids in synaptic transmission in the Schaffer collateral-commissural pathway of the rat hippocampus. *J. Physiol.* **334**: 33–46.

Crick F. (1988) *What Mad Pursuit.* Basic Books, New York.

Gaffan (1991) Spatial organisation of episodic memory. *Hippocampus* **1**: 262–264.

Morris RGM. (1996) Learning, memory and synaptic plasticity: cellular mechanisms, network architecture and the recording of attended experience. In: *The Lifespan Development of Individuals: Behavioral, Neurobiological, and Psychosocial Perspectives* (ed. D. Magnusson). Cambridge University Press, Cambridge, pp. 139–161.

2

The role of amino acid receptors in synaptic plasticity

R. Anwyl

2.1 Introduction

Two main types of synaptic plasticity will be discussed in this chapter, long-term potentiation (LTP) and long-term depression (LTD) (for recent comprehensive reviews, see Bliss and Collingridge, 1993; Linden, 1994). Both LTP and LTD have been shown to occur at excitatory glutamatergic synapses, at which the presumed neurotransmitter agent L-glutamate binds postsynaptically to three major classes of L-glutamate receptors, namely α-amino-3-hydroxy-5-methyl-4-isoxazole proprionic acid (AMPAR)/kainate-R, N-methyl-D-aspartate (NMDAR) and metabotropic (mGluR). LTP has also been shown to occur at inhibitory synapses at which γ-aminobutyric acid (GABA) is the neurotransmitter.

2.2 LTP of AMPA receptor-mediated transmission

2.2.1 LTP of NMDA receptor-dependent AMPA receptor-mediated transmission

Present evidence indicates that AMPAR do not have to be activated in order to induce LTP of NMDAR-dependent synaptic transmission. Kauer *et al.* (1988b) showed in CA1 that, following blockade of AMPAR with 6-cyano-7-nitroquinoxaline-2,3-dione (CNQX), high frequency stimulation or low frequency stimulation paired with membrane depolarization induced a full amplitude LTP of population and intracellular excitatory postsynaptic potentials (EPSPs), the LTP being revealed following washout of the CNQX. It can be concluded from this

---

Cortical Plasticity, edited by M.S. Fazeli and G.L. Collingridge.
© 1996 BIOS Scientific Publishers Ltd, Oxford.

study that sodium entry, or possible second messenger stimulation, associated with AMPAR activation, is not required for LTP induction.

There is strong evidence that the expression of LTP is mediated, at least in part, via AMPAR. Thus, an NMDAR-dependent increase in AMPAR sensitivity of CA1 cells to iontophoretically applied AMPA or quisqualate was found to occur following the induction of LTP (Davies *et al.*, 1988) (*Figure 2.1*). Such an increase in AMPAR functioning could occur by an increase in the number of functional AMPAR, for example by a conversion of inactive to active AMPAR following high frequency stimulation (Malinow, 1994), or by an increase in AMPAR unitary conductance. The increase in AMPAR functioning may be generated by phosphorylation of the receptors by protein kinase A (PKA) (Greengard *et al.*, 1991; Wang *et al.*, 1991), by calcium/calmodulin-dependent protein kinase II (CaMKII) (McGlade-McCulloh, 1993) or by protein kinase C (PKC) (McGlade-McCulloh, 1993). It should be noted that both control and long-term potentiated EPSPs are reduced to the same extent by CNQX, which demonstrates that an identical AMPAR type mediates the control and potentiated EPSPs (Davies *et al.*, 1988). [See Chapter 5 for a detailed consideration of possible post-translational changes in AMPAR (and NMDAR) that may be involved in LTP and LTD.]

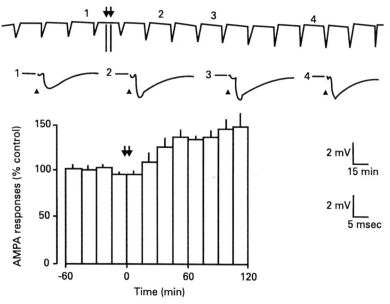

Figure 2.1. Increase in AMPAR sensitivity following LTP induction in CA1 hippocampus. The top trace shows the responses generated by iontophoretic ejection of AMPA at 15-min intervals. Tetanic stimulation was applied at the time indicated by the arrows. The traces of field EPSPs are shown before (1) and following (2–4) tetanic stimulation. The lower graph shows the mean amplitude of the AMPA responses. Tetanic stimulation resulted in immediate synaptic potentiation and a gradual increase in the peak amplitude of the iontophoretic AMPA responses following tetanic potentiation. Reproduced from Reymann *et al.* (1990) by permission of Oxford University Press.

The necessity for activation of NMDAR to occur for the induction, but not the maintenance, of LTP of NMDAR-dependent AMPAR-mediated synaptic transmission was first demonstrated by Collingridge *et al.* (1983a,b) at Schaffer collateral–commissural synapses with CA1 pyramidal cells. These authors demonstrated that, in the slice preparation, iontophoresis of the NMDAR antagonist 2-amino-5-phosphonopentanoate (AP5) on to the synaptic area being stimulated reversibly blocked the induction of LTP, but did not prevent the maintenance of LTP. The essential role of NMDAR in the induction of LTP was confirmed using competitive NMDAR antagonists both *in vitro* in CA1 by Harris *et al.* (1984) who showed that there was a high correlation between a series of phosphonopentanoic homologs and their ability to block LTP, and also *in vivo* in dentate gyrus by Morris *et al.* (1986). Noncompetitive antagonists of the NMDAR, such as MK-801, phencyclidene, σ-opiates and cyclazocine also blocked the induction but not the maintenance of LTP *in vitro* and *in vivo* in CA1 and dentate gyrus (Abraham and Mason, 1988; Coan and Collingridge, 1987; Maren *et al.*, 1991; Schwartzwelder *et al.*, 1988; Stringer and Guyenet, 1983; Stringer *et al.*, 1983a,b; Zhang and Levy, 1992) at doses which noncompetitively inhibited responses to NMDA (Lacey and Henderson, 1983). Antagonists at the allosteric glycine site on the NMDAR also prevent LTP induction *in vitro* (Bashir *et al.*, 1990; Izumi *et al.*, 1990; Oliver *et al.*, 1990) and *in vivo* (Thiels *et al.*, 1992). NMDAR-dependent LTP in the hippocampus, in addition to that occurring in CA1, has now been shown to occur at lateral and medial perforant pathways to hippocampal dentate gyrus granule cells and the commissural–associational pathway to hippocampal CA3 (Colino and Malenka, 1993; Errington *et al.*, 1987; Morris *et al.*, 1986).

NMDAR-dependent LTP was first reported in the neocortex by Artola and Singer (1987), and has now been described in several regions of the neocortex (for review, see Tsumoto, 1992). Very strong evidence has been presented that essentially very similar or identical mechanisms of NMDAR-dependent LTP to that in the hippocampus occur in the neocortex (Artola and Singer, 1993; Kirkwood *et al.*, 1993). The corticostriatal pathway to the striatum has also been shown to generate an NMDAR-dependent LTP, but only under experimental conditions when the NMDAR component of synaptic transmission is enhanced in magnesium-free media (Calabresi *et al.*, 1992b).

The generally accepted mechanisms for the role of NMDAR insuch NMDAR-dependent LTP are as follows. At a very low stimulation rate (<0.01 Hz), the transmitter L-glutamate binds to both the AMPAR and NMDAR. The EPSP elicited at potentials close to the resting level is mediated mainly by the AMPAR, with opening of the NMDAR channel being inhibited due to the voltage-dependent block by magnesium from the extracellular media. Activation of the NMDAR makes an increasing contribution to the EPSPs as the membrane potential is depolarized due to relief of the magnesium block and, at membrane potentials depolarized from –30 mV, the NMDAR mediates a large component of the EPSPs. Thus, the NMDAR acts as a molecular coincidence detector, with the necessity for the binding of L-glutamate to enable activation of NMDAR to occur in a temporally coincident manner with depolarization of the membrane. The activation of NMDAR sufficiently to induce LTP can be achieved in two main ways, firstly by coupling low frequency-evoked EPSPs with intense postsynaptic depolarization so that EPSPs activate NMDAR (Gustafsson et al., 1987; Kelso et al., 1986) and, secondly, by stimulating afferent fibers at high frequency so that summation of the EPSPs results in substantial depolarization and thereby NMDAR activation (Collingridge et al., 1983b). The depolarization evoked during high frequency stimulation is generated by summation of both NMDAR and non-NMDAR EPSPs, extracellular potassium and reduction of inhibitory postsynaptic potentials (IPSPs). The NMDAR component of the depolarization becomes apparent at frequencies of 5 Hz (a similar frequency to the threshold for LTP induction), and increases with higher stimulation frequencies (Collingridge et al., 1988; Herron et al., 1986; Wigstrom and Gustaffson, 1985). High frequency afferent stimulation by itself is not sufficient for induction of LTP, as holding the postsynaptic membrane potential at sufficiently hyperpolarized levels to prevent NMDAR activation during high frequency stimulation prevents the induction of LTP (Kelso et al., 1986; Malinow and Miller, 1986).

The direct application of NMDA in normal extracellular media only induces a short-term potentiation lasting up to about 30 min rather than full LTP (Collingridge et al., 1983a; Kauer et al., 1988a; McGuinness et al., 1991a). This inability to induce full LTP solely with application of NMDA is probably due in part to inappropriate NMDAR activation, because excess activation of NMDAR can lead to impairment of LTP induction (see below) and also because of the absence of activation of mGluR (see below). Under certain conditions, exogenous NMDA application has been shown to induce LTP, for example, in a high calcium medium (Thibault et al., 1989). However, the high calcium medium probably led to a bypassing of mGluR activation, in a way similar to the induction of LTP by application of high calcium alone (Turner et al., 1982). Perfusion of NMDA following an initial application of quisqualate in the presence of high potassium was also able to induce a relatively long-lasting potentiation, the quisqualate activating mGluR (Izumi et al., 1987). Application of glutamate has been shown to induce LTP (Cormier et al., 1993), the glutamate activating both NMDAR and mGluR. Such LTP induced by application of glutamate did not require presynaptic stimulation, as it occurred even during presynaptic block of evoked EPSPs with adenosine and tetrodotoxin (Cormier et al., 1993). The occurrence of a postsynaptic locus for potentiation of glutamate-activated currents was established definitely by Zilberter et al. (1990), who demonstrated that repetitive stimulation with applied glutamate or NMDA (plus glycine) induced a marked increase in the response to glutamate in isolated hippocampal neurons.

Activation of NMDAR is required for the induction of NMDAR-dependent LTP to allow calcium influx into the postsynaptic cell. Thus, the NMDAR has a high calcium permeability as revealed by measurement of calcium by imaging studies (MacDermott et al., 1986), and high frequency stimulation has been shown to increase the intracellular calcium concentration in synaptically activated dendritic spines by calcium influx via NMDAR (Alford et al., 1993; Muller and Connor, 1991). The calcium influx via NMDAR appears to be insufficient for induction of NMDAR-dependent LTP. Thus, the block of calcium release from intracellular stores with dantrolene (Obenhaus *et al.*, 1989), or depletion of intracellular calcium stores with thapsigargin (Harvey and Collingridge, 1992) led to the inhibition of LTP induction. In experiments measuring high frequency-induced dendritic calcium transients, it was shown that calcium entry through the NMDA receptor accounted for approximately 35% of the total signal, while calcium-induced calcium release from intracellular stores accounted for the remaining 65% of the signal (Alford *et al.*, 1993). Calcium influx via P- and T-voltage-gated calcium channels (Ito, 1995) may also be required for LTP induction.

The timing of activation of the NMDAR has to be very precise if it is to lead to the induction of LTP. Untimely activation of the NMDAR with respect to high frequency stimulation can actually inhibit the induction of LTP. In particular, activation of NMDAR prior to high frequency stimulation, by administering magnesium-free media (Coan *et al.*, 1989), NMDAR agonists (Izumi *et al.*, 1992) or subthreshold stimulation (Huang *et al.*, 1992) resulted in the inhibition of the induction of LTP. The physiological significance of such an inhibitory effect of NMDAR is not known at present.

Postsynaptic mGluR appear to play few direct roles in low frequency-evoked synaptic transmission, as antagonism of the mGluR did not affect the EPSPs evoked by test low frequency stimulation in CA1 (Bashir *et al.*, 1993a). However, high frequency trains of stimuli have been shown to induce an mGluR-mediated slow EPSP in CA3 neurons following mossy fiber stimulation (Gerber *et al.*, 1993) and also, as shown by Batchelor *et al.* (1994), in Purkinje cells following high frequency parallel fiber stimulation. The generation of mGluR-mediated EPSPs at high, but not low, frequency afferent stimulation may be a result of the mGluR being located perisynaptically, and therefore only being activated if the L-glutamate concentration in the synaptic cleft becomes relatively high during intense synaptic activity (Baude *et al.*, 1993). The activation of postsynaptic mGLuR only during intense synaptic activity (as for the NMDAR, although because of different mechanisms) would make them very suitable receptors for an involvement in synaptic plastic processes which only result if the synaptic transmission occurs above a basal level.

An early indication for an involvement of mGLuR in the induction of LTP was found using the mGluR antagonists AP3 and AP4 and the mGluR agonist aminocyclopentane-1*S*,3*R* dicarboxylate (ACPD). Thus, Reymann and Matthies (1989) found that perfusion of D- or L-AP4, with application starting 10 min prior to induction, started to inhibit LTP of the population EPSP in CA1 at a time beginning 1–1.5 h after induction, and completely inhibited LTP after 2.5–3 h (Reymann and Matthies, 1989). Studies from the same group also demonstrated that L-AP3 inhibited late LTP at a time about 1 h from the induction of LTP (Behnisch and Reymann, 1991). D,L-AP3 was shown in other studies to inhibit the induction of both short-term potentiation at 1 min following stimulation, and LTP at 20 min following stimulation (Izumi *et al.*, 1991). However, certain of the effects of the D,L-form of AP3 may be due to a direct antagonism on NMDA receptors by the D-form, as 1 mM D,L-AP3 strongly inhibited NMDA-evoked currents (Izumi *et al.*, 1991). The inhibitory effect of D,L-AP3 on LTP was confirmed as being caused by the L- but not the D-isomer (Behnisch *et al.*, 1991).

Studies with ACPD also supported the involvement of the mGluR in LTP. In the presence of ACPD comprising a 50:50 mixture of active and inactive isomer (10 µM of a racemic form of ACPD was used; for simplicity, the concentration of the active isomer is given), short-term potentiation at 1–5 min post-tetanically and LTP of the field EPSP in CA1 were greatly enhanced compared with control slices (McGuinness *et al.*, 1991b). This was confirmed in later studies by Behnisch and Reymann (1992). Moreover, tetanically induced short-term potentiation lasting 30 min was converted into LTP in the presence of 25 µM *trans*-ACPD (Aniksztejn *et al.*, 1992). ACPD was also shown to induce a slow onset LTP (peak attained in 90 min) in CA1, although the induction mechanisms of such LTP were found to be NMDAR independent (Bortolotto and Collingridge, 1993, 1995; see also Chinestra *et al.*, 1994). ACPD can also induce a much faster onset LTP (peak attained in <10 min) providing that NMDAR and the mGluR are activated coincidently. Thus, in the dentate gyrus, application of ACPD combined with membrane depolarization sufficient to induce NMDAR activation rapidly led to the induction of LTP (O'Connor *et al.*, 1995) (*Figure 2.2*).

The introduction of the highly selective mGluR antagonist (RS)-α-methyl-4-carboxyphenylglycine (MCPG) (Eaton *et al.*, 1993) enabled the role of mGluR in synaptic plasticity to be understood in much more depth (Bashir *et al.*, 1993b; Bortolotto *et al.*, 1994) (*Figures 2.3* and *2.4*). MCPG reversibly blocked LTP, although not short-term potentiation, in CA1, thus demonstrating the absolute requirement for the activation of mGluR by synaptically released L-glutamate in the induction of LTP (Bashir *et al.*, 1993b). In further studies, MCPG was used to clarify and expand the difference in functioning between NMDAR and mGluR in the induction of LTP. Thus, activation of mGluR was shown to stimulate a molecular switch which led to a long-lasting (several hours) intracellular cascade involving the production of a second messenger (probably PKC in an autophosphorylated state) (Bortolotto *et al.*, 1994). High frequency-induced LTP was not induced by activation of mGluR in isolation, but only by the co-activation of mGluR and NMDAR. However, the mGluR and NMDAR did not have to be co-activated in a temporal coincident manner. Providing that the activation of mGluR, and therefore of the resulting long-lasting activation of intracellular

---

Figure 2.2. ACPD-induced LTP of the AMPAR-mediated component of the EPSC in the dentate gyrus in the submerged slice preparation is dependent on activation of NMDAR. Upper graph: perfusion of ACPD (10 µM, 2 min) induces a transient depression followed by an LTP of the patch clamp recording of the AMPAR component (measured at 3 msec) of the EPSC in the associational–commissural layer of the dentate gyrus. The membrane was held at –70 mV, and depolarizing pulses to –30 mV were applied for 300 msec at 2-min intervals throughout the experiment. The original traces show the EPSC prior to and following the induction of the potentiation. Lower graph: the NMDAR antagonist AP5 prevents the induction of LTP, but not the transient depression following perfusion of ACPD. Reproduced from O'Connor *et al.* (1995). Reprinted with permission from *Nature* **367**: 557–559. © 1995 Macmillan Magazines Limited.

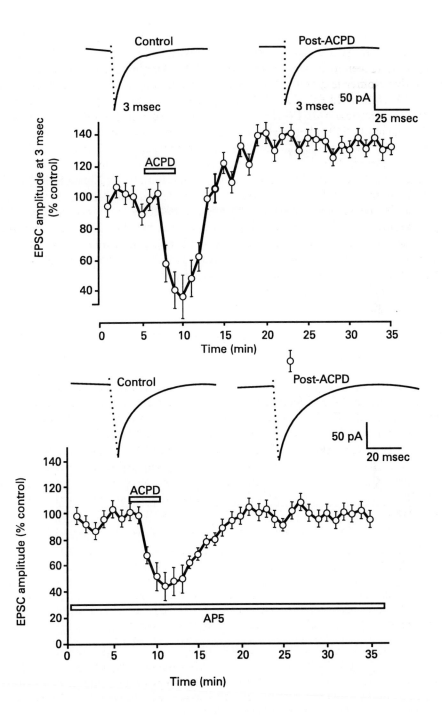

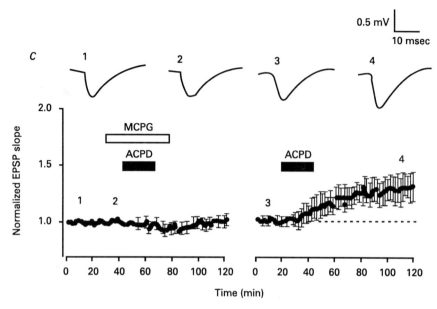

Figure 2.3. MCPG block of mGluR agonist-induced LTP in CA1 hippocampus. Lower graphs: activation of mGluR with the selective mGluR agonist ACPD induces a slow onset LTP of the field EPSP which reaches a peak in about 60 min (right trace). Such LTP was reversibly blocked by preincubation with the mGluR antagonist MCPG (500 µM). Upper traces (1–4) show original records of field EPSPs at the times shown on the graphs. Reproduced from Bashir *et al.* (1993a) Metabotropic receptors contribute to the induction of long-term depression in the CA1 region of the hippocampus. *Eur. J. Pharmacol.* **239**: 265–266, with permission from Elsevier Science.

messengers, was sufficiently strong, the mGluR activation could precede that of the NMDAR. Thus, MCPG only blocked the induction of LTP at nonconditioned synapses, that is at which no activation of the mGluR had occurred previously. If synapses had been conditioned, that is activation of mGluRs had occurred previously during high frequency stimulation or by application of ACPD, then MCPG did not block the induction of LTP (Bortolotto *et al.*, 1994). In addition, prolonged low frequency stimulation (900 pulses at 2 Hz) applied to a conditioned state of the synapses led to deconditioning, a turning off of the molecular switch, a state in which MCPG did block LTP induced by high frequency stimulation (Bortolotto *et al.*, 1994). MCPG subsequently has been confirmed to block the induction of LTP in the dentate gyrus *in vitro* (Yue *et al.*, 1995) and *in vivo* (Richter-Levin *et al.*, 1994; Riedel and Reymann, 1993). Several studies were unable to demonstrate an MCPG block of LTP induction in nonstimulated slices in CA1 (Chinestra *et al.*, 1994; Manzoni *et al.*, 1994a; Wang *et al.*, 1995). Such inability of MCPG to block the induction of LTP in the nonstimulated slice preparation was shown subsequently to be due to the synapses existing in a conditioned state following the preparation of the *in vitro* slices. Thus, following low frequency-induced deconditioning,

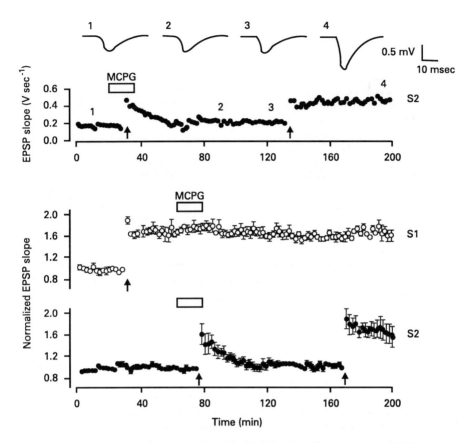

Figure 2.4. MCPG block of tetanically induced LTP in CA1 hippocampus. Middle graph: tetanic stimulation applied to the pathway S1 induces LTP, with MCPG (500 μM) applied after tetanic stimulation failing to block the maintenance of LTP. Lower graph: tetanic stimulation applied to pathway S2 in the presence of the mGluR antagonist MCPG induces only a short-term potentiation lasting about 20–30 min, with induction of LTP being blocked. Following washout of MCPG, tetanic stimulation induces LTP. Reproduced from Bashir *et al.* (1993a) Metabotropic receptors contribute to the induction of long-term depression in the CA1 region of the hippocampus. *Eur. J. Pharmacol.* **239**: 265–266, with permission from Elsevier Science. Upper graph and traces: data from a single experiment with the protocol identical to the lower graph.

MCPG was able to block the induction of LTP (Wang *et al.*, 1995). The essential role of mGluR in the induction of LTP was substantiated further by the generation of mutant mice deficient in mGluR1. High frequency stimulation was unable to induce full amplitude LTP in CA1 in such mice in one study (Alba *et al.*, 1994), although no inhibition of LTP in CA1 or the dentate gyrus was observed in mutant mice deficient in mGluR1 in another study (Conquet *et al.*, 1994).

The activation of $GABA_A R$-mediated IPSPs and the resulting hyperpolarization /conductance increase will normally oppose membrane depolarization and the resulting activation of NMDAR. Thus, in many regions of the central nervous system, the EPSP evoked by low frequency afferent stimulation is followed rapidly by a biphasic IPSP mediated by $GABA_A R$. The IPSPs reduce the amplitude and time course of the EPSP, and therefore prevent the NMDAR contributing to the EPSP, an action which will be made particularly effective because of the slow kinetics of the NMDAR-mediated responses. Such $GABA_A R$ reduction of the synaptic activation of NMDAR was shown clearly by studies in which the bicuculline- and picrotoxin-induced increases of the amplitude and duration of the low frequency-induced EPSP was partially reversed by the NMDAR antagonist AP5 (Dingledine et al., 1986; Herron et al., 1985). Consequently, blocking activation of $GABA_A R$ would be expected to enhance activation of NMDAR during high frequency stimulation and therefore enhance the induction of LTP. Such an action was demonstrated in studies by Wigstrom and Gustaffson (1983, 1985) in which picrotoxin lowered the afferent stimulation amplitude threshold required for LTP induction, although the peak amplitude of LTP was not enhanced.

$GABA_B R$ also play an important role in the induction of LTP. When GABA is released at CA1 synapses following stimulation of the GABAergic axons, $GABA_B$ autoreceptors located on GABAergic terminals are activated, resulting in inhibition of GABA release (Davies et al., 1990). This inhibition of GABA release via activation of GABAergic autoreceptors is frequency dependent, being absent at frequencies below 0.2 Hz and becoming greater with increased frequency (Davies et al., 1990). Thus, during induction of LTP by high frequency stimulation, GABA release is decreased, facilitating postsynaptic depolarization and thereby activation of NMDAR. This phenomenon was demonstrated clearly by the use of $GABA_B$ antagonists in the studies of Davies et al. (1991), with such antagonists inhibiting the induction of LTP by increasing the postsynaptic $GABA_A R$-mediated inhibition. Consistent with such a mechanism, $GABA_B R$ antagonists did not inhibit LTP if $GABA_A R$ inhibition was blocked with antagonists (Davies et al., 1991). Moreover, $GABA_B$ agonists such as baclofen facilitated the induction of LTP (Olpe and Karlsson, 1990). The facilitation of LTP by antagonists of $GABA_B R$ such as phaclofen and CGP 35348 is probably mediated by blocking postsynaptic $GABA_B R$ (Olpe and Karlsson, 1990).

The induction of NMDAR-dependent LTP of AMPAR-mediated synaptic transmission can be summarized as follows (*Figures 2.5* and *2.6*). The initial EPSP of the train is mainly mediated by activation of AMPAR. Little current flow occurs through the NMDAR because the postsynaptic $GABA_A$- and $GABA_B$-mediated IPSPs ensure that the depolarizing EPSP is maintained at a relatively low amplitude and does not therefore attain the potential at which the magnesium block of the NMDAR is strongly relieved. Such a clamping effect of the rapidly activated IPSPs will particularly affect NMDAR EPSPs because of their slow rise time (10–20 msec). Moreover, only very weak activation of the mGluR occurs. During successive IPSPs, rapid depolarization caused by several factors occurs, including summation of AMPAR-mediated EPSPs and a decline in the $GABA_A$-mediated IPSPs due to feedback inhibition on presynaptic $GABA_B R$ and to a depolarizing shift of the chloride equilibrium potential. The

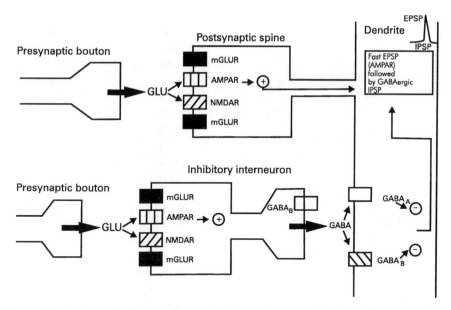

Figure 2.5. A schematic diagram illustrating the involvement of amino acid receptors in the mediation of synaptic transmission occurring at low frequency in CA1 hippocampus. Release of L-glutamate by presynaptic afferent stimulation activates only the postsynatic AMPA subtype of glutamate receptors (AMPAR) on pyramidal cell dendritic spines. Low frequency afferent stimulation does not activate postsynaptic NMDAR or mGluR, the former due to a magnesium block of the NMDAR channel, and the latter because of their possible perisynaptic location. Afferent stimulation also activates inhibitory interneurons in a similar way to that of the pyramidal cells, resulting in release of the inhibitory transmitter GABA which activates $GABA_A$ and $GABA_B$ receptors. The resulting IPSPs cause a reduction of the duration/amplitude of the fast AMPAR-mediated EPSP. Modified from Bliss and Collingridge (1993). Reprinted with permission from *Nature* **361**: 31–39. © 1993 Macmillan Magazines Limited.

depolarization allows current flow through the NMDAR, evoking NMDAR-mediated EPSPs which will also summate and generate further depolarization in a positive feedback process. This summation of NMDAR EPSPs will occur non-linearly with depolarization due to the nonlinear block of the NMDAR by magnesium. The combination of these mechanisms for depolarization will result in the high frequency afferent stimulation causing very strong activation of NMDAR, and thence a large calcium influx and second messenger stimulation. It should be noted that the properties of co-operativity and associativity which are essential features of this type of LTP are readily understandable by the voltage-sensing NMDA-dependent mechanism. The high frequency stimulation and subsequent release of high levels of glutamate will also result in a strong activation of mGluR and thence of the stimulation of intracellular messengers such as inositol trisphosphate and PKC.

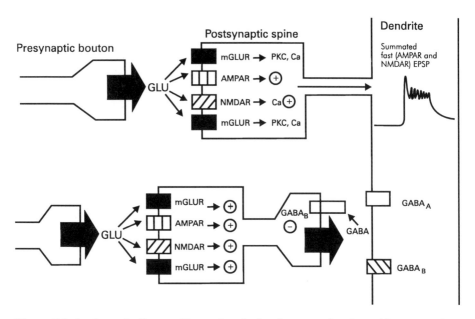

Figure 2.6. A schematic diagram illustrating the involvement of amino acid receptors in the mediation of synaptic transmission during high frequency afferent stimulation. The release of L-glutamate is now increased over an extended period, resulting in activation of AMPAR, NMDAR and mGluR. Summation of AMPAR- and NMDAR-mediated EPSP is enhanced by a reduction of postsynaptic GABAergic transmission due to the activation of presynaptic GABA$_B$R. Intracellular biochemical processes necessary for the induction of long-term potentiation are stimulated by the influx of calcium via the NMDAR and by activation of mGluR. Modified from Bliss and Collingridge (1993). Reproduced from *Nature* **361**: 31–39. © 1993 Macmillan Magazines Limited.

2.2.2 LTP of NMDA receptor-independent AMPA receptor-mediated synaptic transmission

The most widely studied form of NMDAR-independent LTP is that occurring at the mossy fiber–CA3 synapse in the hippocampus, which was first described by Alger and Teyler (1976). Strong evidence has accumulated that the induction of LTP at the mossy fiber–CA3 synapse is not mediated via NMDAR: AP5 did not block the induction of LTP (Harris and Cotman, 1986; Ito and Sugiyama, 1991; Williams and Johnston, 1988; Zalutsky and Nicoll, 1990); the induction of LTP did not show depolarization-dependent associativity, a property usually associated with NMDA-dependent LTP (Nicoll *et al.*, 1988); and, in addition, strong postsynaptic depolarization did not alter the induction of LTP, experimental manipulation that would be expected to strongly affect NMDAR participation (Nicoll *et al.*, 1988). The NMDAR independence is in agreement with the morphological distribution of NMDAR, the region of mossy fiber terminals being devoid of NMDAR (Monaghan and Cotman, 1986).

There is now increasing evidence that mossy fiber–CA3 LTP is presynaptically induced and maintained. Thus, the lack of effect of postsynaptic depolarization on the induction of LTP (Nicoll et al., 1988; Zalutsky and Nicoll, 1990), and the inability of the postsynaptic injection of a calcium chelator to block LTP (Zalutsky and Nicoll, 1990) strongly argues against a postsynaptic location of the LTP induction. Moreover, LTP was inhibited when synaptic transmission was blocked presynaptically by the use of calcium-free media (Castillo et al., 1994; Ito and Sugiyama, 1991). The reduction of paired-pulse whole cell currents following LTP is also strongly indicative that the maintenance of the LTP is presynaptic (Zalutsky and Nicoll, 1990).

Recent evidence is in favor of activation of presynaptic mGluR being necessary for the induction of LTP at mossy fiber–CA3 synapses. Thus, Ito and Sugiyama (1991) have demonstrated that the mGluR agonist ibotenate directly induced LTP following blockade of fast EPSPs by CNQX and AP5, the LTP being expressed after washout of the CNQX and AP5. Such ibotenate-induced LTP could be evoked even in the absence of external calcium (Ito and Sugiyama, 1991). Moreover, the mGluR antagonists D,L-AP3 and AP4 inhibited LTP induction (Ito and Sugiyama, 1991). Compelling recent evidence for the involvement of mGluR in the induction of mossy fiber–CA3 LTP has come from the findings that the more selective mGluR antagonist MCPG blocked high frequency-induced LTP in mossy fiber–CA3 (Bashir et al., 1993b). Furthermore, gene-targeted mutant mice deficient in mGluR displayed a greatly reduced LTP at mossy fiber–CA3 synapses (Conquet et al., 1994). Morphological evidence showing the location of mGluR in the mossy fiber pathway is in agreement with such a conclusion (Conquet et al., 1994).

The induction of LTP in the dorsolateral septal nucleus demonstrates many similarities to the induction of LTP at the mossy fiber–CA3 synapse. Thus, induction of LTP is independent of NMDAR, being resistant to AP5 (Zheng and Gallagher, 1992a), but does occur upon activation of mGluR. The LTP was induced by perfusion of ACPD (Zheng and Gallagher, 1992b), and high frequency-induced LTP and the ACPD-induced LTP were blocked by L-AP3 and L-AP4, leaving a residual short-term potentiation lasting 10–20 min (Zheng and Gallagher, 1990, 1992b). The postsynaptic injection of the nonhydrolyzable analog of GTP, GTPγS, and also of BAPTA, blocked LTP induction, demonstrating the requirement for a postsynaptic G-protein and an increase in intracellular calcium for LTP induction (Zheng and Gallagher, 1992b).

2.3 LTP of NMDA receptor-mediated synaptic transmission

LTP of NMDAR-mediated synaptic transmission was first demonstrated by Bashir et al. (1990) in CA1 hippocampus. NMDA-mediated EPSPs/excitatory postsynaptic currents (EPSCs), isolated from AMPAR-mediated EPSPs/EPSCs with AMPAR antagonists, exhibited LTP following high frequency stimulation. LTP of such isolated NMDAR EPSPs/EPSCs subsequently was substantiated in CA1 (Berretta et al., 1991; Lin et al., 1993) and dentate gyrus (O'Connor et al., 1994; Xie et al., 1992). Stimulatory frequencies from 10 to 200 Hz were all capable of inducing LTP of NMDAR-mediated synaptic transmission (Bashir *et al.*, 1991; O'Connor *et al.*, 1994; Xie *et al.*, 1992). No change in the kinetics of the NMDAR-mediated EPSCs occurred following the induction of LTP, demonstrating that the LTP of the NMDAR EPSC was not generated by a decrease of desensitization or by an alteration in single channel kinetics (O'Connor *et al.*, 1994). The LTP of isolated NMDAR EPSCs was NMDAR dependent, being blocked if the high frequency stimulation was applied in the presence of AP5 (Bashir *et al.*, 1990; O'Connor *et al.*, 1994). Moreover, the LTP of such isolated NMDAR EPSCs was dependent on the activation of mGluR, being blocked by the mGluR antagonists AP3 and MCPG, and being induced by the mGluR agonist ACPD (O'Connor *et al.*, 1994) (*Figures 2.7* and *2.8*).

LTP of the NMDAR-mediated EPSC has also been studied in parallel with the AMPAR-mediated EPSC in the dual-component AMPAR/NMDAR-mediated

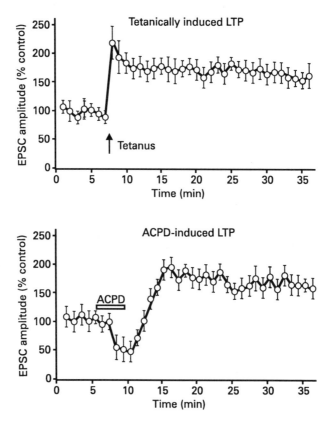

Figure 2.7. LTP of the isolated NMDAR-mediated component of the EPSC. Upper graph: tetanic stimulation induces LTP of the isolated NMDAR EPSC in the dentate gyrus. The AMPAR component of the EPSC was blocked using CNQX. Lower graph: perfusion of ACPD (10 μM, 2 min) induces a transient depression followed by LTP of the isolated NMDAR EPSC. Reproduced from O'Connor *et al.* (1994). Reproduced from *Nature* **367**: 557–559. © 1994 Macmillan Magazines Limited.

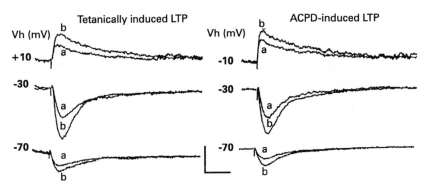

Figure 2.8. LTP of NMDAR EPSC at different holding potentials. Traces of isolated NMDAR EPSCs at different holding potentials (−70, −30 and +10 mV) recorded prior to (traces a) and following (traces b) either tetanic stimulation (left series of traces) or perfusion of ACPD (right series of traces). Horizontal bar 50 msec, vertical bar 30 pA. Reproduced from O'Connor *et al.* (1994). Reproduced from *Nature* **367**: 557–559. © 1994 Macmillan Magazines Limited.

EPSC. This is possible because of the much slower time course of the NMDAR component compared with the AMPAR component. Certain early studies have either failed to observe induction of LTP of NMDAR-mediated synaptic transmission (Kauer *et al.*, 1988b; Perkel and Nicoll, 1993) or found that such LTP was relatively minor (up to 30%) compared with LTP of the AMPAR-mediated transmission (Aszetely *et al.*, 1992; Muller *et al.*, 1988, 1992). However, it has now become clear that LTP of NMDAR-mediated transmission is of a large amplitude and similar in magnitude to that of AMPAR-mediated transmission. Thus, high frequency stimulation was found to generate LTP of the NMDAR component which was of a very similar, and frequently identical, amplitude to that of the AMPAR-mediated component in both CA1 (Clarke and Collingridge, 1995) and the dentate gyrus (O'Connor *et al.*, 1995). The similar enhancement of both AMPAR and NMDAR could result either from presynaptic mediation of LTP maintenance or, alternatively, from changes in postsynaptic receptor functioning such as an increase in receptor number or an increase in channel conductance that are a similar magnitude for both the AMPAR and NMDAR.

2.4 LTD of AMPA receptor-mediated transmission

2.4.1 Introduction

Homosynaptic and heterosynaptic LTD/depotentiation (DP) of AMPAR-mediated synaptic transmission have been observed in several regions of the brain, most notably the hippocampus, neocortex, striatum and cerebellum. Homosynaptic LTD and DP can be induced in the hippocampus in CA1 and the dentate gyrus by relatively prolonged stimulation at low frequencies (1–10 Hz) *in vivo* (Barrionuevo *et al.*, 1980; Staubli and Lynch, 1990), and *in vitro* slices (Dudek and Bear, 1992; Fujii *et al.*, 1991; Mulkey and Malenka, 1992). In the neocortex and the corticostriatal input to the striatum, homosynaptic LTD has usually been induced by high frequency stimulation (~100 Hz) resembling that required for LTP induction (Artola *et al.*, 1990; Calabresi *et al.*, 1992a; Hirsch and Crepel, 1990; Kato, 1993), although recently it has become evident that LTD in the neocortex can be induced with low frequency stimulation identical to that used for LTD induction in the hippocampus if stimulation is given to the direct afferent pathway to the postsynaptic recording site rather than the more usual indirect stimulation (Kirkwood *et al.*, 1993). Heterosynaptic LTD has been observed in a number of studies in the hippocampus, particularly the dentate gyrus and CA3 (Linden, 1994). The LTD in the cerebellum is a conjunctive LTD recorded from Purkinje cells. Such cells receive two distinct sets of excitatory afferents, a single climbing fiber (CF) from the inferior olive, and a large number of parallel fibers (PFs) from granule cells. AMPAR-mediated Purkinje cell EPSPs evoked by PF stimulation undergo LTD following conjunctive activation of PFs and a CF (Ito, 1989).

2.4.2 Role of AMPA receptors

Evidence for the essential participation of AMPAR in the induction of conjunctive LTD in the cerebellum was first made by experiments investigating the induction of LTD by the pairing of depolarization and application of quisqualate, consequently activating AMPAR and mGluR. LTD was blocked if AMPAR activation was prevented with the AMPAR antagonist CNQX (Linden *et al.*, 1991). In addition, LTD (induced in control experiments by pairing depolarization-induced calcium spikes with PF-mediated EPSPs) was blocked in the presence of CNQX (Hemart *et al.*, 1995). Linden *et al.* (1993) established that AMPAR activation was normally required in order to allow sodium influx, as substitution of sodium by either permeant or impermeant ions of the AMPAR channel prevented the induction of LTD. It was speculated that the rise in internal sodium resulting from increased sodium influx could increase intracellular calcium via the sodium–calcium exchange mechanism, or directly stimulate intracellular enzyme activity such as phospholipase C or phospholipase A_2 (Linden *et al.*, 1993). This requirement for the activation of AMPAR in the induction of LTD may be the explanation for the finding that postsynaptic depolarization is essential for induction of LTD. Thus, prevention of the CF-evoked depolarization of the Purkinje cells by activation of inhibitory stellate neurons (Ekerot and Kano, 1985) or by strong hyperpolarization of the Purkinje cells (Crepel and Jaillard, 1991) during PF stimulation blocked the induction of LTD. Moreover, the effects of stimulation of the CF input could be substituted by direct postsynaptic depolarization of the Purkinje cell (Crepel and Jaillard, 1990, 1991; Konnerth *et al.*, 1992).

The expression of cerebellar LTD involving a depression of postsynaptic AMPAR-mediated EPSPs was supported by experiments showing that PF-evoked EPSPs could be substituted by AMPAR-mediated test pulses evoked by exogenous application of AMPAR agonists such as AMPA or glutamate (Crepel and Krupa, 1988; Linden *et al.*, 1991).

Activation of AMPAR is not required for the induction of LTD in the neocortex. Thus LTD was induced even when stimulation was given in the presence of the AMPAR antagonist CNQX, the LTD being observed upon washout of the CNQX (Kato, 1993).

2.4.3 Role of NMDA receptors

The role of NMDAR in the induction of homosynaptic LTD/DP in both the hippocampus and neocortex is controversial at present, with contrasting results from studies investigating the effect of AP5 on the induction of LTD/DP. AP5 was found to block the induction of homosynaptic LTD/DP in certain studies in CA1 hippocampus (Debanne *et al.*, 1994; Dudek and Bear, 1992; Fujii *et al.*, 1991; Mulkey and Malenka, 1992; O'Dell and Kandel, 1994). However, such an AP5 block of homosynaptic LTD/DP was not substantiated in later studies, in either CA1 (Bashir and Collingridge, 1994; Bolshakov and Siegelbaum,1994; Yang *et al.*, 1994) or the dentate gyrus (O'Mara *et al.*, 1995). The difference in these results is not due to an altered NMDA dependence of LTD/DP with age, as both Mulkey and Malenka (1992) and Bolshakov and Siegelbaum (1994) used young animals, while Fujii *et al.* (1991), Dudek and Bear (1992), Bashir and Collingridge (1994) and O'Mara *et al.* (1995) used adult animals. A possible explanation for the differing effects of AP5 is that LTD can be induced either dependently or independently of NMDAR. An alternative explanation is that the LTD/DP is actually independent of NMDAR activation, but requires a threshold level of depolarization which is generated in part by activation of NMDAR. Thus, in certain studies, the blocking action of AP5 on LTD induction may have been generated by a reduction in the depolarization to a level below that necessary for LTP induction. There is substantial evidence that depolarization is required for homosynaptic LTD induction, derived from experiments in which postsynaptic hyperpolarization during the applied afferent stimulation prevents the induction of LTD in CA1 hippocampus (Mulkey and Malenka, 1992), and the visual cortex (Artola *et al.*, 1990). The depolarization required for the induction of LTD may be necessary for calcium influx from the extracellular medium. The induction of homosynaptic LTD/DP does require an elevation of the intracellular calcium above the resting level, as intracellular injection of BAPTA abolished the induction of LTD (Mulkey and Malenka, 1992). However, although evidence has been presented for a calcium influx via voltage-gated calcium channels being necessary for homosynaptic LTD induction (Bolshakov and Siegelbaum, 1994), it was not substantiated (Bashir and Collingridge, 1994).

The involvement of NMDAR in homosynaptic LTD induction in the neocortex is also disputable. Thus, AP5 was found to inhibit the induction of LTD in one study (Kirkwood *et al.*, 1993), but not in several others (Artola *et al.*, 1990; Hirsch and Crepel, 1991; Kato, 1993; Walsh, 1993). Further evidence supporting an NMDAR-independent LTD induction comes from studies in which LTD was unmasked if the induction of LTP was blocked with AP5 (Hirsch and Crepel, 1991; Kato, 1994). AP5 also did not block LTD induction in the striatum (Calabresi *et al.*, 1992a; Lovinger *et al.*, 1993; Walsh, 1993) or nucleus accumbens

(Pennartz et al., 1993). NMDAR are not involved in LTD induction in the cerebellum, both the PF and CF synapses on to Purkinje cells lacking NMDAR in the adult.

Heterosynaptic depression, including the short-lasting depression observed in CA1, and the heterosynaptic LTD observed in the dentate gyrus, requires the activation of NMDAR (Artola and Singer, 1993; Bashir and Collingridge, 1994; Desmond et al., 1991; Linden, 1994). The short-lasting heterosynaptic depression in CA1 was demonstrated to be generated by adenosine, released following the synaptic activation by glutamate of NMDAR (Manzoni et al., 1994b). The heterosynaptic LTD could be induced by a similar mechanism.

2.4.4 Role of mGlu receptors

Substantial evidence has been found for an essential role of mGluR in the induction of LTD/DP in the hippocampus, neocortex, striatum and cerebellum. Thus, the induction of homosynaptic LTD in CA1 hippocampus was blocked by the mGluR antagonist L-AP3 (Stanton et al., 1991; Yang et al., 1994) and MCPG (Bashir and Collingridge, 1994; Bashir et al., 1993b; Bolshakov and Siegebaum 1994; O'Mara et al., 1995; Yang et al., 1994). Moreover, ACPD induced LTD/DP if accompanied either by a train of postsynaptic depolarizing pulses in CA1 (Bolshakov and Siegelbaum, 1994), or in the presence of the $GABA_AR$ antagonist picrotoxin in the dentate gyrus (O'Mara et al., 1995). In the neocortex, the mGluR agonists quisqualate (Kato, 1993) and also ACPD coupled with tetanization in the presence of AP5 (Kato, 1994) induced LTD. Moreover, block of G-protein functioning prevented LTD (Kato, 1993). In the striatum, L-AP3 impaired the induction of LTD (Calabresi et al., 1992b). mGluR activation may be required for the induction of LTD/DP in order to release calcium from intracellular stores (Kato, 1993), or to stimulate other second messengers.

In the cerebellum, mGluR must be co-activated with AMPAR in order to induce conjunctive LTD, the mGluR being synaptically activated by stimulation of the PFs (Hartell, 1994). Initial experiments on the role of mGluR and LTD established that LTD was not induced if AMPAR were activated alone, even if accompanied by depolarization. Thus, repeated test applications of AMPA evoked depolarizing potentials which remained at a constant amplitude (Ito and Karachot 1990a,b). Moreover, conjunctive AMPA application with Purkinje cell depolarization did not evoke LTD of the AMPA-induced inward current (Linden et al., 1991). Positive evidence for a role of mGluR in the induction of LTD was shown in experiments in which LTD was induced by conjunctive application of the mGluR agonist quisqualate with CF stimulation (Kano and Kato, 1987). In addition, LTD of AMPAR-mediated currents was induced if glutamate, quisqualate or AMPA plus ACPD responses were evoked conjunctively with Purkinje cell depolarization (Linden et al., 1991). LTD of quisqualate-induced potentials could be induced by AMPA following a conditioning quisqualate application, or by AMPA plus ACPD if the potentials were evoked sufficiently closely together (Ito and Karachot, 1990a,b). The application of ACPD in conjunction with depolarization-induced activation of Purkinje cell calcium channels was also found to induce LTD of PF EPSPs (Daniel et al., 1992; Hemart et al., 1995).

Very convincing evidence for an essential role of mGluR in the induction of cerebellar LTD has been established recently with the use of mGluR antagonists and also with mutant animals. Thus, LTD was blocked by MCPG (Hartell, 1994) and by L-AP3 (Linden et al., 1991). Moreover, mutant mice deficient in mGluR1 had impaired LTD (Conquet et al., 1994). LTD was also blocked by pertussis toxin, illustrating the requirement for activation of a G-protein-linked receptor such as mGluR (Ito and Karachot, 1990c; Linden et al., 1991). Activation of mGluR alone is not sufficient to induce LTD. Thus, conjunction of quisqalate potentials and Purkinje cell depolarization in the presence of CNQX failed to elicit LTD following washout of the CNQX (Linden et al., 1991).

The essential role of mGluR in the induction of LTD is likely to be due to the linking of such receptors to the stimulation of intracellular kinases and/or the release of calcium from intracellular stores (Crepel and Krupa, 1989; Linden and Connor, 1991a,b). The rise in intracellular calcium is generated, at least in part, by calcium influx via voltage-dependent calcium channels opened during the CF-evoked plateau depolarization (Crepel and Jaillard, 1990, 1991; Hirano, 1990; Konnerth et al., 1992; Sakurai, 1990). Activation of mGluR alone did not normally generate a sufficient rise in the intracellular calcium concentration to evoke LTD (Crepel et al., 1991). However, it is possible that calcium influx from the external media via voltage-gated calcium channels may summate with calcium release from intracellular stores generated following mGluR activation.

2.4.5 Role of GABA receptors

Several recent reports have shown that afferent stimulation in the presence of $GABA_AR$ or $GABA_BR$ agonists induces homosynaptic LTD. In the neocortex, high frequency stimulation applied in the presence of $GABA_AR$ agonists induced LTD (Kato and Yoshimura, 1993). Moreover, in hippocampal CA1, low frequency stimulation combined with either $GABA_AR$ or $GABA_BR$ agonists induced LTD (Yang et al., 1994). In these latter experiments, the exogenous GABA limited the postsynaptic depolarization at the soma to less than 10 mV. These reports are surprising in view of the requirement of depolarization for the induction of LTD (Mulkey and Malenka, 1992). Perhaps the requirements for GABAR activation are associated with activation of intracellular mediators rather than a reduction of membrane depolarization.

2.5 LTD of NMDA receptor-mediated synaptic transmission

LTD of isolated NMDAR-mediated transmission has been observed in the dentate gyrus following the pairing of low frequency stimulation (10 Hz) with either membrane hyperpolarization (Xie et al., 1992) or mild (but not strong) depolarization (Lin et al., 1993), and also following high frequency stimulation (100 Hz) in the striatum (Calabresi *et al.*, 1992a; Lovinger *et al.*, 1993) and in the nucleus accumbens (Kombian and Malenka, 1994). The induction of LTD of NMDAR-mediated transmission in CA1 has been shown to be dependent on activation of mGluR, with L-AP3 and MCPG abolishing induction (Yi *et al.*, 1995).

2.6 LTD of GABA$_A$-mediated synaptic transmission

GABA$_A$-mediated IPSPs are reduced following high frequency stimulation (Liu *et al.*, 1993; Stelzer *et al.*, 1994). In a detailed study in CA1 involving intracellular dendritic recording, orthodromic IPSPs, GABA-mediated conductance during the IPSPs, spontaneous IPSPs and iontophoretically evoked GABA potentials were all strongly reduced following high frequency afferent stimulation (Stelzer *et al.*, 1994). In this study, the induction of the LTD was independent of AMPAR and GABA$_B$R activation (Stelzer *et al.*, 1994). The LTD induction may be mediated via activation of mGluR, as ACPD was found to induce LTD of the GABA$_A$-mediated transmission in CA1, an effect blocked by prior activation of intracellular G-proteins (Liu *et al.*, 1993). The role of NMDAR in the LTD of the GABA$_A$-mediated transmission is controversial at present, with AP5 blocking such LTD in one study (Stelzer *et al.*, 1994), but not in a further study (Liu *et al.*, 1993).

References

Abraham WC, Mason SE. (1988) Effects of the NMDAR channel antagonist CPP and MK 801 on hippocampal field potentials and long-term potentiation in anaesthetised rats. *Brain Res.* **462**: 40–46.
Alba A, Chen C, Herrup K, Rosenmund C, Stevens CF, Tonegawa S. (1994) Reduced hippocampal long-term potentiation and context-specific deficit in associative learning in mGluR1 mutant mice. *Cell* **79**: 365–375.
Alford S, Frengueli BG, Schofield JG, Collingridge GL. (1993) Characterization of calcium signals induced in hippocampal CA1 neurons by the synaptic activation of NMDA receptors. *J. Physiol.* **469**, 693–716.
Alger BE, Teyler TJ. (1976) Long-term and short-term plasticity in the CA1, CA3 and dentate regions of the rat hippocampal slice. *Brain Res.* **110**: 463–480.
Aniksztejn L, Otani S, Ben-Ari Y. (1992) Quisqualate metabotropic receptors modulate NMDA currents and facilitate induction of long-term potentiation through protein kinase C. *Eur. J. Neurosci.* **4**: 500–505.
Artola A, Singer W. (1987) Long-term potentiation and NMDA receptors in rat visual cortex. *Nature* **330**: 649–652.
Artola A, Singer W. (1993) Long-term depression of excitatory synaptic transmission and its relationship to long-term potentiation. *Trends Neurosci.* **16**: 480–487.
Artola A, Brocher, S, Singer W. (1990) Different voltage-dependent thresholds for inducing long-term depression and long-term potentiation in slices of rat visual cortex. *Nature* **347**: 69–72.
Aszetely F, Wigstrom H, Gustaffson B. (1992) The relative contribution of NMDA channels in the expression of long-term potentiation in the hippocampal CA1 region. *Eur. J. Neurosci.* **4**: 681–690.
Barrionuevo G, Schottler F, Lynch G. (1980) The effects of repetitive low frequency stimulation on control and potentiated synaptic responses in the hippcampus. *Life Sci.* **27**: 2385–2389.
Bashir ZI, Collingridge GL. (1994) An investigation of depotentiation of long-term potentiation in the CA1 region of the hippocampus. *Exp. Brain. Res.* **100**: 437–443.

Bashir ZI, Alford S, Davies SN, Randall AD, Collingridge GL. (1990) Long-term potentiation of NMDA receptor-mediated synaptic transmission in the hippocampus. *Nature* **349**: 156–158.

Bashir ZI, Jane DE, Sunter DC, Watkins JC, Collingridge GL. (1993a) Metabotropic receptors contribute to the induction of long-term depression in the CA1 region of the hippocampus. *Eur. J. Pharmacol.* **239**: 265–266.

Bashir ZI, Bortolotto ZA, Davies CH *et al.* (1993b) Induction of LTP in the hippocampus needs synaptic activation of glutamate metabotropic receptors. *Nature* **363**: 347–350.

Bashir ZI, Tam B, Collingridge GL. (1994) Activation of the glycine site in the NMDA receptor is necessary for the induction of LTP. *Neurosci. Lett.* **108**: 261–266.

Batchelor AM, Madge DJ, Garthwaite J. (1994) Synaptic activation of metabotropic glutamate receptors in the parallel fibre–Purkinje cell pathway in rat cerebellar slices. *Neuroscience* **63**: 911–915.

Baude A, Nusser Z, Roberts JDB, Mulvihill E, McIlhinney RAJ, Somogyi P. (1993) The metabotropic glutamate receptor (mGluR1) is concentrated at perisynaptic membrane of neuronal subpopulations as detected by immunogold reaction. *Neuron* **11**: 771–787.

Behnisch T, Reymann KG. (1993) Co-activation of metabotropic glutamate and N-methyl-D-aspartate receptors is involved in mechanisms of long-term potentiation. *Neuroscience* **51**: 37–47.

Behnisch T, Fjodorow K, Reymann KG. (1991) L-2-amino-3-phosphonoproprionate blocks late synaptic long-term potentiation. *Neuro Rep.* **2**: 386–388.

Berretta N, Berton F, Bianchi R, Brunelli M, Capogna M, Francesconi W. (1991) Long-term potentiation of NMDA receptor-mediated EPSP in guinea-pig hippocampal slices. *Eur. J. Neurosci.* **3**: 850–854.

Bliss TVP, Collingridge GL. (1993) A synaptic model of memory: long-term potentiation in the hippocampus. *Nature* **361**: 31–39.

Bortolotto ZA, Collingridge GL. (1993) Characterization of LTP by the activation of glutamate metabotropic receptors in area CA1 of the hippocampus. *Neuropharmacology* **32**: 1–9.

Bortolotto ZA, Collingridge GL. (1995) On the mechanism of long-term potentiation induced by 1S,3R-aminocyclopentane-1,3-dicarboxylic acid (ACPD) in rat hippocampal slices. *Neuropharmacology* **34**: 543–551.

Bortolotto ZA, Bashir ZI, Davies CH, Collingridge GL. (1994) A molecular switch activated by metabotropic glutamate receptors regulates induction of long-term potentiation. *Nature* **362**: 740–743.

Bolshakov VY, Siegelbaum SA. (1994) Postsynaptic induction and presynaptic expression of hippocampal long-term depression. *Science* **264**: 1148–1152.

Calabresi P, Pisano A, Mercuri NB, Bernadi G. (1992a) Long-term synaptic depression in the striatum: physiological and pharmacological characterization. *J. Neurosci.* **12**: 4224–4233.

Calabresi P, Pisano A, Mercuri NB, Bernadi G. (1992b) Long-term potentiation in the striatum is unmasked by removing the voltage-dependent magnesium block of NMDA receptor channels. *Eur. J. Neurosci.* **4**: 929–935.

Castillo PE, Weisskopf MC, Nicoll RA. (1994) The role of Ca channels in hippocampal mossy fibre synaptic transmission and long-term potentiation. *Neuron* **12**: 261–269.

Chinestra P, Aniksztejn L, Diabira D, Ben-Ari Y. (1994) Major differences between long-term potentiation and ACPD-induced slow onset potentiation in hippocampus. *J. Neurophysiol.* **70**: 2684–2689.

Clarke KA, Collingridge GL. (1995) Synaptic potentiation of dual-component excitatory postsynaptic currents in the rat hippocampus. *J. Physiol.* **482**: 39–52.

Coan EJ, Collingridge GL. (1987) Effects of phencyclidine, SKF 10,047 and related psychotomimetic agents on N-methyl-D-aspartate mediated synaptic responses in rat hippocampal slices. *Br. J. Pharmacol.* **91**: 547–556.

Coan EJ, Irving, AJ and Collingridge GL. (1989) Low frequency activation of the NMDA receptor system can prevent the induction of LTP. *Neurosci. Lett.* **105**: 205–210.

Colino A, Malenka RJ. (1993) Mechanisms underlying induction of long-term potentiation in rat medial and lateral perforant paths *in vitro*. *J. Neurophysiol.* **69**: 1150–1159.

Collingridge GL, Bliss TVP. (1987) NMDA receptors – their role in long-term potentiation. *Trends Neurosci.* **10**: 288–293.

Collingridge GL, Kehl SJ, McLennan, H. (1983a) The antagonism of amino-acid-induced excitations of rat hippocampal CA1 neurons *in vitro J. Physiol.* **334**: 19–31.

Collingridge GL, Kehl SJ, McLennan H. (1983b) Excitatory amino acids in synaptic transmission in the Schaffer collateral–commissural pathway of the rat hippocampus. *J. Physiol.* **334**: 33–46.

Collingridge GL, Herron CE, Lester RAJ. (1988) Frequency-dependent N-methyl-D-aspartate receptor-mediated synaptic transmission in rat hippocampus. *J. Physiol.* **399**: 301–312.

Conquet F, Bashir ZI, Davies CH et al. (1994) Motor deficit and impairment of synaptic plasticity in mice lacking mGluR1. *Nature* **372**: 237–243.

Cormier RJ, Mauk MD, Kelly PT. (1993) Glutamate iontophoresis induces long-term potentiation in the absence of evoked presynaptic activity. *Neuron* **10**: 907–919.

Crepel F, Jaillard D. (1990) Protein kinases, nitric oxide and long-term depression of synapses in the cerebellum. *NeuroReport* **1**: 133–136.

Crepel F, Jaillard D. (1991) Pairing of pre- and postsynaptic activities in cerebellar Purkinje cells induces long-term changes in synaptic efficacy *in vitro*. *J. Physiol.* **432**: 123–141.

Crepel F, Krupa M. (1988) Activation of protein kinase C induces a long-term depression of glutamate sensitivity of cerebellar Purkinje cells. An *in vitro* study. *Brain Res.* **458**: 397–440.

Crepel F, Daniel H, Hemart N, Jaillard D. (1991) Effect of ACPD and AP3 on parallel-fibre-mediated EPSPs of Purkinje cells in cerebellar Purkinje cells *in vitro*. *Exp. Brain Res.* **86**: 402–406.

Daniel H, Hemart N, Jaillard D, Crepel F. (1992) Coactivation of metabotropic glutamate receptors and voltage-gated calcium channels induces long-term depression in cerebellar Purkinje cells *in vitro*. *Exp. Brain Res.* **90**: 327–331.

Davies CH, Davies SN, Collingridge GL. (1990) Paired-pulse depression of monosynaptic GABA-mediated inhibitory postsynaptic responses in rat hippocampus. *J. Physiol.* **424**: 513–531.

Davies CH, Starkey SJ, Pozza MF, Collingridge GL. (1991) $GABA_B$ autoreceptors regulate the induction of LTP. *Nature* **349**: 609–611.

Davies SN, Lester RA, Reymann KG, Collingridge GL. (1988) Temporally distinct pre- and postsynaptic mechanisms maintain long-term potentiation. *Nature* **338**: 500–503.

Debanne D, Gahwiler BH, Thompson SM. (1994) Asynchronous pre- and postsynaptic activity induces associative long-term depression in area CA1 of the rat hippocampus *in vitro*. *Proc. Natl Acad. Sci. USA* **91**: 1148–1152.

Desmond NL, Colbert CM, Zhang DX, Levy WB. (1991) NMDA receptor antagonists block the induction of long-term depression in the hippocampal dentate gyrus of the anaesthetized rat. *Brain Res.* **552**: 93–98.

Dingledine R, Haynes MA, King AL. (1986) Involvement of N-methyl-D-aspartate receptors in epileptiform bursting in the rat hippocampal slice. *J. Physiol.* **380**: 175–189.

Dudek SM, Bear MF. (1992) Homosynaptic long-term depression in area CA1 of hippocampus and effects of N-methyl-D-aspartate receptor blockade. *Proc. Natl Acad. Sci. USA* **89**: 4363–4367.

Eaton SA, Jane DE, Jones PLSJ et al. (1993) Competitive antagonism at metabotropic glutamate receptors by (S)-4-carboxyphenylglycine and (RS)-methyl-4-carboxyphenylglycine. *Eur. J. Pharmacol. Mol. Pharmacol.* **244**: 195–197.

Ekerot C-F, Kano M. (1985) Long-term depression of parallel fibre synapses following stimulation of climbing fibres. *Brain Res.* **342**: 357–360.

Errington ML, Lynch MA, Bliss TVP. (1987) Long-term potentiation in the dentate gyrus: induction and increased glutamate release are blocked by D-(-)aminophosphonovalerate. *Neuroscience* **20**: 279–284.

Fujii S, Saito K, Miyakawa H, Ito K, Kato H. (1991) Reversal of long-term potentiation (depotentiation) induced by tetanus stimulation of the input to CA1 neurons of guinea pig hippocampal slices. *Brain Res.* **555**: 112–122.

Gerber U, Luthi A, Gahwiler BH. (1993) Inhibition of a slow synaptic response by a metabotropic glutamate receptor antagonist in hippocampal CA3 pyramidal cells. *Proc. R. Soc. Lond.* **254**: 169–172.

Greengard P, Jen J, Nairn JC, Stevens CF. (1991) Enhancement of the glutamate response by cAMP-dependent protein kinase in hippocampal neurons. *Science* **253**: 1135–1138.

Gustafsson B, Wigstom H, Abraham WC, Huang YY. (1987) Long-term potentiation in the hippocampus using depolarizing current pulses as the conditioning stimulus to single volley synaptic potentials. *J. Neurosci.* **7**: 774–780.

Harris EW, Cotman CW. (1986) Long-term potentiation of guinea-pig mossy fiber responses is not blocked by N-methyl-D-aspartate antagonists. *Neurosci. Lett.* **70**: 132–137.

Harris EW, Ganong AH, Cotman CW. (1984) Long-term potentiation in the hippocampus involves activation of N-methyl-D-aspartate receptors. *Brain Res.* **323**: 132–137.

Hartell NA. (1994) Induction of cerebellar long-term depression requires activation of glutamate metabotropic receptors. *NeuroReport* **5**: 913–916.

Harvey J, Collingridge GL. (1992) Thapsigargin blocks the induction of long-term potentiation in rat hippocampal slices. *Neurosci. Lett.* **139**: 197–200.

Hemart N, Daniel H, Jaillard D, Crepel F. (1995) Receptors and second messengers involved in long-term depression in rat cerebellar slices *in vitro*; a reappraisal. *Eur. J. Neurosci.* **7**: 45–53.

Herron CE, Williamson R, Collingridge GL. (1985) A selective N-methyl-D-aspartate antagonist depresses epileptiform activity in rat hippocampal slices. *Neursci. Lett.* **61**: 255–260.

Herron CE, Lester RAJ, Coan EJ, Collingridge GL. (1986) Frequency dependent involvement of NMDA receptors in the hippocampus: a novel synaptic mechanism. *Nature* **322**: 265–268.

Hirsch JC, Crepel F. (1990) Use-dependent changes in synaptic efficacy in rat prefrontal neurons *in vitro*. *J. Physiol.* **427**: 31–49.

Hirsch JC, Crepel F. (1991) Blockade of NMDA receptors unmasks a long-term depression in synaptic efficacy in rat prefrontal neurons *in vitro*. *Exp. Brain Res.* **85**: 621–624.

Huang Y-Y, Colino A, Selig DK, Malenka RC. (1992) The influence of prior synaptic activty on the induction of long-term potentiation. *Science* **255**: 730–733.

Ito I, Sugiyama H. (1991) Roles of glutamate receptors in long-term potentiation at hippocampal mossy fiber synapses. *NeuroReport* **2**: 333–336.

Ito K. (1995). Voltage-gated calcium channel blockers, ω-AgaIVA and Ni, suppress the induction of θ-burst induced long-term potentiation in guinea-pig hippocampal CA1 neurons. *Neurosci. Lett.* **183**: 112–115.

Ito M. (1989) Long-term depression. *Annu. Rev. Neurosci.* **12**: 85–102.

Ito M, Karochet L. (1990a) Long-term desensitization of quisqualate-specific glutamate receptors in Purkinje cells investigated with wedge recording from rat cerebellar slices. *Neurosci. Res.* **7**: 168–171.

Ito M, Karachot L. (1990b) Receptor subtypes involved in, and time course of, the long-term desensitization of glutamate receptors in cerebellar granule cells. *Neurosci. Res.* **8**: 303–307.

Izumi Y, Miyakawa H, Ho K, Kato H. (1987) Quisqualate and N-methyl-D-aspartate (NMDA) receptors in induction of long-term potentiation using conditioning solutions. *Neurosci. Lett.* **83**: 201–206.

Izumi Y, Clifford DB, Zorumski CF. (1990) Glycine antagonists block the induction of long-term potentiation of CA1 of rat hippocampal slices. *Neurosci. Lett.* **112**: 251–256.

Izumi Y, Clifford DB, Zorumski CF. (1991) 2-Amino-3-phosphono-proprionate blocks the induction and maintenance of long-term potentiation in rat hippocampal slices. *Neurosci. Lett.* **122**: 187–190.

Izumi Y, Clifford DB, Zorumski CF. (1992) Low concentrations of N-methyl-D-aspartate inhibit the induction of long-term potentiation in rat hippocampal slices. *Neurosci. Lett.* **122**: 187–190.

Kano M, Kato M. (1987) Quisqualate receptors are specifically involved in cerebellar synaptic plasticity. *Nature* **325**: 276–279.

Kato N. (1993) Dependency of long-term depression on postsynaptic metabotropic glutamate receptors in visual cortex. *Proc. Natl Acad. Sci. USA* **90**: 3650–3654.

Kato N. (1994) Long-term depression requiring *t*-ACPD activation and NMDA receptor blockade. *Brain Res.* **665**: 158–160.

Kato N, Yoshimura H. (1993) Tetanization during $GABA_A$ receptor activation induces long-term depression in visual cortex slices. *Neuropharmacology* **32**: 511–513.

Kauer JA, Malenka RC, Nicoll RA. (1988a) NMDA application potentiates synaptic transmission in the hippocampus. *Nature* **334**: 250–252.

Kauer JA, Malenka RC, Nicoll R.A. (1988b) A persistent postsynaptic modification mediates long-term potentiation in the hippocampus. *Neuron* **1**: 911–917.

Kelso SR, Ganong AH, Brown TH. (1986) Hebbian synapses in hippocampus. *Proc. Natl Acad. Sci. USA* **83**: 5326–5330.

Kirkwood A, Dudek SD, Gold JT, Aizenman CD, Bear MF. (1993) Common forms of synaptic plasticity in hippocampus and neocortex *in vitro*. *Science* **260**: 1518–1521.

Kombian SB, Malenka RC. (1994) Simultaneous LTP of non-NMDA and LTD of NMDA receptor-mediated responses in the nucleus accumbens. *Nature* **368**: 242–246.

Konnerth A, Dreeson J, Augustine GJ. (1992) Brief dendritic signals initiate long-lasting synaptic depression in cerebellar Purkinje cells. *Proc. Natl Acad. Sci. USA* **89**: 7051–7055.

Lacey MG, Henderson G. (1983) Antagonism of N-methyl-D-aspartate acid excitation of rat hippocampal pyramidal neurones *in vitro* by phencyclidine applied in known concentration. *Soc. Neurosci. Abstr.* **9**: 260.

Lin J-H, Way L-J, Gean P-W. (1993) Pairing of pre- and postsynaptic activities in hippocampal CA1 neurons induces long-term modifications of NMDA receptor-mediated synaptic potential. *Brain Res.* **603**: 117–120.

Linden DJ. (1994) Long-term synaptic depression in the mammalian brain. *Neuron* **12**: 457–472.

Linden DJ, Connor JA. (1991a) Participation of postsynaptic protein kinase C in cerebellar long-term depression in culture. *Science* **254**: 1656–1659.

Linden DJ, Connor JA. (1991b) Long-term depression of glutamate currents in cultured cerebellar Purkinje neurons does not require nitric oxide signalling. *Eur. J. Neurosci.* **4**: 10–15.

Linden DJ, Dickenson MH, Smeyne M, Connor JA. (1991) A long-term depression of AMPA currents in cultured cerebellar Purkinje neurons. *Neuron* **7**: 81–89.

Linden DJ, Smeyne M, Connor, JA. (1993) Induction of cerebellar long-term depression in culture requires postsynaptic action of sodium ions. *Neuron* **11**: 1093–1100.

Liu Y, Disterhoft JF, Slater NT. (1993) Activation of metabotropic glutamate receptors induces long-term depression of GABAergic inhibition in hippocampus. *J. Neurophysiol.* **69**: 1000–1004.

Lomo T. (1966) Frequency potentiation of excitatory synaptic activity in the dentate area of the hippocampal formation. *Acta Physiol. Scand.* **68** (Suppl.): 128.

MacDermott A, Mayer ML, Westbrook GL, Smith SJ, Barker JL. (1986) NMDA-receptor activation increases cytoplasmic calcium concentration in cultured spinal cord neurons. *Nature* **321**: 519–522.

Malinow R. Miller JP. (1986) Postsynaptic hyperpolarization during conditioning reversibly blocks induction of long-term potentiation. *Nature* **320**: 529–530.

Malinow R. (1994) LTP: desparately seeking resolution. *Science* **266**: 1195–1196.

Manzoni OJ, Weisskopf MG, Nicoll RA. (1994a) MCPG antagonizes metabotropic glutamate receptors, but not long-term potentiation, in the hippocampus. *Eur. J. Neurosci.* **6**: 1050–1054.

Manzoni OJ, Manabe T, Nicoll RA. (1994b) Release of adenosine by activation of NMDA receptors in the hippocampus. *Science* **265**: 2098–2101.

Maren S, Baudry M, Thompsen RF. (1991) Differential effects of ketamine and MK-801 on the induction of long-term potentiation. *NeuroReport* **2**: 239–242.

Mayer ML, Westbrook GL. (1987) Permeation and block of N-methyl-D-aspartate receptor channels by divalent cations in mouse cultures of central neurone. *J. Physiol.* **394**: 501–527.

McGlade-McCulloh E, Yamamoto H, Tan S-E, Brickley DA, Soderling TR. (1993) Phosphorylation and regulation of glutamate receptors by calcium/calmodulin-dependent protein kinase II. *Nature* **362**: 640–642.

McGuinness N, Anwyl R, Rowan MJ. (1991a) Inhibition of an N-methyl-D-aspartate induced short-term potentiation in the rat hippocampal slice. *Brain Res.* **562**: 335–338.

McGuinness N, Anwyl R, Rowan MJ. (1991b) *Trans*-ACPD enhances long-term potentiation in the hippocampus. *Eur. J. Pharmacol.* **197**: 231–232.

Monaghan DT, Cotman CW. (1986) Anatomical organization of NMDA, Rainate and quisqualate receptors. In: *Excitatory Amino Acids* (eds PJ Roberts, J Storm-Mathisen, HF Bradford). Macmillan, London, pp. 279–299.

Morris RGM, Anderson E, Lynch GS, Baudry M. (1986) Selective impairment of learning and blockade of LTP by NMDA receptor antagonist, AP5. *Nature* **319**: 774–776.

Mott DD, Lewis DV, Ferrari CM, Wison WA, Schwartzwelder HS. (1990) *Neurosci. Lett.* **113**: 222–226.

Mulkey RM, Malenka RC. (1992) Mechanisms underlying induction of homosynaptic long-term depression in area CA1 of the hippocampus. *Neuron* **9**: 967–975.

Muller W, Connor JA. (1991) Dendritic spines as individual compartments for synaptic calcium responses. *Nature* **354**: 73–76.

Muller D, Joly M, Lynch G. (1988) Contributions of quisqualate and NMDA receptors to the induction and expression of LTP. *Science* **242**: 1694–1697.

Muller D, Arai A, Lynch G. (1992) Factors governing the potentiation of the NMDA receptor-mediated responses in hippocampus. *Hippocampus* **2**: 29–38.

Nicoll RA, Kauer JA, Malenka RC. (1988) The current excitement in long-term potentiation. *Neuron* **1**: 97–103.

Obenhaus A, Mody I, Baimbridge KG. (1989) Dantrolene-Na (Dantrium) blocks induction of long-term potentiation in hippocampal slices. *Neurosci. Lett.* **98**: 172–178.

O'Connor J, Rowan MJ, Anwyl R. (1994) Long-lasting enhancement of NMDA receptor-mediated synaptic transmission by metabotropic glutamate receptor activation. *Nature* **367**: 557–559.

O'Connor J, Rowan MJ, Anwyl R. (1995) Tetanically induced LTP involves a similar increase in the AMPA and NMDA receptor components of the excitatory postsynaptic current – investigations of the involvement of mGlu receptors. *J. Neurosci.* **15**: 2013–2020.

O'Dell TJ, Kandel ER. (1994) Low frequency stimulation erases LTP through an NMDA receptor-mediated activation of protein phosphatases. *Learn. Mem.* **1**: 129–139.

Oliver MW, Kessler M, Larson J, Schottler F, Lynch G. (1990) Glycine site associated with the NMDA receptor modulates long-term potentiation. *Synapse* **5**: 265–270.

Olpe H, Karlsson, G. (1990) The effects of baclofen and two $GABA_B$ antagonists on long-term potentiation. *Naunyn-Schmiedeberg's Arch. Pharmacol.* **342**: 194–197.

O'Mara SM, Rowan MJ, Anwyl R. (1995) Metabotropic glutamate receptor-induced long-term depression and depotentiation in the dentate gyrus of the rat hippocampus *in vitro*. *Neuropharmacology* **34**: 983–989.

Pennartz CMA, Ameeran RF, Groenewegen HJ, Lopez da Silva FH. (1993) Synaptic plasticity in an *in vitro* slice preparation of the rat nucleus accumbens. *Eur. J. Neurosci.* **5**: 107–117.

Perkel DJ, Nicoll RA. (1993) Evidence for all or none regulation of neurotransmitter release: implication for long-term potentiation. *J. Physiol.* **471**: 481–500.

Reymann KG, Matthies H. (1989) 2-Amino-4-phosphonobutyrate selectively eliminates late phase of long-term potentiation in rat hippocampus. *Neurosci. Lett.* **98**: 166–171.

Reymann KG, Davies SN, Matthies H, Kase H, Collingridge GL. (1990) Activation of a K252l-sensitive protein kinase is necessary for a postsynaptic phase of long-term potentiation in area CA1 of rat hippocampus. *Eur. J. Neurosci.* **2**: 481–486.

Richter-Levin G, Errington ML, Maegawa H, Bliss TVP. (1993) Activation of metabotropic glutamate receptors is necessary for long-term potentiation in the dentate gyrus and for spatial learning. *Neuropharmacology* **33**: 853–857.

Riedel G, Reymann KG. (1993) An antagonist of the metabotropic glutamate receptor prevents LTP in the dentate gyrus of freely moving rats. *Neuropharmacology* **32**: 929–931.

Sakurai M. (1990) Calcium is an intracellular mediator of the climbing fiber in induction of cerebellar long-term potentiation. *Proc. Natl Acad. Sci. USA* **87**: 3383–3385.

Schwartswelder HS, Ferrari C, Anderson WW, Wilson WA. (1988) The drug MK-801 attenuates the development, but not the expression, of long-term potentiation and stimulus train-induced bursting in hippocampal slices. *Neuropharmacology* **28**: 441–445.

Stanton PK, Chattarji S, Sejnowski TJ. (1991) 2-Amino-3-phosphoproprionic acid, an inhibitor of glutamate stimulated phosphoinositide turnover, blocks induction of homosynaptic long-term depression, but not potentiation, in rat hippocampus. *Neurosci. Lett.* **127**: 61–66.

Staubli U, Lynch G. (1990) Stable depression of potentiated synaptic responses in the hippocampus with 1–5 Hz stimulation. *Brain Res.* **513**: 113–121.

Stelzer A, Simon G, Kovacs G, Rabindra R. (1994) Synaptic disinhibition during maintenance of long-term potentiation in the CA1 hippocampal subfield. *Proc. Natl Acad. Sci. USA* **91**: 3058–3062.

Stringer JL, Guyenet PG. (1983) Elimination of long-term potentiation in the hippocampus by phencyclidine and ketamine. *Brain Res.* **258**: 159–164.

Stringer JL, Greenfield LJ, Hackett, JT, Guyenet, PG. (1983a) Blockade of long-term potentiation by phencyclidene and σ-opiates in the hippocampus *in vitro* and *in vivo*. *Brain Res.* **280**: 127–138.

Stringer JL, Hackett JT, Guyenet PG. (1983b) Long-term potentiation blocked by phencyclidine and cyclazocine *in vitro*. *Eur. J. Pharmacol.* **98**: 381–388.

Thibault O, Joly M, Muller D, Schottler F, Dudek S, Lynch G. (1989) Long-lasting physiological effects of bath applied N-methyl-D-aspartate. *Brain Res.* **476**: 170–173.

Thiels E, Weisz DJ, Berger TW. (1992) *In vivo* modulation of N-methyl-D-aspartate receptor-dependent long-term potentiation by the glycine modulatory site. *Neuroscience* **46**: 501–509.

Tsumoto T. (1992) Long-term potentiation and long-term depression in the neocortex. *Prog. Neurobiol.* **39**: 209–228.

Turner RW, Baimbridge KG, Miller JJ. (1982) Calcium induced long-term potentiation in the hippocampus. *Neuroscience* **7**: 1411–1416.

Walsh JP. (1993) Depression of excitatory synaptic input in rat striatal neurons. *Brain Res.* **608**: 123–128.

Wang L-Y, Salter MW, McDonald JF. (1991) Regulation of kainate receptors by cAMP-dependent protein kinase and phosphatases. *Science* **253**: 1132–1135.

Wang Y, Rowan MJ, Anwyl R. (1995) (RS)-α-Methyl-4-carboxyphenylglycine inhibits long-term potentiation only following the application of low frequency stimulation in the rat dentate gyrus *in vitro*. *Neurosci. Lett.* **197**: 207–210.

Wigstrom H, Gustaffson B. (1983) Facilitated induction of hippocampal long-lasting potentiation during the blockade of inhibition. *Nature* **301**: 603–604.

Wigstrom H, Gustaffson B. (1984) A possible correlate of the postsynaptic condition for long-lasting potentiation in the guinea-pig hippocampus. *Neurosci. Lett.* **44**: 327–332

Williams S, Johnston D. (1988) Muscarinic depression of long-term potentiation in CA3 hippocampal neurons. *Science* **242**: 84–87.

Xie X, Berger TW, Barrionuevo, G. (1992) Isolated NMDA receptor-mediated synaptic responses express both LTP and LTD. *J. Neurophysiol.* **67**: 1009–1013.

Yang XD, Connor JA, Faber DS. (1994) Weak excitation and simultaneous inhibition induce long-term depression in hippocampal CA1 neurons. *J. Neurophysiol.* **71**: 1586–1590.

Yi P-L, Chang F-G, Tsai J-J, Hung C-H, Gean P-W. (1995) The involvement of metabotropic glutamate receptors in long-term depression of N-methyl-D-aspartate receptor-mediated synaptic potential in the rat hippocampus. *Neurosci. Lett.* **185**: 207–210.

Zalutsky RA, Nicoll RA. (1990) Mossy fiber long-term potentiation shows specificity but no apparent cooperativity. *Science* **248**: 1619–1624.

Zhang DX, Levy WB. (1992) Ketamine blocks the induction of LTP at the lateral entorhinal cortex-dentate gyrus synapses. *Brain Res.* **593**: 124–127.

Zheng F, Gallagher JP. (1992a) Metabotropic glutamate receptor agonists potentiate a slow afterdepolarization in CNS neurons. *NeuroReport* **3**: 622–624.

Zheng F, Gallagher JP. (1992b) Metabotropic glutamate receptors are required for the induction of long-term potentiation. *Neuron* **9**: 163–172.

Zilberter YI, Uteshev V, Sokolova SN, Motin LG, Eremjan HH. (1990) Potentiation of glutamate-activated currents in isolated hippocampal neurons. *Neuron* **5**: 597–602.

3

Second messengers in LTP and LTD

Erik D. Roberson, Joey D. English and J. David Sweatt

3.1 Introduction

In this chapter, we will examine the roles of various second messenger systems in the induction and expression of long-term potentiation (LTP) and long-term depression (LTD). A few amino acid receptors are capable of modulating synaptic transmission directly, for example by affecting potassium or calcium channel function via a G-protein intermediate. However, the vast majority of the effects of amino acid receptors result from the production of intracellular second messengers. Thus, downstream of amino acid receptor stimulation, dissecting the molecular mechanisms of LTP and LTD becomes a matter of second messengers and their effectors.

Our appreciation of the tremendous complexity of intracellular signaling pathways has increased at least as fast as our understanding of these systems (*Figure 3.1*). There are many examples of a second messenger diverging to act on multiple effectors and of several second messengers converging to act on a single effector. Thus, to facilitate a more mechanistic treatment of their roles, our discussion will, for the most part, be organized around the messengers' effectors, that is the kinases, proteases and phosphatases which they activate. In cases where the effectors are poorly understood, such as the putative retrograde messengers, we will focus on the messengers themselves. In each section, we will first briefly examine the biochemical properties and regulation of the system, then discuss its role in LTP or LTD in area CA1 of the hippocampus, where the biochemistry of the phenomena has been most thoroughly studied.

Special note should be made at the outset of the pervasive role of the calcium ion. One of the few points of consensus in the fields of LTP and LTD is that calcium plays a central role in triggering the induction of these processes. That being the case, a chapter on second messengers in LTP and LTD is essentially a chapter on calcium, even when the focus is on additional messengers. Calcium is perhaps the best example of a divergent second messenger able to activate multiple downstream

---

Cortical Plasticity, edited by M.S. Fazeli and G.L. Collingridge.
© 1996 BIOS Scientific Publishers Ltd, Oxford.

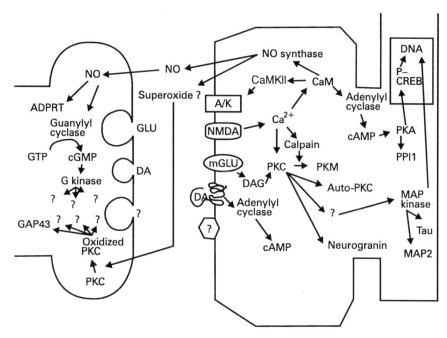

Figure 3.1. Overview of biochemical mechanisms contributing to the induction and expression of LTP in area CA1 of hippocampus. A number of biochemical mechanisms likely to be involved in NMDA receptor-dependent LTP in area CA1 of rat hippocampus are included. Although in this figure mechanisms are assigned solely to either the pre- or postsynaptic compartment, the majority of the mechanisms could contribute to LTP through actions in either locus.

effectors, and in fact many targets of calcium are enzymes which catalyze the production of other second messengers, for example the calcium/calmodulin-dependent form of adenylyl cyclase, which makes cAMP. Thus, it should be understood that calcium is likely to play a prominent role in the production or action of many, if not all, of the second messenger systems discussed.

3.2 Kinases

One of the major advances in our understanding of the molecular mechanisms of LTP was the discovery by Malinow *et al.* in 1988 that protein kinase inhibitors block both its induction and expression. Protein kinases, enzymes which covalently attach phosphate groups on to the side chains of serine, threonine and tyrosine, are attractive candidates for roles in LTP, because phosphorylation represents an effective but reversible means for modulating protein function and because many protein kinases are regulated by second messengers. In this section, we will examine the roles of two calcium-activated kinases, protein kinase C (PKC) and calcium/calmodulin-dependent protein kinase II (CaMKII), the cAMP-dependent protein kinase, members of the tyrosine kinase family, and a possible role for the p42 mitogen-activated protein kinase (MAPK).

3.2.1 Calcium-activated protein kinases

One major effect of the calcium influx necessary for LTP is the activation of calcium-dependent protein kinases, especially PKC and CaMKII. Both PKC and CaMKII are activated transiently by second messengers and also can enter an autonomously activated state independent of second messengers. The role of these calcium-regulated protein kinases in both LTP induction and maintenance has been studied using a variety of techniques, and the evidence for the involvement of these protein kinases in the LTP generally is of three types: effects of inhibitors, effects of activators and direct measurement of enzymatic activity.

Mechanisms of transient activation. The 'classical' PKC isozymes (cPKCs) (Nishizuka, 1992) are designated α, βI/βII and γ. The physiologically relevant second messenger activators of these cPKC isozymes have been studied extensively and are similar. cPKCs require calcium and phospholipid, especially phosphatidylserine (PS), for their activation (*Figure 3.2*). The affinity of cPKCs for calcium is greatly increased by the presence of diacylglycerols (DAGs) such that cPKCs can be activated by DAGs without increased calcium levels. Tumor-promoting phorbol esters mimic the action of DAGs and activate cPKCs at physiological calcium levels. Calcium binding causes translocation of PKC to the plasma membrane as a step in its activation. Transient translocation, however, does not necessarily imply activation.

The cPKCs consist of a regulatory and a catalytic domain within one subunit. The regulatory domain binds phorbol esters, DAG and calcium, and contains a pseudosubstrate domain which is thought to bind the catalytic site and inhibit PKC activity in the absence of second messenger activators. The catalytic domain contains the active site for phosphotransferase activity.

The remaining PKC isozymes known to occur in the nervous system are the 'new' PKCs (nPKCs), PKC δ and ε, and an 'atypical' PKC (aPKC), PKC ζ. These isoforms do not require calcium for their activation. PKC δ and ε are activated by PS and DAG or phorbol esters in the absence of calcium. PKC ζ is active in the absence of calcium, phospholipid and DAG, but is stimulated somewhat by PS.

The CaMKII enzyme is a heteromultimer comprised of two types of individual subunits, 10–12 of which together form one single CaMKII holoenzyme. The two types of subunit are designated α and β, and can combine in various ratios to give active CaMKII molecules. Each single-subunit molecule contains both an active site and a calmodulin-binding site. The enzyme is activated by calcium through the action of the calcium-binding protein calmodulin, which when occupied by calcium binds to a CaMKII subunit and elicits exposure of the catalytic domain of the enzyme (*Figure 3.2*). Once calcium returns to resting levels, calmodulin dissociates and enzymatic activity returns to basal levels.

PKC and CaMKII in LTP induction. Because of the transient nature of second messenger-dependent activation of PKC and CaMKII, this type of activation is not considered likely to play a role in the sustained expression of LTP. Rather, work has focused on the possibility that transient activation plays a role in the induction of LTP.

Kinase inhibitor studies. A variety of cell-permeable protein kinase inhibitors have been used to implicate a role for protein kinases in LTP induction. Several inhibitors known to block protein kinase C activity *in vitro* block the induction of LTP while leaving a short-term, decremental potentiation lasting from 20 to 90 min (*Figure 3.4b*). These include mellitin (Lovinger *et al.*, 1987), polymyxin B (Lovinger *et al.*, 1987), H-7 (Colley *et al.*, 1990; Malinow *et al.*, 1988), k-252a (Matthies *et al.*,

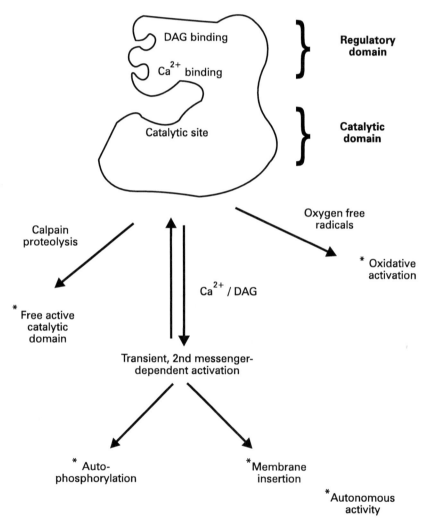

Figure 3.2. Mechanisms of PKC activation in LTP. Mechanisms for the generation of transiently or autonomously activated PKC are highlighted. See text for a detailed description.

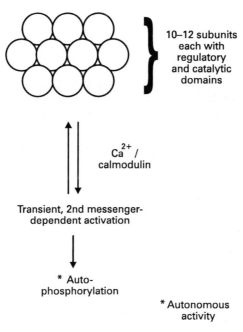

Figure 3.3. Mechanisms for CaMKII activation in LTP. Mechanisms for the generation of transiently or autonomously activated CaMKII are highlighted. See text for a detailed description.

1991), k-252b (Reymann et al., 1990), sphingosine (Malinow et al., 1988) and staurosporine (Denny et al., 1990b). The interpretation of these experiments is limited, because all of these inhibitors block kinases other than PKC and may have nonspecific effects on other enzymes or ion channels.

Cell-permeable inhibitors of CaMKII can also inhibit LTP induction. KN-62, a CaMKII inhibitor, blocks LTP induction in a manner similar to PKC inhibitors (Ito et al., 1991). Also, inhibitors of calmodulin such as trifluoperazine and calmidazolium (Malenka et al., 1989) can block LTP induction. While these agents can inhibit other processes, they generally implicate CaMKII or calmodulin-dependent processes in LTP induction.

More convincing evidence for involvement of PKC and CaMKII in LTP induction has come from studies using inhibitor peptides. Postsynaptic injection of the PKC inhibitor peptide PKC_{19-31} blocks induction of LTP (Malinow et al., 1989; Wang and Feng, 1992), as does the CaMKII inhibitor peptide $CaMKII_{273-302}$ (Malinow et al., 1989). In these experiments, modified peptides, which are much less potent inhibitors of their respective kinases, do not block LTP (Malinow et al., 1989). Two calmodulin-binding peptides, CBP and CBP3, also block LTP induction when injected postsynaptically (Malenka et al., 1989).

The use of genetically altered mice lacking certain protein kinases has provided another means to study kinase involvement in LTP that is analogous to kinase inhibition. Mice lacking the γ isoform of PKC exhibit deficits in LTP induction, though these deficits can be overcome by conditioning stimuli (Abeliovich et al.,

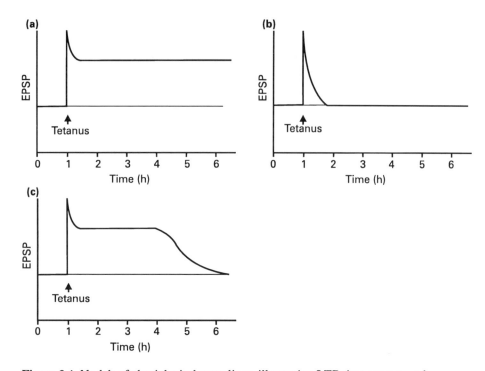

Figure 3.4. Models of physiological recordings, illustrating LTP time courses under various conditions. (a) Diagram of control physiological data typically obtained from extracellular or intracellular recordings from hippocampal slices subjected to LTP-inducing tetanic stimulation. (b) Illustration of the short-lasting potentiation observed when tetanic stimulation is delivered to slices preincubated with protein kinase inhibitors. (c) Illustration of the blockade of late-phase LTP observed when protein synthesis or PKA activity is blocked before delivery of tetanic stimulation.

1993). Also, mice lacking the α-subunit of CaMKII are deficient in LTP, though LTP is not eliminated in all cases (Silva *et al.*, 1992). The ability to induce LTP under some conditions in the PKCγ and α-CaMKII knockout mice, however, does not rule out contributions from other isoforms of these kinases. Although these mice exhibited apparently normal synaptic transmission, the possibility of subtle effects of genetic alterations on neuronal development confounds the interpretation of these experiments.

Kinase activator studies. The tumor-promoting phorbol esters are potent activators of PKC (Castagna *et al.*, 1982). Application of active phorbol esters causes a robust increase in synaptic transmission in the hippocampus (Malenka *et al.*, 1986). Activation of PKC with phorbol esters, however, does not result in long-lasting potentiation, as synaptic responses return to baseline following washout of the drug (Muller *et al.*, 1988). The increase in synaptic transmission caused by PKC activation appears to be due to an increase in presynaptic neurotransmitter release (Malenka *et al.*, 1986, 1987). Injection of active PKC into postsynaptic neurons also results in synaptic potentiation (Hu *et al.*, 1987). These experiments indicate that increased PKC activity either pre- or postsynaptically can cause synaptic potentiation, though a transient increase in PKC activity is not sufficient to induce lasting LTP.

Direct assays of kinase activity. Several studies have indicated that both PKC and CaMKII are activated in response to LTP-inducing stimulation. Klann *et al.* (1993) demonstrated PKC activation in LTP induction using assays of kinase activity in tissue extracts. In addition, a role for PKC activity in LTP induction has also been implicated in studies using antibodies that recognize various PKC isoforms. Sacktor *et al.* (1993) have demonstrated a transient translocation of several PKC isoforms to membrane fractions immediately following tetanic stimulation, suggesting that transient second messenger-stimulated PKC activity occurs during LTP induction. Fukunaga *et al.* (1993) have shown that CaMKII is also activated in LTP induction; they observed an increase in both second messenger-independent and second messenger-activated CaMKII activity 5 min following tetanic stimulation.

Autonomous activation. In LTP, a brief period of stimulation elicits a long-lasting change in neuronal function. Therefore, it is important to understand at the biochemical level how a transient signal can be translated into a persistent effect in neurons. None of the second messengers associated with LTP are believed to be produced in a sustained manner. However, with many kinases, transient activation by a second messenger can lead to generation of a novel form of the kinase that is subsequently active independent of the continued presence of the second messenger. The idea of persistent activation of protein kinases is appealing in the context of a mechanism for the maintenance of LTP, because transient activation of a cell surface receptor, with a corresponding short-lived increase in a second messenger, can lead to long-term changes in effector enzyme activity and thus perhaps to long-term changes in synaptic efficacy (Lisman, 1985; Schwartz, 1993).

Both CaMKII and PKC have been implicated in the maintenance of LTP. We will review evidence that autonomously active CaMKII and PKC contribute to the maintenance of LTP in an early phase (E-LTP) lasting over the first few hours after induction.

Autonomous activation of PKC. Several lines of evidence indicate a role for PKC activation in the maintenance of E-LTP in the hippocampus: PKC inhibitors can reverse E-LTP; both presynaptic and postsynaptic PKC activation lead to increased synaptic efficacy; and PKC can be persistently activated and this autonomous activation is known to occur during E-LTP. In this section, we will briefly overview mechanisms for the generation of autonomously active PKC, and then discuss the data implicating autonomously active PKC in E-LTP.

As discussed above, PKC activation in response to typical second messenger activation is transient, lasting only as long as the messenger is present. However, several scenarios have now been identified that can result in an increase in PKC activity that is independent of typical second messenger activators (*Figure 3.2*):

(i) The best characterized mechanism for generating an increase in autonomous PKC activity is proteolytic activation (Inoue *et al.*, 1978). Proteolytic activation of PKC occurs via cleavage at specific sites between the regulatory domain and catalytic domain, resulting in a freed 45–50 kDa catalytic fragment (known as PKM) which is active in the absence of calcium and phospholipid. Calcium treatment of hippocampal homogenate causes an increase in autonomous PKC activity, an effect which is blocked by inhibiting calcium-activated proteases (Sessoms *et al.*, 1992–1993) (see Section 3.3).

(ii) Autophosphorylation of PKC in the presence of calcium and phospholipid is known to occur (Mochly and Koshland, 1987). Autophosphorylation of PKC lowers its K_m for calcium in a manner analogous to the effect of DAG, and a subtle increase in activity occurs even in the absence of calcium. Autophosphorylated PKC also has a higher affinity for phorbol ester binding. At the present time, it remains to be determined directly if autophosphorylation of PKC can lead to its autonomous activation.

(iii) In the presence of phospholipid vesicles, calcium induces PKC binding to membranes and causes a subpopulation of PKC to become inserted into the membrane (Bazzi and Nelsestuen, 1988). The 'membrane-inserted' PKC is active in the absence of calcium, DAG, phorbol esters and additional phospholipids, and is not readily dissociated from the membrane by washing with EGTA (Bazzi and Nelsestuen, 1988), which distinguishes this type of activation from the membrane association observed with transient activation.

(iv) Under certain conditions, an increase in calcium- and phospholipid-independent PKC activity can be achieved by oxidation of PKC (Gopalakrishna and Anderson 1989). Although oxidation of PKC by H_2O_2 can lead to oxidative inactivation of PKC, mild oxidation in the presence of ATP can lead to generation of calcium- and phospholipid-independent PKC (Gopalakrishna and Anderson, 1989). Oxidation-induced persistent PKC activation has been demonstrated in both purified PKC preparations (Gopalakrishna and Anderson, 1989) and in hippocampal homogenates (Palumbo *et al.*, 1992). The physiological relevance of oxidative modification of PKC and the stability of oxidatively modified PKC are unknown. Oxygen free radicals such as nitric oxide, however, are generated in the central nervous system and act as second messengers. The possibility that transient oxygen free radical production may activate PKC is an attractive one.

Kinase inhibitor studies. Inhibitor studies have provided evidence that continued, autonomous activity of protein kinases is in fact required for E-LTP. The kinase inhibitor H-7, which blocks catalytic activity of PKC by competitive inhibition with ATP, reversibly blocks the expression of E-LTP (Hidaka and Kobayashi, 1992; Malinow *et al.*, 1988). Another ATP-competitive kinase inhibitor, staurosporine, also decreases established LTP (Matthies *et al.*, 1991), although a contradictory report has been published (Denny *et al.*, 1990b). On the other hand, the PKC inhibitor sphingosine, which competes not with ATP but with second messenger activators (Hidaka and Kobayashi, 1992), blocks only LTP induction but not the expression of E-LTP (Malinow *et al.*, 1988). These reports are consistent with the hypothesis that LTP maintenance depends on continued, activator-independent kinase activity. Injection of the selective PKC inhibitor peptide PKC_{19-31} immediately following LTP induction blocks expression of LTP (Wang and Feng, 1992). Postsynaptic injection of PKC_{19-31} plus the PKC inhibitor polymyxin B either 75 min or 3 h following LTP induction also blocks LTP expression (Wang and Feng, 1992). These studies support the hypothesis that the continued activity of postsynaptic PKC is required for E-LTP. Another report implicates presynaptic protein kinase activity in E-LTP maintenance (Malinow *et al.*, 1989).

Kinase substrate studies. The phosphorylation state of one identified PKC substrate, GAP43 (F1, B50), is altered in *post hoc* biochemical assays in association with LTP (Akers and Routtenberg, 1985; Akers *et al.*, 1986; Lovinger *et al.*, 1985; Nelson *et al.*, 1989). Direct labeling of hippocampal slices *in situ* with radioactive inorganic phosphate has confirmed that the phosphorylation of GAP43 is increased for at least 1 h following LTP induction (Gianotti *et al.*, 1992). Also, increased GAP43 phosphorylation associated with LTP has been demonstrated using antibodies that recognize GAP43 phosphorylated at the PKC phosphorylation site (Leahy *et al.*, 1993). GAP43 is a particularly interesting PKC substrate as it is nervous tissue-specific (Aloyo *et al.*, 1982), located predominantly in the presynaptic terminal (van Lookeren Campagne *et al.*, 1989; van Hooff *et al.*, 1986) and its phosphorylation is correlated with neurotransmitter release (Dekker *et al.*, 1989a,b).

The phosphorylation of another PKC substrate has also been demonstrated. A 17-kDa PKC substrate that resembles neurogranin (RC3) protein (Baudier *et al.*, 1989, 1991; Represa *et al.*, 1990; Watson *et al.*, 1990) shows increased phosphorylation in *post hoc* and back-phosphorylation assays in association with LTP (Chen *et al.*, 1993; Klann *et al.*, 1992). Neurogranin is structurally and functionally similar to GAP43 but, in contrast to GAP43, it is a predominantly postsynaptic PKC substrate (Baudier *et al.*, 1991).

These data demonstrating increased PKC substrate phosphorylation during E-LTP are consistent with the hypothesis of persistent, autonomous PKC activation. An alternative interpretation is that a transient activation of PKC leads to a stable phosphorylation of GAP43 and neurogranin. This possibility has been ruled out in the case of neurogranin, as the increased phosphorylation is not maintained when LTP induction is followed by application of a kinase inhibitor (Chen *et al.*, 1993).

Kinase assay studies. Several recent studies have measured the activity of protein kinases directly in *post-hoc* assays following the induction of LTP. In the first such study, Klann *et al.* have shown that a lasting increase in autonomous PKC activity is associated with LTP (Klann *et al.*, 1991). This finding supports previous studies using kinase inhibitors to show that PKC activity is necessary for LTP maintenance. The mechanism of the persistent PKC activation during E-LTP appears to involve increased phosphorylation of PKC (Klann *et al.*, 1993; Schwartz, 1993).

A role for PKC activity in E-LTP has also been implicated in studies using antibodies that recognize various PKC isoforms. Sacktor *et al.* (1993) have demonstrated that an increase in the active, catalytic fragment of PKC isoform ζ (PKM ζ) is associated with LTP 30 min following LTP induction. This increase in PKM ζ is interpreted as an increase in second messenger-independent PKC activity because this fragment can autophosphorylate in the absence of PKC activators (Sacktor *et al.*, 1993). In addition, Powell *et al.* (1994) have demonstrated proteolytic activation of PKC in N-methyl-D-aspartate (NMDA) receptor-independent LTP.

Autonomous activation of CaMKII. The data implicating a role for CaMKII in LTP have been reviewed recently by Lisman (1994), so we will provide only a cursory overview of this area.

CaMKII, upon activation by calcium and calmodulin, can autophosphorylate by an intersubunit, intraholoenzyme reaction. The autophosphorylated enzyme is autonomously active, that is it is active independently of the continued presence of calcium and calmodulin (*Figure 3.3*). This mechanism sets up a self-perpetuating CaMKII activation that is independent of subunit turnover and can theoretically last the lifetime of the neuron. Interestingly, it has been hypothesized that this mechanism in and of itself may be sufficient for LTP, independent of any changes in gene expression (Lisman, 1985).

Direct assays of CaMKII activity have implicated a role for its persistent activation in LTP (Fukunaga *et al.*, 1993). An increase in both second messenger-independent and second messenger-activated CaMKII activity occurs 5 min following LTP induction and lasts for at least 60 min into LTP maintenance (Fukunaga *et al.*, 1993). These results suggest that CaMKII may also play a role in the expression of E-LTP.

3.2.2 Cyclic AMP-dependent protein kinase

The cAMP system. cAMP is the original second messenger, discovered by Earl Sutherland in his studies of glycogen metabolism. The formation of cAMP from ATP is catalyzed by adenylyl cyclase. There are many isoforms of adenylyl cyclase, all of which are stimulated by G-proteins. An adenylyl cyclase isoform which is also stimulated synergistically by calcium and calmodulin is prominently expressed in the hippocampus (Xia *et al.*, 1991).

The main effector of cAMP is the cAMP-dependent protein kinase, PKA. The PKA holoenzyme is a heterotetramer comprised of two catalytic (C) subunits, which have phosphotransferase activity, and two regulatory (R) subunits, which have cAMP binding activity and an autoinhibitory domain which blocks C subunit activity. In the absence of cAMP, PKA exists in the inactive holoenzyme state. Binding cAMP induces a conformational change in the R subunit and causes C subunit to be released. The free C subunit is enzymatically active.

PKA and LTP. Chetkovich *et al.* (1991) discovered that levels of cAMP increase immediately following an LTP-inducing stimulus. The production of cAMP occurs downstream of the calcium trigger for LTP, as it depends both on extracellular calcium and activation of NMDA receptors (Chetkovich *et al.*, 1991). Calcium stimulates production of cAMP in conjunction with calmodulin by activating the calmodulin-sensitive form of adenylyl cyclase found in the hippocampus (Chetkovich and Sweatt, 1993). Dopamine receptor-stimulated G-proteins may also contribute, as antagonists of these receptors block the production of cAMP in LTP (Frey *et al.*, 1993). The rise in cAMP levels is transient, returning to baseline after 10–20 min (Chetkovich and Sweatt, 1993).

cAMP made immediately after the induction of LTP activates PKA (Roberson and Sweatt, 1993). Interestingly, though, the activity of PKA is not required to support LTP during the initial phases of LTP; application of PKA inhibitors during the first hour after tetanization interferes only with a late phase of LTP (L-LTP) which develops after 3–4 h (*Figure 3.4c*) (Frey *et al.*, 1993; Matthies and Reymann, 1993).

There are as yet no data to support the idea the PKA is persistently activated in an autonomous form like PKC and CaMKII. So what happens between PKA activation early in LTP and the beginning of L-LTP? The L-LTP phase is also blocked by inhibitors of transcription and translation, suggesting that perhaps this phase of LTP depends on the synthesis of a new gene product (Nguyen *et al.*, 1994). Thus, a link between PKA and changes in gene expression has been sought. In many cell types, PKA induces transcription of genes by activating a transcription factor which interacts with the cAMP response elements (CREs) found upstream of certain genes (Montminy and Bilezikjian, 1987). This factor, the CRE-binding protein (CREB), has a single site for phosphorylation by PKA and, when phosphorylated, it stimulates transcription of the downstream gene (Yamamoto *et al.*, 1988). It is hypothesized that PKA phosphorylates CREB during the early stages of LTP, initiating a cascade of changes in gene expression which eventually produces the L-LTP. This hypothesis has not been tested directly, but transgenic mice lacking CREB have abnormal LTP and deficiencies in long-term learning tasks (Bourtchuladze *et al.*, 1994).

3.2.3 Protein tyrosine kinases

Neuronal tyrosine kinases. Two general families of protein tyrosine kinases (PTKs) have been described (Cantley *et al.*, 1991). One is a family of receptors for cellular growth factors such as platelet-derived growth factor (PDGF), nerve growth factor (NGF) and brain-derived neurotrophic factor (BDNF). These transmembrane receptors contain a cytosolic PTK domain which is activated upon growth factor binding. The second is a family of cytosolic, nonreceptor kinases, with the best characterized members being the *src*-related PTKs. Though widely studied, the regulatory mechanisms of this PTK family remain poorly understood.

Both PTK families classically have been considered as important regulators of cellular proliferation and differentiation in mitotic cells (Cantley et al., 1991). It is therefore interesting to note that many PTKs are expressed at high levels in postmitotic neuronal cells and that PTK activity is observable in such cells (Cotton and Brugge, 1983; Ellis et al., 1988). In addition, many neuronal proteins have been identified as PTK substrates; these include synaptic vesicle proteins such as synaptophysin (Pang et al., 1988), neurotransmitter receptors such as the nicotinic acetylcholine receptor (Huganir et al., 1984), glutamate receptors (Moss et al., 1993), and voltage-gated potassium channels (Huang et al., 1993). These observations suggest that PTKs expressed in postmitotic neurons may have novel roles in regulating synaptic plasticity.

LTP studies.

Kinase inhibitor studies. Perhaps the best evidence for involvement of PTKs in LTP induction comes from studies with membrane-permeant PTK inhibitors. O'Dell et al. (1991b) have demonstrated that general PTK inhibitors such as genestein and lavendustin A block LTP induction but do not affect maintenance of established LTP. Interpretation of these studies, however, is limited. As all members of these PTK families share a highly conserved catalytic domain, such general inhibitors have broad spectrum effects. Thus, these studies cannot identify which PTK family or particular PTK contributes to LTP induction.

One approach to studying the involvement of particular PTKs in LTP is the use of transgenic mice. This variation of the 'kinase inhibitor study' has been employed to investigate the role of several *src*-related, nonreceptor cytosolic PTKs. In particular, LTP has been studied in mice containing a selective knockout of either p59 *fyn*, p60 *src*, p56 *lyn*, or p123 *yes*. Analysis of LTP in these transgenic animals revealed that only *fyn* mutant animals are deficient in LTP induction (Grant et al., 1992). Specifically, the threshold for induction is increased and, in many cases, LTP is absent altogether. These results, however, should be interpreted cautiously. As PTKs are involved in cellular determination and proliferation, animals deficient in certain PTKs may have subtle developmental anomalies which affect the establishment of properly functioning neurons. In fact, histological examination of the hippocampus in *fyn* mutant mice revealed that anatomical defects are present. Furthermore, recent evidence suggests that *fyn* kinase is involved in axonal myelination (Umemori et al., 1994). Thus, the deficiency in LTP in the *fyn* mutant mice might not involve mechanisms underlying normal LTP induction.

Kinase activator studies. Recent work has demonstrated that neuronal growth factor receptors might play a role in synaptic plasticity. Application of BDNF or NT-3, ligands of neuronal *trk* receptors, leads to long-lasting increase in synaptic strength (Kang and Schuman, 1995). Although the biochemical mechanisms by which synaptic transmission is enhanced are unknown, the initial work suggests that a presynaptic enhancement of neurotransmitter release might be involved. In addition, it is unclear at present how growth factor-induced potentiation relates to LTP, and thus whether or not the activation of these receptor PTKs contributes to tetanus-induced LTP.

3.2.4 Mitogen-activated protein kinases

Another intriguing yet poorly understood kinase family with potentially important neuronal functions is the mitogen-activated protein kinase (MAPK) family. Like protein tyrosine kinases, these proline-directed serine/threonine kinases have been studied classically as mediators of cellular proliferation in mitotic cells, yet are expressed at high levels in postmitotic neuronal cells (Crews and Erikson, 1993; Fiore et al., 1993b). Interestingly, MAPKs have been identified as components of two main signaling cascades: one initiated by activation of receptor PTKs and another initiated by activation of PKC. As described in Sections 3.2.1 and 3.2.3, both of these systems are involved in the induction of LTP.

MAPKs are expressed in pyramidal neurons and have been localized to the soma and dendritic tree, suggesting a role in postsynaptic function (Fiore et al., 1993b). Also, MAPKs are activated via covalent modification, allowing for autonomous or persistent activity (Murphy et al., 1994). Finally, MAPK substrates regulate many processes thought to be involved in modulating synaptic strength. These include second messenger generation [activation of cytosolic phospholipase A_2, which preferentially liberates arachidonic acid (Clark et al., 1991; Lin et al., 1993), cytoskeleton modulation [regulation of the actin-binding properties of microtubule-associated protein 2, a dendritic-specific protein (Brugg and Matus, 1991; Cáceres et al., 1983)] and translation/transcription (reviewed by Blenis, 1993; Roberts, 1992). Taken together, the above observations suggest that MAPKs may play a role in regulating synaptic plasticity.

Several lines of evidence suggests that one MAPK in particular, the 42-kDa MAPK isoform (p42 MAPK), might be a component of the biochemical machinery supporting LTP. NMDA receptor simulation leads to p42 MAPK activation in hippocampal, but not cortical, primary neuronal cultures, while metabotropic glutamate (mGlu) receptor stimulation leads to p42 MAPK activation in cortical cultures (Bading and Greenberg, 1991; Fiore et al., 1993a). Both NMDA receptor and mGlu receptor stimulation also increase p42 MAPK activity in slices of the adult hippocampus (J.D. English and J.D. Sweatt, unpublished observations). As NMDA receptors and mGlu receptors potentially signal via PKC activation, it is interesting to note that PKC stimulation leads to p42 MAPK activation in area CA1 of the adult hippocampus (Stratton et al., 1989; J.D. English and J.D. Sweatt, unpublished observations). Overall, these data link p42 MAPK activation to NMDA receptors, mGlu receptors and PKC, all thought to be components of the biochemical machinery involved in LTP induction. The primary caveat of such experiments is that, for the most part, they have utilized primary neuronal cultures; as MAPKs are involved in cellular differentiation, it is possible that their regulation during development differs from their regulation in determined neurons.

3.3 Proteases

Proteases were one of the first major classes of second messenger effectors hypothesized to play a role in LTP (Lynch and Baudry, 1984). Because of the importance of calcium in the induction of LTP, calcium-dependent thiol proteases, known as calpains, have been studied most extensively. Two forms of calpain are expressed in brain, calpain I which is active at micromolar calcium concentrations and calpain II which is active at millimolar calcium concentrations.

Calpain inhibitors block the induction of LTP (del Cerro *et al.*, 1990b; Denny *et al.*, 1990a). Delivery of an LTP-inducing stimulus in the presence of these drugs produces potentiation which decreases to the baseline over approximately 30 min. Application of the inhibitors does not affect established LTP. These observations suggest that calpain-mediated proteolysis may be required for induction, but not the expression of E-LTP.

The calpain substrates relevant to LTP are not known, but two candidates have received particular attention. Spectrin is a cytoskeletal protein and calpain substrate whose degradation is associated with morphological changes in other cell types. Calpain-mediated cleavage of spectrin may underlie some of the structural modifications of the synapse which apparently occur during LTP (Chang and Greenough, 1984). The other likely substrate is PKC (see Section 3.2.1).

3.4 Retrograde messengers

Although it has been the subject of intense study, the location of the changes responsible for the expression of LTP has not been determined. Clearly though, the induction of LTP seems to hinge on a postsynaptic event (calcium influx), and there is evidence that at least part of the expression of LTP is presynaptic under certain conditions (see Section 3.6.1). Any presynaptic component of LTP expression must depend on a message from the postsynaptic cell signaling that LTP should be induced, and the hunt for such 'retrograde messengers' has begun. (See also Chapter 4.)

In the absence of any evidence of a retrograde form of normal synaptic transmission, interest has focused on small molecules that could be produced in the postsynaptic cell and diffuse across the synaptic cleft and into the presynaptic cell to exert their actions. In this section, we will survey the evidence supporting a retrograde messenger role for two gases, nitric oxide (NO) and carbon monoxide (CO), and two lipid molecules, arachidonic acid (AA) and platelet-activating factor (PAF).

It should be kept in mind that, while these messengers are often studied in the context of the retrograde messenger hypothesis, many of the data supporting their roles in LTP do not distinguish between their serving inter- and intracellular roles. It is reasonable to assume that whether they act on other cells or not, any of these messengers will have many effects within the cell in which they are produced.

3.4.1 Nitric oxide

Nitric oxide is a free radical produced by nitric oxide synthase (NOS) during the oxidation of arginine to citrulline. NOS is stimulated by calcium and calmodulin, which provides a means for coupling its activity to the calcium trigger for LTP. NO levels increase after LTP-inducing stimulation, an effect blocked by NOS inhibitors and removal of extracellular calcium (Chetkovich *et al.*, 1993).

Early reports indicated that NOS activity is required for LTP (Haley *et al.*, 1992; O'Dell *et al.*, 1991a; Schuman and Madison, 1991). However, it has become clear that, under many conditions, including higher temperature, stronger stimulation and in older rats, LTP induction does not depend on NOS (Chetkovich *et al.*, 1993; Haley *et al.*, 1993; Williams *et al.*, 1993). To date, transgenic mouse studies have not helped to clarify the picture, as only one of the two forms of NOS present in the hippocampus has been disrupted. In mice with mutations in neuronal NOS, LTP is normal, but is still blocked by NOS inhibitors, presumably because of the remaining endothelial NOS (O'Dell *et al.*, 1994).

The retrograde messenger hypothesis does not necessarily predict that application of the messenger itself will have any effect in the absence of synaptic activity. Rather, a retrograde messenger should be able to complement presynaptic activity which is subthreshold for LTP induction, to produce LTP. A subthreshold tetanus delivered during application of NO gas leads to LTP (Zhuo *et al.*, 1993). More interestingly, it appears that even NO released from a single pyramidal neuron, in which LTP is induced by pairing depolarization with a weak tetanus, is capable of complementing the weak tetanus to produce LTP in neighboring cells (Schuman and Madison, 1994).

It is not known how NO exerts its effects. Most interest has focused on two effectors: soluble guanylyl cyclase, which catalyzes the production of cGMP thereby activating cGMP-dependent protein kinase (PKG), and ADP-ribosyltransferase (ADPRT), which covalently modifies proteins by adding an ADP ribose moiety from NAD^+ and NADH. PKG is homologous to PKA, except that the regulatory and catalytic domains are expressed on the same polypeptide. Levels of cGMP do increase during LTP as a result of NO production (Chetkovich *et al.*, 1993), and PKG activity is required for LTP (Zhuo *et al.*, 1994). ADPRT activity may also be required for LTP (Schuman *et al.*, 1994).

3.4.2 Carbon monoxide

Carbon monoxide (CO) shares several properties with NO. Although it is not a free radical, CO is highly diffusible and is reactive with heme groups; like NO, CO can stimulate the activity of guanylyl cyclase. CO is produced by heme oxygenase, the enzyme which catalyzes the first step in heme degradation. Preventing CO production using heme oxygenase inhibitors blocks the induction of LTP (Stevens and Wang, 1993; Zhuo *et al.*, 1993). However, the relative potency of heme oxygenase inhibitors in blocking LTP appears to correlate better with their ability to block NOS than to block heme oxygenase (Meffert *et al.*, 1994). The application of CO gas paired with a subthreshold tetanus does induce LTP in an NMDA receptor-independent manner (Zhuo et al., 1993).

3.4.3 Arachidonic acid

Arachidonic acid (AA) is a 20-carbon fatty acid which commonly occupies the sn-2 position of membrane phospholipids. It is liberated from phospholipids by the action of phospholipase A_2 (PLA_2) or the combined actions of phospholipase C and diacylglycerol lipase. A number of calcium-sensitive forms of PLA_2 have been described, and phospholipase C activity is regulated by activation of mGlu receptors. AA levels, measured by push–pull perfusion techniques, increase following the induction of LTP (Lynch et al., 1991).

Preventing the production of AA with nordihydroguaiaretic acid (NDGA), an inhibitor of PLA_2, blocks the induction of LTP (Williams and Bliss, 1988; O'Dell et al., 1991a). This experiment must be interpreted with caution, however. PLA_2 activation is the first step in the production of a series of lipid metabolites, including leukotrienes, prostaglandins, thromboxanes and alkyl ethers like PAF (see Section 3.4.4), and therefore the NDGA result may reflect the dependence of LTP on these other species. Another caveat to this experiment is that NDGA also inhibits the activity of lipoxygenase, an enzyme which produces superoxide radicals as a byproduct in converting AA into leukotrienes. Superoxide itself may be required for LTP (Klann and Sweatt, 1994); thus, the fact that NDGA prevents LTP induction could reflect a necessity for AA as a substrate for superoxide generation rather than as a messenger *per se*.

AA is capable of complementing a subthreshold, weak tetanus to produce LTP (Williams et al., 1989), but its ability to do so is not independent of NMDA receptor activation (O'Dell et al., 1991a). During LTP, the action of a retrograde messenger should be downstream of NMDA receptor activation which triggers its production, so application of the retrograde messenger itself should circumvent the need for NMDA receptor stimulation. Apparently, AA acts upstream of the NMDA receptors, perhaps by augmenting neurotransmitter release through activation of PKC.

3.4.4 Platelet-activating factor

Platelet-activating factor (PAF) is an alkyl ether analog of phospatidylcholine, in which the ester bond of the sn-1 fatty acid is replaced by an ether linkage. Originally named for its ability to cause platelet aggregation, PAF is also present in the central nervous system. Unlike other retrograde messenger candidates, PAF acts through a specific cell surface receptor. The PAF receptor is expressed throughout the cortex and in the hippocampus, where it causes increases in intracellular calcium, apparently by mobilization of intracellular stores (Bito et al., 1992). Antagonists of the PAF receptor block the induction of LTP (del Cerro et al., 1990a; Kato et al., 1994), and PAF analogs can complement a subthreshold tetanus to produce LTP even in the presence of NMDA receptor antagonists (Clark et al., 1992).

3.5 Phosphatases

The majority of this chapter has focused on the second messenger systems and effectors involved in LTP. In this section, we will turn to the newer and less thoroughly studied field of LTD biochemistry. The main effectors of second messengers in LTD appear to be protein phosphatases, enzymes which reverse the action of protein kinases by removing the covalently attached phosphate group.

Multiple different phosphatases are believed to play a role in LTD. Calcineurin (phosphatase 2B) is stimulated by low levels of calcium in concert with calmodulin, and dephosphorylates a limited set of substrates. Protein phosphatases 1 and 2A (PP1 and PP2A) on the other hand have very broad substrate specificity. Neither PP1 nor PP2A is regulated directly by second messengers. However, PP1 is inhibited by so-called inhibitor protein 1 (I1) when I1 is phosphorylated by PKA. I1 is one of the few substrates of calcineurin. Thus, activation of calcineurin by calcium leads to dephosphorylation of I1, and thus to activation of PP1.

There is good evidence implicating calcineurin and PP1 or PP2A in LTD. Bath application or postsynaptic injection of calcineurin or PP1 and PP2A inhibitors block LTD induction (Mulkey *et al.*, 1993, 1994). In addition, cell-permeable PP1 and PP2A inhibitors block the expression of LTD (Mulkey *et al.*, 1993). PP1 and PP2A inhibitors also block depotentiation (i.e. reversal) of LTP (O'Dell and Kandel, 1994).

LTP and LTD, despite being opposite in sign, have several striking similarities. Both require NMDA receptor activation, postsynaptic calcium influx, calmodulin and possibly mGlu receptor activation (Dudek and Bear, 1992; Mulkey and Malenka, 1992). Given these similarities, how is synaptic depression induced? LTD is apparently induced by smaller calcium transients than those required for LTP (Mulkey and Malenka, 1992). Although the details of the model are yet to be proven, it is likely that the lower levels of calcium are sufficient to activate calcineurin and begin a phosphatase cascade without activating the kinases discussed in Section 3.2. However, when an LTP-inducing stimulus drives calcium even higher, activation of I1 by PKA causes a shutdown of the phosphatase cascade.

The concept of phosphatase involvement in LTD is interesting given the involvement of protein kinases in LTP induction and maintenance. While LTP appears to be maintained (at least in the early stages) by increased substrate phosphorylation, LTD may be a manifestation of decreased phosphorylation of protein kinase substrates. In the case of depotentiation, presumably the dephosphorylated proteins are those whose phosphorylation is increased by the kinase activation underlying LTP. As described in Section 3.2.2, both CaMKII and PKC are hypothesized to undergo persistent activation by autophosphorylation during LTP. If phosphatase activation is a trigger for depotentiation, then this mechanism may allow for a turn-off of autonomously activated kinases and a true reversal of LTP. In the case of 'pure' LTD, that is when LTP has not been induced previously, the issue is less clear: either there is ongoing phosphorylation of relevant phosphatase substrates before LTD, or LTD is in reality depotentiation of pre-existing LTP.

3.6 Application of biochemical techniques to the study of LTP and LTD

Although our understanding of the biochemical events which underlie LTP and LTD certainly remains incomplete, much progress has been made in identifying some of the fundamental biochemical components. In addition to their face value in helping our understanding of the molecular mechanisms of LTP, these insights allow for novel experimental approaches, combining our current understanding of LTP biochemistry with the techniques available in cellular and molecular biology to address more general questions about LTP. In this section, we will consider biochemical approaches by which the questions of the locus of the long-term changes which support LTP and the relationship of hippocampal LTP and LTD with learning events and memory formation in the intact animal can be examined.

3.6.1 Pre vs. post

One of the most controversial areas in LTP research concerns the locus/loci of LTP expression. Though it is generally agreed that postsynaptic calcium flux is an important trigger for LTP induction, it is unclear whether the changes that underlie the maintenance of LTP occur presynaptically and/or postsynaptically. To date, this question has been addressed primarily through electrophysiological experiments that examined presynaptic release probabilities or postsynaptic glutamate sensitivity before and after LTP induction. Such experiments have had vastly different conclusions: some argue for presynaptic changes, some for postsynaptic changes, some for both (Bekkers and Stevens, 1990; Kauer *et al.*, 1988; Manabe and Nicoll, 1994; Stevens and Wang, 1994; see Chapter 4).

Another way to approach this question experimentally is to take advantage of the current insights into the biochemical components involved in LTP. The components of particular interest are proteins which are localized exclusively to one side of the synapse. Candidates include kinase substrates such as GAP43 (presynaptic) and neurogranin (postsynaptic), synaptic vesicle proteins or other presynaptic proteins involved in transmitter release, cytoskeletal proteins such as MAP2 (postsynaptic), neurotransmitter receptors and enzymes such as p42 MAPK (postsynaptic). Biochemical evidence which demonstrates a persisting phosphorylation, modification or activation of such proteins during LTP would provide direct evidence for a persisting change occurring either pre- or postsynaptically.

Experiments designed around this rationale have provided evidence for both presynaptic and postsynaptic biochemical events which persist into the early maintenance phase of LTP. As mentioned in Section 3.2.1, several laboratories have demonstrated that the phosphorylation of GAP43, a PKC substrate localized presynaptically, increases for at least 1 h into LTP. In addition, *post hoc* and back-phosphorylation assays show that the phosphorylation state of neurogranin, a PKC substrate localized postsynaptically, is also persistently increased during LTP maintenance. Together, these observations suggest that persisting biochemical changes occur on both sides of the synapse during LTP maintenance.

3.6.2 LTP, LTD and learning and memory formation

LTP and LTD are widely studied as model systems for the cellular and molecular basis of information storage in the mammalian central nervous system (see Chapter 11). To date, however, there is little direct evidence that hippocampal synapses actually undergo LTP or LTD during learning episodes or memory formation. The question remains: what are the biochemical and cellular events which underlie learning and memory and how similar are these events to those studied in the models? In this section, we will first examine the limitations of one recent experimental approach to this question, the use of transgenic mice to determine whether inhibiting LTP is correlated with inhibiting learning and memory. We will then discuss fundamentally different molecular and biochemical approaches to determine whether what happens during LTP also happens during learning and memory.

Knockout mice. One popular approach to investigating a connection between LTP and learning and memory has been the study of transgenic mice. The overall rationale of these studies is to examine the effect on hippocampal LTP as well as the impact upon the animal's performance in a hippocampal-dependent learning task (e.g. the Morris water maze or fear conditioning). Here, a correlation between a deficit in LTP and a deficit in behavioral learning tasks would suggest a role for LTP in learning.

Unfortunately, such studies have not been without problems. The mice studied often had developmental deficits which resulted in abnormal anatomical structure or abnormal behavior, for example the *fyn* mutant (see Section 3.2.3). Furthermore, it can be argued that the physiological and behavioral responses examined in these mice simply reflects the overall compensation that the nervous system makes to the chronic absence of a particular component. That is, these responses do not necessarily give insight into the role of a particular protein in the normal response (which would require that all other components be equivalent to the normal, wild-type situation), but rather demonstrate how the nervous system responds to its persisting absence. Such responses might differ subtly or substantially from those which underlie synaptic plasticity and learning and memory in wild-type animals.

The use of regulated knockouts, that is transgenic mice designed to allow for the expression of the protein in question during development and the transient suppression of its expression for an acute experimental analysis of physiological and behavioral effects, would greatly ameliorate the interpretation of transgenic studies. Such a design would reduce the confounding problems of potential developmental abnormalities and compensational responses to chronic loss. Thus, this technique would allow for a better investigation of how physiology and behavior are affected when one putative component of the LTP machinery cannot contribute to the otherwise 'normal' cascades.

Other approaches. Most of the approaches used to link LTP and LTD with learning and memory formation to date have been inhibition studies attempting to correlate a block of LTP/LTD induction with a block of learning and memory formation. An alternative approach to such inhibitor studies is to correlate the events that occur during LTP and LTD to the events that occur during learning and memory in normal mice. This can now be done simply by designing probes for the biochemical changes which have been demonstrated to occur during LTP and LTD and then using them to determine whether such changes also occur in hippocampus as a result of an animal's learning episode/memory formation. One attractive class of probes is antibodies specific for the phosphorylated form of particular proteins. Candidates include kinases which autophosphorylate (e.g. PKC and CaMKII) and kinase substrates (e.g. neurotransmitter receptors, neurogranin, transcription factors).

It is widely assumed that changes in gene expression are required to support LTP and learning and memory formation. To date, support for this idea has come largely from studies in which transcriptional and translational processes are pharmacologically inhibited and the effect upon LTP or learning and memory formation determined. Molecular techniques now allow for an alternative approach for investigating this question. That is, it is becoming possible to generate transgenic mice which carry a 'reporter construct' for monitoring transcription factor-dependent gene transcription. Such a construct contains a transcription factor-specific recognition sequence (e.g. the CRE sequence recognized by CREB) which would regulate the expression of the gene for an easily detectable marker enzyme (e.g. β-galactosidase). Changes in gene expression initiated by a particular transcription factor could thus be monitored *in vitro* during LTP and *in situ* following a learning episode.

References

Abeliovich A, Chen C, Goda Y, Silva AJ, Stevens CF, Tonegawa S. (1993) Modified hippocampal long-term potentiation in PKC gamma-mutant mice. *Cell* **75**: 1253–1262.

Akers RF, Routtenberg A. (1985) Protein kinase C phosphorylates a 47 M_r protein (F1) directly related to synaptic plasticity. *Brain Res.* **334**: 147–151.

Akers RF, Lovinger DM, Colley PA, Linden DJ, Routtenberg A. (1986) Translocation of protein kinase C activity may mediate hippocampal long-term potentiation. *Science,* **231**: 587–589.

Aloyo YJ, Zwiers H, Gispen WH. (1982) B-50 protein kinase C in rat brain. *Prog. Brain Res.* **54**: 303–315.

Bading H, Greenberg ME. (1991) Stimulation of protein tyrosine phosphorylation by NMDA receptor activation. *Science* **253**: 912–914.

Baudier J, Bronner C, Kligman D, Cole RD. (1989) Protein kinase C substrates from bovine brain. Purification and characterization of neuromodulin, a neuron-specific calmodulin-binding protein. *J. Biol. Chem.* **264**: 1824–1828.

Baudier J, Deloume JC, Dorsselaer AV, Black D, Matthes HWD. (1991) Purification and characterization of a brain-specific protein kinase C substrate, neurogranin (p17). *J. Biol. Chem.* **266**: 229–237.

Bazzi MD, Nelsestuen GL. (1988) Constitutive activity of membrane-inserted protein kinase C. *Biochem. Biophys. Res. Commun.* **152**: 336–343.

Bekkers JM, Stevens CF. (1990) Presynaptic mechanism for long-term potentiation in the hippocampus. *Nature* **346**: 724–729.

Bito H, Nakamura M, Honda Z, Izumi T, Iwatsubo T, Seyama Y, Oguura A, Kudo Y, Shimizu T. (1992) Platelet-activating facctor (PAF) receptor in rat brain: PAF mobilizes intracellular calcium in hippocampal neurons. *Neuron* **9**: 285–294.

Blenis J. (1993) Signal transduction via MAP kinases: proceed at your own RSK. *Proc. Natl Acad. Sci. USA* **90**: 5889–5892.

Bourtchuladze R, Frenguelli B, Blendy J, Cioffi D., Schutz G, Silva AJ. (1994) Deficient long-term memory in mice with a targeted mutation of the cAMP response element binding protein. *Cell* **79**: 59–68.

Brugg B, Matus A. (1991) Phosphorylation determines the binding of microtubule-associated protein 2 (MAP2) to microtubules in living cells. *J. Cell Biol.* **114**: 735–743.

Cáceres A, Payne MR, Binder LI, Steward O. (1983) Immunocytochemical localization of actin and microtubule associated protein MAP2 in dendritic spines. *Proc. Natl Acad. Sci. USA* **80**: 1738–1742.

Campagne MvanL, Oestreicher AB, Henegouwen PM, Gispen WH. (1989) Ultrastructural and immunocytochemical localization of B-50/GAP43, a protein kinase C substrate, in isolated presynaptic nerve terminals and neuronal growth cones. *J. Neurocytol.* **18**: 479–489.

Cantley LC, Auger KR, Carpenter C, Duckworth B, Graiani A, Kapeller R, Soltoff S. (1991) Oncogenes and signal transduction. *Cell* **64**: 281–302.

Castagna M, Takai Y, Kaibuchi K, Sano K, Kikkawa U. (1982) Direct activation of calcium-activated, phospholipid-dependent protein kinase by tumor-promoting phorbol esters. *J. Biol. Chem.* **257**: 7847–7851.

Chang FLF, Greenough WT. (1984) Transient and enduring morphological correlates of synaptic activity and efficacy in the rat hippocampal slice. *Brain Res.* **309**: 35–46.

Chen SJ, Klann E, Sweatt JD. (1993) Maintenance of LTP is associated with an increase in the phosphorylation of neurogranin protein. *Soc. Neurosci. Abstr.* **19**: 1707.

Chetkovich DM, Sweatt, JD. (1993) NMDA receptor activation increases cyclic AMP in area CA1 of the hippocampus via calcium/calmodulin stimulation of adenylyl cyclase. *J. Neurochem.* **61**: 1933–1942.

Chetkovich DM, Gray R, Johnston D, Sweatt JD. (1991) *N*-methyl-D-aspartate receptor activation increases cAMP and voltage-gated calcium activity in area CA1 of the hippocampus. *Proc. Natl Acad. Sci. USA* **88**: 6467–6471.

Chetkovich DM, Klann E, Sweatt JD. (1993) Nitric oxide synthase-independent long-term potentiation in area CA1 of hippocampus. *NeuroReport* **4**: 919–922.

Clark GD, Hapel LT, Zorumski CF, Bazan NG. (1992) Enhancement of hippocampal excitatory synaptic transmission by platelet-activating factor. *Neuron* **9**: 1211–1216.

Clark JD, Lin L-L, Kriz RW, Ramesha CS, Sultzman LA, Lin AY, Milona N, Knopf JL. (1991) A novel arachidonic acid-selective cytosolic PLA2 contains a calcium-dependent translocation domain with homology to PKC and GAP. *Cell* **65**: 1043–1051.

Colley PA, Sheu F-S, Routtenberg A. (1990) Inhibition of protein kinase C blocks two components of LTP persistence, leaving initial potentiation intact. *J. Neurosci.* **10**: 3353–3360.

Cotton PC, Brugge JS. (1983) Neural tissues express high levels of the cellular *src* gene product pp60 (c-src). *Mol. Cell. Biol.* **3**: 1157–1162.

Crews CM, Erikson RL. (1993) Extracellular signals and reversible protein phosphorylation: what to MEK of it all. *Cell* **74**: 215–217.

Dekker LV, Graan PN, Oestreicher AB, Versteeg DHG, Gispen WH. (1989a) Inhibition of noradrenaline release by antibodies to B-50 (GAP43). *Nature* **342**: 74–76.

Dekker LV, Graan, PNED, Versteeg DHG, Oestreicher AB, Gispen WH. (1989b) Phosphorylation of B-50 (GAP43) is correlated with neurotransmitter release in rat hippocampal slices. *J. Neurochem.* **52**: 24–30.

del Cerro S, Arai A, Lynch G. (1990a) Inhibition of long-term potentiation by an antagonist of platelet-activating factor receptors. *Behav. Neural Biol* **54**: 213–217.

del Cerro S, Larson J, Oliver MW, Lynch G. (1990b) Development of hippocampal long-term potentiation is reduced by recently introduced calpain inhibitors. *Brain Res.* **530**: 91–95.

Denny JB, Polan-Curtain J, Ghuman A, Wayner MJ, Armstrong DL. (1990a) Calpain inhibitors block long-term potentiation. *Brain Res.* **534**: 317–320.

Denny JB, Polan-Curtain J, Rodriguez S, Wayner MJ, Armstrong DL. (1990b) Evidence that protein kinase M does not maintain long-term potentiation. *Brain Res.* **534**: 201–208.

Dudek SM, Bear MF. (1992) Homosynaptic long-term depression in area CA1 of hippocampus and effect of NMDA receptor blockade. *Proc. Natl Acad. Sci. USA* **89**: 4363–4367.

Ellis PD, Bissoon N, Gurd JW. (1988) Synaptic protein tyrosine kinase: partial characterization and identification of endogenous substrates. *J. Neurochem.* **51**: 611–620.

Fiore RS, Murphy TH, Sanghera JS, Pelech SL, Baraban JM. (1993a) Activation of p42 mitogen-activated protein kinase by glutamate receptor stimulation in rat primary cortical cultures. *J. Neurochem.* **61**: 1626–1633.

Fiore RS, Bayer VE, Pelech SL, Posada J, Cooper JA, Baraban JM. (1993b) p42 mitogen-activated protein kinase in brain: prominent localization in neuronal cell bodies and dendrites. *Neuroscience.* **55**: 463–472.

Frey U, Huang Y-Y, Kandel ER. (1993) Effects of cAMP simulate a late stage of LTP in hippocampal CA1 neurons. *Science* **260**: 1661–1664.

Fukunaga K, Stoppini L, Miyamoto E, Muller D. (1993) Long-term potentiation is associated with an increased activity of calcium/calmodulin-dependent protein kinase II. *J. Biol. Chem.* **268**: 7863–7867.

Gianotti C, Nunzi MG, Gispen WH, Corradetti R. (1992) Phosphorylation of the presynaptic protein B-50 (GAP-43) is increased during electrically induced long-term potentiation. *Neuron* **8**: 843–848.

Gopalakrishna R, Anderson WB. (1989) Calcium- and phospholipid-independent activation of protein kinase C by selective oxidative modification of the regulatory domain. *Proc. Natl Acad. Sci. USA* **86**: 6758–6762.

Grant SGN, O'Dell TJ, Karl KA, Stein PL, Soriano P, Kandel ER. (1992) Impaired long-term potentiation, spatial learning, and hippocampal development in *fyn* mutant mice. *Science* **258**: 1903–1910.

Haley JE, Wilcox GL, Chapman, PF. (1992) The role of nitric oxide in hippocampal long-term potentiation. *Neuron* **8**: 211–216.

Haley JE, Malen PL, Chapman, PF. (1993) Nitric oxide synthase inhibitors block long-term potentiation induced by weak but not strong tetanic stimulation at physiological brain temperature in rat hippocampal slices. *Neurosci. Lett.* **160**: 85–88.

Hidaka H, Kobayashi R. (1992) Pharmacology of protein kinase inhibitors. *Annu. Rev. Pharmacol. Toxicol.* **32**: 377–397.

Hu GY, Hvalby O, Walaas SI, Albert KA, Skjeflo P, Andersen P, Greengard P. (1987) Protein kinase C injection into hippocampal pyramidal cells elicits features of long term potentiation. *Nature* **328**: 426–429.

Huang XY, Morielli AD, Peralta EG. (1993) Tyrosine kinase-dependent suppression of a potassium channel by the G protein-coupled m1 muscarinic acetylcholine receptor. *Cell* **75**: 1145–1156.

Huganir RL, Miles K, Greengard P. (1984) Phosphorylation of the nicotinic acetylcholine receptor by an endogenous tyrosine-specific protein kinase. *Proc. Natl Acad. Sci. USA* **81**: 6968–6972.

Inoue M, Kishimoto A, Takai Y, Nishizuka Y. (1978) Studies on a cyclic nucleotide-independent protein kinase and its proenzyme in mammalian tissues: II. Proenzyme and its activation by calcium-dependent protease from brain. *J. Biol. Chem.* **252**: 7610–7616.

Ito I, Hidaka H, Sugiyama H. (1991) Effects on KN-62, a specific inhibitor of calcium/calmodulin-dependent protein kinase II, on long-term potentiation in the rat hippocampus. *Neurosci. Lett.* **121**: 119–121.

Kang H, Schuman EM. (1995) Long-lasting neurotrophin-induced enhancement of synaptic transmission in the adult hippocampus. *Science* **267**: 1658–1662.

Kato K, Clark GD, Bazan NG, Zorumski CF. (1994) Platelet-activating factor as a potential retrograde messenger in CA1 hippocampal long-term potentiation. *Nature* **367**: 175–179.

Kauer JA, Malenka RC, Nicoll RA. (1988) A persistent postysynaptic modification mediates long-term potentiation in the hippocampus. *Neuron* **1**: 911–917.

Klann E, Sweatt JD. (1994) Induction of LTP is associated with an increase in PKC activity mediated by nitric oxide and superoxide. *Soc. Neurosci. Abstr.* **20**: 446.

Klann E, Chen SJ, Sweatt JD. (1991) Persistent protein kinase activation in the maintenance phase of long-term potentiation. *J. Biol. Chem.* **266**: 24253–24256.

Klann E, Chen SJ, Sweatt JD. (1992) Increased phosphorylation of a 17-kDa protein kinase C substrate (P17) in long-term potentiation. *J. Neurochem.* **58**: 1576–1579.

Klann E, Chen SJ, Sweatt JD. (1993). Mechanism of protein kinase C activation during the induction and maintenance of long-term potentiation probed using a selective peptide substrate. *Proc. Natl Acad. Sci. USA* **90**: 8337–8341.

Leahy JC, Luo Y, Kent CS, Meiri KF, Vallano ML. (1993) Demonstration of presynaptic protein kinase C activation following long-term potentiation in rat hippocampal slices. *Neuroscience.* **52**: 563-574.

Lin L-L, Wartmann M, Lin AY, Knopf JL, Seth A, Davis RJ. (1993) cPLA2 is phosphorylated and activated by MAP kinase. *Cell* **72**: 269-278.

Lisman JE. (1985) A mechanism for memory storage insensitive to molecular turnover: a bistable autophosphorylating kinase. *Proc. Natl Acad. Sci. USA* **82**: 3055-3057.

Lisman JE. (1994) The CaMKII hypothesis for storage of synaptic memory. *Trends Neurosci.* **17**: 406-412.

Lovinger DM, Akers RF, Nelson RB, Barnes CA, McNaughton BL, Routtenberg A. (1985) A selective increase in phosporylation of protein F1, a protein kinase C substrate, directly related to three day growth of long term synaptic enhancement. *Brain Res.* **343**: 137-143.

Lovinger DM, Wong KL, Murakami K, Routtenberg A. (1987) Protein kinase C inhibitors eliminate hippocampal long-term potentiation. *Brain Res.* **436**: 177-183.

Lynch G, Baudry M. (1984) The biochemistry of memory: a new and specific hypothesis. *Science* **224**: 1057-1063.

Lynch MA, Clements MP, Voss KL, Bramham CR, Bliss TVP. (1991) Is arachidonic acid a retrograde messenger in long-term potentiation? *Biochem. Soc. Trans.* **19**: 391-396.

Malenka RC, Madison DV, Nicoll RA. (1986) Potentiation of synaptic transmission in the hippocampus by phorbol esters. *Nature* **321**: 175-177.

Malenka RC, Ayoub GS, Nicoll RA. (1987) Phorbol esters enhance transmitter release in rat hippocampal slices. *Brain Res.* **403**: 198-203.

Malenka RC, Kauer JA, Perkel DJ, Mauk MD, Kelly PT, Nicoll RA, Waxham MN. (1989) An essential role for postsynaptic calmodulin and protein kinase activity in long-term potentiation. *Nature* **340**: 554-557.

Malinow R, Madison DV, Tsien RW. (1988) Pesistent protein kinase activity underlying long-term potentiation. *Nature* **335**: 820-824.

Malinow R, Schulman H, Tsien RW. (1989) Inhibition of postsynaptic PKC or CaMKII blocks induction but not expression of LTP. *Science* **245**: 862-866.

Manabe T, Nicoll RA. (1994) Long-term potentiation: evidence against an increase in transmitter release probability in the CA1 region of the hippocampus. *Science* **265**: 1888-1892.

Matthies H, Reymann KG. (1993) Protein kinase A inhibitors prevent the maintenance of hippocampal long-term potentiation. *NeuroReport* **4**: 712-714.

Matthies H Jr, Behnisch T, Kase H, Matthies H, Reymann KG. (1991) Differential effects of protein kinase inhibitors on pre-established long-term potentiation in rat hippocampal neurons *in vitro*. *Neurosci. Lett.* **121**: 259-262.

Meffert MK, Haley JE, Schuman EM, Schulman H, Madison DV. (1994) Inhibition of hippocampal heme oxygenase, nitric oxide synthase, and long-term potentiation by metalloporphyrins. *Neuron* **13**: 1225-1233.

Mochly RD, Koshland DE. (1987) Domain structure and phosphorylation of protein kinase C. *J. Biol. Chem.* **262**: 2291-2297.

Montminy MR, Bilezikjian LM. (1987) Binding of a nuclear protein to the cyclic-AMP response element of the somatostatin gene. *Nature* **328**: 175-178.

Moss SJ, Blackstone CD, Huganir RL. (1993) Phosphorylation of recombinant non-NMDA glutamate receptors on serine and tyrosine residues. *Neurochem. Res.* **18**: 105-110.

Mulkey RM, Malenka RC. (1992) Mechanisms underlying induction of homosynaptic long-term depression in area CA1 of the hippocampus. *Neuron* **9**: 967-975.

Mulkey RM, Herron CE, Malenka RC. (1993) An essential role for protein phosphatases in hippocampal long-term depression. *Science* **261**: 1051-1055.

Mulkey RM, Endo S, Shenolikar S, Malenka RC. (1994) Involvement of a calcineurin/inhibitor-1 phosphatase cascade in hippocampal long-term depression. *Nature* **369**: 486-488.

Muller D, Turnbull J, Baudry M, Lynch G. (1988) Phorbol ester-induced synaptic facilitation is different than long-term potentiation. *Proc. Natl Acad. Sci. USA* **85**: 6997-7000.

Murphy TH, Blatter LA, Bhat RV, Fiore RS, Wier WG, Baraban JM. (1994) Differential regulation of calcium/calmodulin-dependent protein kinase II and p42 MAP kinase activity by synaptic transmission. *J. Neurosci.* **14**: 1320–1331.

Nelson RB, Linden DJ, Hyman C, Pfenninger KH, Routtenberg A. (1989) The two major phosphoproteins in growth cones are probably identical to two protein kinase C substrates correlated with persistence of long-term potentiation. *J. Neurosci.* **9**: 381–389.

Nguyen PV, Abel T, Kandel ER. (1994) Requirement for a critical period of transcription for a late phase of LTP. *Science* **265**: 1104–1107.

Nishizuka Y. (1992) Intracellular signaling by hydrolysis of phospholipids and activation of protein kinase C. *Science* **258**: 607–614.

O'Dell TJ, Kandel ER. (1994) Low-frequency stimulation erases LTP through an NMDA receptor-mediated activation of protein phosphatases. *Learn. Mem.* **1**: 129–139.

O'Dell TJ, Hawkins RD, Kandel ER, Arancio O. (1991a) Tests of the roles of two diffusible substances in long-term potentiation: evidence for nitric oxide as a possible early retrograde messenger. *Proc. Natl. Acad. Sci. USA* **88**: 11285–11289.

O'Dell TJ, Kandel ER, Grant SGN. (1991b) Long-term potentiation in the hippocampus is blocked by tyrosine kinase inhibitors. *Nature* **353**: 558–560.

O'Dell TJ, Huang PL, Dawson TM, Dinerman, JL, Snyder SH, Kandel ER, Fishman MC. (1994) Endothelial NOS and the blockade of LTP by NOS inhibitors in mice lacking neuronal NOS. *Science* **265**: 542–546.

Palumbo EJ, Sweatt JD, Chen S-J, Klann E. (1992) Oxidation-induced persistent activation of protein kinase C in hippocampal homogenates. *Biochem. Biophys. Res. Commun.* **187**: 1439–1445.

Pang DT, Wang JKT, Valtorta F, Benfenati F, Greengard P. (1988) Protein tyrosine phosphorylation in synaptic vesicles. *Proc. Natl Acad. Sci. USA* **85**: 762–766.

Powell CM, Johnston D, Sweatt JD. (1994) Autonomously active PKC in the maintenance phase of NMDA receptor independent long-term potentiation. *J. Biol. Chem.* **269**: 27958–27963.

Represa A, Deloume JC, Sensenbrenner M, Ben-Ari Y, Baudier J. (1990) Neurogranin: immunocytochemical localization of a brain-specific protein kinase C substrate. *J. Neurosci.* **10**: 3782–3792.

Reymann KG, Davies SN, Matthies H, Kase H, Collingridge GL. (1990) Activation of a K-252b-sensitive protein kinase is necessary for a post-synaptic phase of long-term potentiation in area CA1 of rat hippocampus. *Eur. J. Neurosci.* **2**: 481–486.

Roberson ED, Sweatt JD. (1993) Cyclic AMP-dependent protein kinase is activated during the induction of long-term potentiation. *Soc. Neurosci. Abstr.* **19**: 1708.

Roberts TM. (1992) A signal chain of events. *Science* **360**: 534–535.

Sacktor TC, Osten P, Valsamis H, Jiang X, Naik MU, Sublette E. (1993) Persistent activation of the zeta isoform of protein kinase C in the maintenance of long-term potentiation. *Proc. Natl Acad. Sci. USA* **90**: 8342–8346.

Schuman EM, Madison DV. (1991) A requirement for the intercellular messenger nitric oxide in long-term potentiation. *Science* **254**: 1503–1506.

Schuman EM, Madison DV. (1994) Locally distributed synaptic potentiation in the hippocampus. *Science* **263**: 532–536.

Schuman EM, Meffert MK, Schulman H, Madison DV. (1994) An ADP-ribosyltransferase as a potential target for nitric oxide action in hippocampal long-term potentiation. *Proc. Natl Acad. Sci. USA* **91**: 11958–11962.

Schwartz JH. (1993) Cognitive kinases. *Proc. Natl Acad. Sci. USA* **90**: 8310–8313.

Sessoms JS, Chen S-J, Chetkovich DM, Powell CM, Roberson ED, Sweatt JD, Klann E. (1992–1993) Calcium-induced persistent protein kinase activation in rat hippocampal homogenates. *Sec. Mess. Phosphoprot.* **14**: 109–126.

Silva AJ, Stevens CF, Tonegawa S, Wang Y. (1992) Deficient hippocampal long-term potentiation in calcium–calmodulin kinase II mutant mice. *Science* **257**: 201–206.

Stevens CF, Wang Y. (1993) Reversal of long-term potentiation by inhibitors of heme oxygenase. *Nature* **364**: 147–149.

Stevens CF, Wang Y. (1994) Changes in reliability of synaptic function as a mechanism for plasticity. *Nature* **371**: 652–653.

Stratton KR, Worley PF, Huganir RL, Baraban JM. (1989) Muscarinic agonists and phorbol esters increase tyrosine phosphorylation of a 40-kilodalton protein in hippocampal slices. *Proc. Natl Acad Sci. USA* **86**: 2489–2501.

Umemori H, Sato S, Yagi T, Aizawa S, Yamamoto T. (1994) Initial events of myelination involve fyn tyrosine kinase signalling. *Nature* **367**: 572–576.

van Hoof CO, DeGraan GPN, Boonstra J, Oestreicher AB, Schmidt MMH, Gispen WH. (1986) Nerve growth factor enhances the level of the protein kinase C substrate B-50 in pheochromocytoma PC12 cells. *Biochem. Biophys. Res. Commun.* **139**: 644–651.

Wang JH, Feng DP. (1992) Postsynaptic protein kinase-C essential to induction and maintenance of long-term potentiation in the hippocampal CA1 region. *Proc. Natl Acad. Sci. USA* **89**: 2576–2580.

Watson JB, Battenberg KK, Wong FE, Bloom FE, Sutcliffe JG. (1990) Subtractive cDNA cloning of RC3, a rodent cortex-enriched mRNA encoding a novel 78 residue protein. *J. Neurosci.* **26**: 397–408.

Williams JH, Bliss TVP. (1988) An *in vitro* study of the effect of lipoxygenase and cyclo-oxygenase inhibitors of arachidonic acid on the induction and maintenance of long-term potentiation in the hippocampus. *Neurosci. Lett.* **107**: 301–306.

Williams JH, Errington MA, Lynch MA, Bliss TVP. (1989) Arachidonic acid induces a long-term activity-dependent enhancement of synaptic transmission in the hippocampus. *Nature* **341**: 739–742.

Williams JH, Li Y-G, Nayak A, Errington ML, Murphy KPSJ, Bliss TVP. (1993) The suppression of long-term potentiation in rat hippocampus by inhibitors of nitric oxide synthase is temperature and age dependent. *Neuron* **11**: 877–884.

Xia Z, Refsdal CD, Merchant KM, Dorsa DM, Storm DR. (1991) Distribution of mRNA for the calmodulin-sensitive adenylate cyclase in rat brain: expression in areas associated with learning and memory. *Neuron* **6**: 431–443.

Yamamoto KK, Gonzalez GA, Biggs WH III, Montminy MR. (1988) Phosphorylation-induced binding and transcriptional efficacy of nucelar factor CREB. *Nature* **334**: 494–498.

Zhuo M, Small SA, Kandel ER, Hawkins RD. (1993) Nitric oxide and carbon monoxide produce activity-dependent long-term synaptic enhancement in hippocampus. *Science* **260**: 1946–1950.

Zhuo M, Hu Y, Schultz C, Kandel ER, Hawkins RD. (1994) Role of guanylyl cyclase and cGMP-dependent protein kinase in long-term potentiation. *Nature* **368**: 635–639.

Further reading

Bliss TVP, Collingridge GL. (1993) A synaptic model of memory: long-term potentiation in the hippocampus. *Nature* **361**: 31–39.

Cohen P. (1989) The structure and regulation of protein phosphatases. *Annu. Rev. Biochem.* **58**: 453–508.

Frank DA, Greenberg ME. (1994) CREB: a mediator of long-term memory from mollusks to mammals. *Cell* **79**: 5–8.

Garthwaite J. (1991) Glutamate, nitric oxide and cell–cell signalling in the nervous system. *Trends Neurosci.* **14**: 60–67.

Gasic GP, and Hollman M. (1992) Molecular neurobiology of glutamate receptors. *Annu. Rev. Physiol.* **54**: 507–536.

Huang KP. (1989) The mechanism of protein kinase C activation. *Trends Neurosci.* **12**: 425–432.

Johnston D, Williams S, Jaffe D, Gray R. (1993) NMDA-receptor-independent long-term potentiation. *Annu. Rev. Physiol.* **54**: 489–512.

Linden DJ. (1994) Long-term synaptic depression in the mammalian brain. *Neuron* **12**: 457–472.

Lisman JE. (1994) The CaMKII hypothesis for storage of synaptic memory. *Trends Neurosci.* **17**: 406–412.

Mahoney CW, Huang KP. (1994) Molecular and catalytic properties of protein kinase C. In: *Protein Kinase C* (ed. JF Kuo). Oxford University Press, Oxford, pp. 16–63.

Malenka RC. (1994) Synaptic plasticity in the hippocampus: LTP and LTD. *Cell* **78**: 535–538.

Malenka RC, Nicoll RA. (1993) NMDA receptor-dependent synaptic plasticity: multiple forms and mechanisms. *Trends Neurosci.* **16**: 521–527.
Raymond LA, Blackstone CD, Huganir RL. (1993) Phosphorylation of amino acid neurotransmitter receptors in synaptic plasticity. *Trends Neurosci.* **16**: 147–153.
Scott JD. (1991) Cyclic nucleotide-dependent protein kinases. *Pharmacol. Ther.* **50**: 123–145.

The locus of expression of NMDA receptor-dependent LTP in the hippocampus

T.V.P. Bliss and M.S. Fazeli

4.1 Introduction

In this chapter, we consider the following question: what is the locus of expression of long-term potentiation (LTP) at those hippocampal synapses where its induction requires the activation of postsynaptically located N-methyl-D-aspartate (NMDA) receptors, that is at perforant path–granule cell synapses in the dentate gyrus, at associational–commissural synapses in areas CA3 and Schaffer–commissural synapses in area CA1. It will become clear from our survey of the experimental approaches that have been adopted, and the evidence that has been adduced, that the question cannot yet be answered with any certainty. The various possibilities were laid out at the time of the first description of LTP by Bliss and Lømo in 1973 when they concluded that the phenomenon "could take the form of . . . an increase in the amount of transmitter released per synapse, an increase in the sensitivity of the postsynaptic junctional membrane, or a reduction in the resistance of the narrow stem by which spines are attached to the parent dendrite". A great deal of data have since been amassed in support of each of these three classes of explanation – presynaptic, postsynaptic and morphological – and their corresponding loci. (Morphological changes could include growth of synapses or an increase in their number; in these cases, the distinction between pre- and postsynaptic changes ceases to be meaningful.) In most instances, the evidence has been such as to favor one or other locus without excluding the others; for example, evidence for an increase in transmitter release does not preclude changes in receptor function. In some experimental designs, to be discussed

---

Cortical Plasticity, edited by M.S. Fazeli and G.L. Collingridge.
© 1996 BIOS Scientific Publishers Ltd, Oxford.

below, the results were interpreted initially as being incompatible with a presynaptic component to LTP. In all such cases, however, alternative explanations have been offered which call for presynaptic involvement. A conservative assessment of current evidence would therefore ascribe the expression of LTP to both pre- and postsynaptic mechanisms, while allowing for the possibility that the proportion of pre- and postsynaptic contributions changes with time.

4.2 Changes in transmitter release

4.2.1 Changes in glutamate efflux

Attempts to localize the mechanisms underlying the maintenance of LTP in the hippocampus began by asking whether its induction gave rise to a persistent increase in transmitter efflux. In the first such experiment, Skrede and Malthe-Sørenssen (1981) delivered tetanic stimulation to the hippocampal slice and detected an enhanced resting and evoked release of preloaded [^{3}H] aspartate. This was followed by a series of experiments by Bliss and his colleagues in which the extracellular glutamate concentration in the dentate gyrus of anesthetized rats was sampled using a push–pull cannula to which were attached metal recording electrodes to allow the measurement of evoked responses. These experiments established that LTP was associated with an increase in the concentration of extracellular glutamate (Bliss *et al.*, 1986; Dolphin *et al.*, 1982; Errington *et al.*, 1987; Lynch *et al.*, 1989a). Moreover, if the induction of LTP was blocked, either by physiological manuevers (Bliss *et al.*, 1986) or by drugs such as the NMDA receptor antagonist, D(-)AP5 (Errington *et al.*, 1987), or the phospholipase A_2 inhibitor nordihydroguaiaretic acid (Lynch *et al.*, 1989b), the increase in glutamate release did not occur. Although these results provided the first sustained body of evidence for a presynaptic component to LTP, their interpretation has not gone unchallenged (Nicolls and Attwell, 1991). One unresolved issue has been the extent to which the detected increase in extracellular glutamate concentration could be explained by a decrease in glutamate uptake rather than by increases in spontaneous or evoked release from potentiated synapses. Attempts to address this problem by the use of glutamate uptake inhibitors have not succeeded because of the rapid onset of epileptiform activity with the use of these drugs *in vivo* (M.L. Errington and T.V.P. Bliss, unpublished results). More recently, a 'dialysis electrode' containing the glutamate-specific enzyme glutamate oxidase has been used to monitor real-time changes in the concentration of extracellular glutamate in the dentate gyrus *in vivo*. Extracellular glutamate diffuses into the lumen of the dialysis tubing where it is oxidized to 2-oxo-glutamate and hydrogen peroxide; the latter is electroactive and can be detected by the current produced at a platinum electrode held at the appropriate redox potential. Increasing the rate of stimulation of the perforant path produces an increase in glutamate signal, as expected if glutamate is the transmitter at perforant path–granule cell synapses. With this technique, both background and stimulus-dependent increases in glutamate concentration could be detected after induction of LTP in the dentate gyrus (Galley et al., 1993). Surprisingly, a similar LTP-related increase in glutamate efflux could not be detected in area CA1, either with the push–pull cannula (Aniksztejn et al., 1989), or with the dialysis electrode (M.L. Errington, P.T. Galley and T.V.P. Bliss, unpublished results). However, the significance of these negative findings in area CA1 is uncertain, since an increase in glutamate efflux with increasing stimulus frequency was not demonstrated; it is possible that uptake mechanisms for glutamate are more active in this region than in the dentate gyrus.

A second approach to the study of glutamate release in LTP is to compare depolarization-induced release from slices of hippocampus, or from hippocampal synaptosomes, in potentiated and control tissue. A relative increase in release of endogenous or radiolabeled glutamate has been reported both in the dentate gyrus (Lynch et al., 1989b) and in areas CA1 (Ghijsen et al., 1992) and CA3 (Feasey et al., 1986), following the induction of NMDA-receptor dependent LTP either *in vitro* or in the anesthetized animal. While transmitter release produced by prolonged periods of depolarization is at best an indirect model of the rapid exocytosis triggered by an action potential at the nerve terminal, the model does have the advantage that it can be used to assess transmitter release at arbitrarily long times after the induction of LTP in the intact animal. The technique was used by Bliss et al. (1987) to show that, relative to control tissue, depolarization-induced release remained elevated in potentiated tissue prepared from the dentate gyrus of animals killed 3 days after the induction of LTP while, if the comparison was made after 1 week, when evoked responses had declined to baseline, no difference was found. This result demonstrates that presynaptic changes can be maintained for as long as LTP itself, that is for several days in the dentate gyrus (see also Meberg et al., 1993).

A promising new technique for detecting changes in transmitter release, the excised patch or sniffer pipette, which was developed originally to study the release of acetylcholine from growth cones (Hume et al., 1983), has recently been applied to the detection of glutamate release in the hippocampus (Maeda et al., 1995a; O'Connor et al., 1995a). A patch electrode armed with an outside-out patch is ripped from the soma of a pyramidal cell, and crammed (the vernacular of the patch-sniffers is not for the faint-hearted) into the synaptic region of the pathway of interest. The idea is that the receptor population of the patch will act as a sensitive bioassay system for synaptically released glutamate. Providing the properties of the patch receptors remain constant, any change in the response to synaptic activation will reflect a change in transmitter release. There are, however, uncertainties about the stability of the patch receptors in an extracellular environment containing an array of proteases, or diffusible messengers whose activity may be modulated by tetanic stimulation and which in turn could modify the properties of the population of patch receptors. Thus, the potentiation of NMDA receptor-mediated responses reported by O'Connor et al. (1995a) using this method may reflect either an increase in transmitter release or an LTP-associated modification of patch receptor properties. An absence of change in responsiveness following induction of LTP or long-term depression (LTD) would provide more secure evidence for a postsynaptic locus. A preliminary report suggests that this may be the case in area CA1 (Maeda et al., 1995b)

4.2.2 An increase in synaptic vesicle turnover detected in a model of LTP in cultured hippocampal neurons

The frequency of spontaneous miniature excitatory postsynaptic currents (EPSCs) in cultured hippocampal neurons can be enhanced for prolonged periods by brief application of 50 μM glutamate (Malgaroli and Tsien, 1992). The increase in frequency occurs without change in the mean amplitude of spontaneous EPSCs, or in the response to applied agonists (Malgaroli et al., 1992), strongly suggesting a presynaptic basis for the effect. As with tetanus-induced LTP, induction of glutamate-induced potentiation is NMDA receptor dependent and is blocked by postsynaptic hyperpolarization (Malgaroli and Tsien, 1992), indicating that similar cellular mechanisms may be involved in the two types of potentiation. Striking visual confirmation that presynaptic mechanisms are engaged in glutamate-induced potentiation has been supplied by Malgaroli et al. (1995) who made fluorescently tagged antibodies to an intraluminal epitope of the synaptic vesicle protein synaptotagmin. Control experiments showed that antibodies were able to gain access to the epitope only when it was exposed to the extracellular domain during exocytosis. Hence, an increase in fluorescence is an indication of an increase in the frequency of vesicle fusion. Active terminals were labeled before and for an hour after the induction of glutamate-induced potentiation of miniature EPSC frequency. Application of glutamate produced a marked increase in fluorescence in labeled terminals, with the least active terminals before potentiation showing the largest increases afterwards (see Section 4.5). The extent to which glutamate-induced potentiation in culture constitutes a valid model for tetanus-induced LTP is debatable; nevertheless, these experiments provide powerful evidence that presynaptic factors contribute significantly to at least one form of NMDA receptor-dependent synaptic plasticity in the hippocampus.

4.3 Changes in postsynaptic responsiveness

4.3.1 Effects of applying glutamate receptor agonists

Postsynaptic changes in LTP, whatever the mechanism, must ultimately be detectable as an increase in the response to a fixed synaptic input. Synaptic inputs, however, are themselves potentially modifiable, and experimental tests of postsynaptic sensitivity must rely instead on exogenous application of an appropriate receptor agonist. Early in vitro experiments failed to identify an increased responsiveness to iontophoretically applied glutamate (Lynch et al., 1976), possibly because the efficiency of glutamate transporters prevented access to synaptic receptors. However, using iontophoretic applications of α-amino-3-hydroxy-5-methyl-4-isoxazole proprionic acid (AMPA), which is not a substrate for glutamate transporters, Davies et al. (1989) were able to demonstrate a delayed increase in AMPA receptor-mediated responses following the induction of LTP; the increase was apparent within 15 min, and reached an asymptotic value within 1–2 h (see Chapter 2, *Figure 2.1*). These results are suggestive, although the degree to which the observed response is generated by extrasynaptic receptors remains unknown, and with the relatively long application times used (up to 25 sec) most synaptic AMPA receptors would have been desensitized. Techniques for applying exogenous receptor agonists over a time course and volume which approximates the release of transmitter at a single synapse are required, and exist in principle. The strategy involves the synthesis of a photolabile 'caged' glutamate, an inactive molecule from which glutamate can be released by brief exposure to UV light (Corrie *et al.*, 1993; Katz and Dalva, 1994). Brief and localized release of transmitter can be achieved by focusing UV light on to the slice, using two-photon excitation in which the slice is illuminated at visible wavelengths, with effective frequency doubling into the UV range occurring only over a highly restricted focal volume. Two-photon excitation has allowed Yuste and Denk (1995) to obtain images of synaptically generated calcium transients in dendritic spines in the hippocampal slice, but rapid, focal uncaging of glutamate has so far not been reported.

4.3.2 Changes in kinetics of receptor-mediated synaptic responses

The amplitude and time course of synaptically induced currents are controlled by the binding and unbinding of agonist to receptor, desensitization of the receptor and, possibly, diffusion, uptake or enzymatic degradation of the transmitter. Phosphorylation of receptors can change their open probability which will affect the amplitude of the response (Raymond *et al.*, 1993) while modifications leading to changes in mean open time (or mean cluster length) will affect the decay time constant (Magleby and Stevens, 1972). In contrast, an increase in transmitter release would be expected to affect the amplitude but not the time course of the response. So how are the time courses of synaptic currents altered in LTP? For NMDA receptor-mediated responses, the answer is very little, either in area CA1 (Asztely *et al.*, 1992; Clark and Collingridge, 1995; Collingridge *et al.*, 1992) or in dentate gyrus (O'Connor *et al.*, 1995a); for the fast AMPA receptor-mediated component, Ambros-Ingersen *et al.* (1991) reported a prolongation of the falling phase, but this result has not been confirmed (Collingridge *et al.*, 1992; Isaacson and Nicoll, 1991). On balance, these observations suggest that post-translational modifications affecting mean open time of the receptor do not occur. Changes which affect open probability remain a possible explanation for the change in amplitude of synaptic current. In neurons in culture, phosphorylation of AMPA receptor subunits on serine and threonine residues leads to an increase in AMPA receptor-mediated current (Soderling *et al.*, 1994) and there is indirect evidence that tyrosine phosphorylation of NMDA receptors also increases the magnitude of synaptic currents (Wang and Salter, 1994). Thus, phosphorylation of AMPA and NMDA receptors may regulate the amplitude of their respective components of the evoked synaptic current, and thus provide a postsynaptic mechanism for LTP. In this context, it is of interest that two groups have found a persistent increase in the phosphorylation of the 2B subunit of the NMDA receptor following the induction of LTP in the dentate gyrus of freely moving or anesthetized rats (Rosenblum *et al.*, 1996; Rostas *et al.*, 1996).

4.4 Changes in synthesis or phosphorylation of pre- and postsynaptic proteins

4.4.1 Presynaptic proteins

Long-term changes in the synthesis or post-translational status of presynaptic proteins provide evidence for a presynaptic involvement in the expression of LTP, whether or not the function of such changes is known. Two laboratories have documented an increase in the phosphorylation of the presynaptic growth-associated protein GAP43 in LTP (Linden *et al.*, 1988; Routtenberg *et al.*, 1985; Schrama *et al.*, 1986). The initial studies employed the indirect method of back-phosphorylation, in which tissue is extracted and then incubated with labeled inorganic phosphate; an increase in endogenous phosophorylation is inferred from a reduction in exogenous labeling. An increase in *in situ* phosphorylation of GAP43 following the induction of LTP in area CA1 has since been demonstrated by Gianotti *et al.* (1992); these post-translational changes can be detected within 10 min, but persist only for 60–90 min (Ramakers *et al.*, 1995). In contrast, changes in the concentration of GAP43 protein have also been observed as long as 3 days after induction of LTP (Meberg *et al.*, 1993). The concentration of a number of synaptic vesicle proteins, including synaptophysin, synapsin and synaptotagminin, have been reported following induction of LTP in the dentate gyrus; the increases were not apparent at 45 min post-induction, but were present at 3 h (Lynch *et al.*, 1994).

4.4.2 Postsynaptic proteins

Although GAP43 and synaptic vesicle proteins are localized to the presynaptic terminal, there must be some doubt in the case of experiments in the dentate gyrus as to whether the changes occur at perforant path terminals, or on the axon terminals of the granule cells, some of which terminate in the hilus and would be included in the dentate tissue dissected for analysis. If this were the case, the change would be in the presynaptic region of the postsynaptic cell, and for the purposes of the present analysis would count as a postsynaptic change. An example of just such an effect is the increase in the expression of the presynaptic membrane protein, syntaxin, in mossy fiber terminals, following the induction of LTP in the perforant path (Smirnova *et al.*, 1993). An increase in mRNA for syntaxin was detected by *in situ* hybridization in granule cell bodies, and the increase in protein was detected immunohistochemically in the terminal zone of mossy fiber axons.

There is a substantial literature linking the rapid transcription of immediate-early genes (IEGs) to LTP in the perforant path (Cole *et al.*, 1989; Dragunow *et al.*, 1989; Wisden *et al.*, 1990; Worley *et al.*, 1993) Their role in the cascade leading to activation of effector genes is discussed in Chapter 6, along with evidence for expression of downstream genes involved in second-messenger cascades (Thomas *et al.*, 1994). For our present purposes, the fact that the IEG *zif268* is activated in granule cells following the induction of LTP, as is another IEG, *c-fos*, in the unanesthetized animal (Abraham *et al.*, 1991), indicates that gene activation is taking place in the postsynaptic cell during a time window of 1–2 h after induction. However, it has not been established that IEG activation is necessary for LTP. Whether or not similar activation of IEGs occurs in presynaptic neurons (e.g. the projection cells of layer II in the entorhinal cortex which give rise to the perforant path) does not seem to have been examined.

Increased phosphorylation of postsynaptic proteins has also been detected. In addition to describing the time course of phosphorylation of GAP43 following the induction of LTP in area CA1, Ramakers *et al.* (1995) established that phosphorylation of the postsynaptic neuronal protein, neurogranin, was enhanced for a period around 60 min after induction. Another example, the tyrosine phosphorylation of NMDAR-2B, has already been mentioned.

We conclude that LTP is associated with increases in postsynaptic gene expression as well as post-translational modification of postsynaptic proteins in hippocampal neurons, and that some of these changes can be long-lasting. However, until we know more about the role that such changes play in synaptic transmission, it would be premature to conclude that they constitute evidence for a postsynaptic mechanism for the maintenance of LTP. Other possibilities include modulation of transmitter release from the axon terminals of the postsynaptic cell, or a role in the synthesis, activation or release of retrograde messengers.

4.5 Evidence from quantal analysis

Of all the techniques available to tackle the question addressed in this chapter, quantal analysis once held the greatest promise. Pioneering attempts were made in the early 1980s to attempt a quantal analysis of LTP using conventional intracellular recording (Voronin, 1983), but the inherent noisiness of recordings made with high impedance sharp electrodes was a serious drawback, making it difficult to detect peaks in the amplitude distribution of evoked responses, despite the development of deconvolution and statistical techniques for fitting noisy data (Redman, 1990). The introduction of whole-cell recording to the hippocampal slice by Edwards *et al.* (1989) brought a dramatic improvement in signal-to-noise ratio, and with it the hope that the locus of LTP would be identified rapidly. To say that hope has not been realized would be something of an understatement. The problems are two-fold. First, unresolved discrepancies in the results obtained from different laboratories ostensibly doing the same experiment. Second, problems of interpretation, arising from attempts to apply an analysis designed for a cell with a single synapse, the muscle fiber, to the hippocampal pyramidal cell possessing as many as 25 000 synapses. Where there is so little agreement, there are few constraints on the imagination, and the most basic tenets of quantal analysis are open to question. For example, an increase in the frequency, as opposed to the amplitude, of spontaneous excitatory synaptic currents is, in the classical model, a straightforward presynaptic event. However, consider a population of synapses with a nonzero probability of release but without functional postsynaptic receptors. These synapses would be effectively silent. A procedure which uncovered or inserted receptors at these synapses would then lead to an increase in the frequency of spontaneous synaptic currents by an entirely postsynaptic mechanism. While the voluminous literature affords comfort to those of a purely presynaptic persuasion, as well as to the postsynaptic fundamentalist, the reader may find it hard to resist the conclusion that the evidence points to both pre- and postsynaptic changes.

Initial results using the whole cell technique provided evidence for a presynaptic component to LTP in area CA1 (Bekkers and Stevens, 1990; Malinow and Tsien, 1990). This conclusion was based on an analysis of the coefficient of variation, CV, of synaptic responses to weak stimuli, collected before and after the induction of LTP. For the simple binomial model of the release process, CV (defined as $CV^2 = $ variance/mean^2) is a function of n, the number of release sites, and p_r, the probability of release (see Korn and Faber, 1991):

$$CV^2 = (1-p_r)/np_r$$

Since both n and p_r are presynaptic parameters, the observed changes in CV after the induction of LTP were interpreted as evidence for a presynaptic change, though changes in postsynaptic parameters were not excluded. In agreement with these data, an increase in p_r without change in quantal amplitude has been reported in a study using minimal stimulation in the hippocampal slice (Stevens and Wang, 1994). The absence of change in quantal amplitude in these experiments appears inconsistent with a postsynaptic contribution. Another standard method of quantal analysis is to estimate the presynaptic parameter, p_r, from the proportion of failures – that is, stimuli which fail to produce a response. Recordings between pairs of connected CA3 and CA1 neurons appeared to support a presynaptic locus, since the probability of failures decreased markedly after the induction of LTP (Malinow, 1991). [An earlier attempt using conventional 'sharp' electrodes to study LTP between pairs of CA3 and CA1 cells had failed, because, for reasons that remain unclear, LTP at the level of the single connection could very rarely be induced (Friedlander *et al.*, 1990).] A particularly clear-cut result was obtained by Bolshakov and Siegelbaum (1994) in paired recordings from young rats. An action potential in a CA3 neuron released at most one quantum of transmitter. LTP was associated with an increase in p_r (reduction in failures), with no change in quantal size (but see *Science*, 271, 1604–1606, 1996, for an exchange of letters between Malinow and Zainen, and Bolshakov and Siegelbaum on whether the addition of new synapses with low quantal size can provide an alternative explanation for these findings).

The conclusions drawn from the above data are vulnerable at a number of levels. First, the experiments were done in dissociated cultures or on hippocampal slices from young animals at room temperature, and were followed for less than 2 h. Second, the conclusions are model dependent. If quantal release at central synapses is not described accurately by a simple binomial model – and it is becoming increasingly apparent that the assumption of a constant release probability for all synapses is wrong (Rosenmund *et al.*, 1993) – then the conclusions may not be valid; for instance, under the more realistic multinomial model, where p_r is allowed to vary across synapses, changes in CV would be predicted with changes in the postsynaptic parameter, quantal size (Faber and Korn, 1992). (Since the most likely presynaptic mechanism for LTP is a change in p_r, the simple binomial model with its assumption of a constant p_r at all release sites amounts to a manifesto for postsynaptic mechanisms.) Moreover, recent experiments, described below, have caused Malinow to reinterpret his 'failures' data in terms of a postsynaptic model in which clusters of latent AMPA receptors are uncovered by the induction process (Liao *et al.*, 1995). Third, other laboratories have concluded that there are changes in quantal size either with or, in one case, without (Foster and McNaughton, 1991) presynaptic changes (reviewed in Bliss and Collingridge, 1993). An increase in quantal size and frequency of spontaneous mEPSCs following the induction of LTP was reported by Manabe *et al.* (1992). Spontaneous release presumably occurs from the entire set of afferent terminals. What is the situation in the small subset of potentiated synapses? In a recent extension of this experiment, the action of Sr^{2+} in prolonging the process of transmitter release at activated synapses was cleverly exploited to allow the inference that quantal size at potentiated synapses is increased (Oliet *et al.*, 1996). There was, however, also evidence for an increase in the frequency of evoked quantal events.

Leaving aside the question of whether different mechanisms sustain potentiation at different times, we may ask whether LTP is the same at all synapses? Larkman *et al.* (1992) raised the intriguing possibility that the locus of expression at a given synapse depends on the existing value of p_r; synapses where p_r is already near its maximal value of 1 will be potentiated predominantly by postsynaptic mechanisms, while for low p_r synapses, LTP is primarily presynaptic (note that this scheme assumes a multinomial model). Additional support for this idea has come from the experiments by Malgaroli *et al.* (1995), who found that the relative increase in the rate of exocytosis after LTP was greater in those boutons which were less active before induction. The data of Bolshakov and Siegelbaum (1995) add an extra twist to the tale: they found that p_r was high (0.9) in 4- to 8-day-old rats, declining to a mean of 0.5 in 2- to 3-week old rats; discounting the possibility of postsynaptic changes, they suggest that the absence of LTP (and, conversely, the pronounced degree of LTD) in very young rats is a reflection of the already near maximal p.

4.5.2 A pharmacological method for estimating p_r

A novel technique for measuring p_r at NMDA receptor-containing synapses has been introduced by Rosenmund *et al.* (1993) and applied to the measurement of p_r at hippocampal synapses by Hessler *et al.* (1993) and Weisskopf and Nicoll (1995). The method exploits the fact that the NMDA receptor antagonist MK801 is an activity-dependent channel blocker; the drug can only gain access to the channel when it is in the open state, that is when the receptor has been activated by binding of transmitter under conditions in which the Mg^{2+} block of the channel is relieved. These conditions can be met by stimulation of an afferent pathway in low Mg^{2+}. For a given stimulus frequency, the rate at which the drug blocks the NMDA receptor-dependent component of the evoked response will then depend on the probability of release, p_r, at the terminal(s) of the stimulated axon(s). The direct approach of isolating the NMDA response and measuring its rate of decay in the same pathway before and after the induction of LTP is not possible because the action of MK801 is irreversible. Nevertheless, by comparing rates of decay in control and potentiated pathways, Manabe and Nicoll (1994) concluded that LTP of the AMPA receptor-mediated response was not associated with a change in release probability at potentiated synapses. More recently, two other groups have examined tetanically induced potentiation of the isolated NMDA component, and have reported an increase in release probability using the MK801 method (Clark and Collingridge, 1996; Kullmann et al., 1996). The reason for the discrepancy is not known: in both sets of experiments, the decay of the NMDA component in MK801 was used to monitor the probability of release, a parameter ostensibly independent of the type of receptor, so the reason for the discrepancy cannot be readily explained by the fact that AMPA receptor-mediated LTP was monitored in one experiment and NMDA receptor-mediated LTP in the other. Note that this approach would not detect a mechanism which involved the growth or uncovering of extra synapses with the same release probability as the original population, since the time course of decay in the presence of MK801 would not be affected.

4.5.3 Silent synapses

The idea that synapses may be functionally silent was introduced by Wall (1977), in the context of signaling in the spinal cord, and the concept has cast a long shadow over quantal analysis elsewhere in the central nervous system (Redman, 1990). What proportion of synapses are silent becomes as salient a question as what proportion of synapses are potentiable. If the answer is most and few respectively, the experience of Sayer et al., who very rarely detected LTP in pairs of connected cells in the hippocampal slice (Friedlander et al., 1990), becomes less perplexing. On the other hand, in the young (but not very young) rat, Bolshakov and Siegelbaum found it relatively easy to find pairs of interconnected cells that were connected by nonsilent synapses, and which could be readily potentiated. A silent synapse could be silent in a number of ways. Release probability could be effectively zero, or p_r could be finite and the postsynaptic membrane devoid of receptors; or both conditions might exist, with silence on both sides of the synapse. Furthermore, postsynaptic silence might apply to the whole population of receptors, or only to selected subtype(s). Two groups (Liao et al., 1995, and Isaac et al., 1995) have produced evidence for the existence of synapses in area CA1 which express NMDA receptors but are silent with respect to AMPA receptors. In the protocol of Liao et al., a cell was voltage clamped near resting membrane potential, and the stimulus intensity reduced to a level which consistently failed to produce a response (100 failures in a row). The cell was then depolarized and stimulation continued at the same weak intensity. At these depolarized voltages, an NMDA receptor-mediated response was often present. This, the authors concluded, showed that the stimulating electrode was activating a fiber with a nonzero release probability, and that the failure to detect a response at normal membrane potentials (where the NMDA response would be small because of the voltage-dependent block by Mg^{2+}) must, therefore, be due to an absence of AMPA receptors. After pairing, an AMPA response appeared with nonzero probability. Thus, in this model, the reduction in failures following LTP is due not to a presynaptic modulation of release probability, but to the insertion of a latent population of AMPA receptors. There is an alternative explanation, however, for which some evidence exists (Kullmann et al., 1996). The NMDA receptor is two orders of magnitude more sensitive to glutamate than the AMPA receptor, and it is possible that the receptor can detect changes in ambient levels of glutamate caused by spill-over of transmitter from neighboring synapses. If this is the source of the NMDA receptor-mediated response seen when the cell is depolarized, then the simplest reason for the persistent failures at normal membrane potentials is that the synapse is silent in the presynaptic sense, that is its probability of release is near zero. It then becomes unnecessary to postulate the existence of latent AMPA receptors: LTP could be accounted for by an increase in p_r. This explanation could be tested by a protocol in which whole-cell attachments were made to a connected pair of CA3 and CA1 cells. Stimuli delivered to the CA3 cell would produce action potentials only in the axon of that cell, and this would limit spill-over to the adjacent *en passant* terminals from the same axon. The test requires that the paired cells be connected weakly initially (p_r near zero). The spill-over hypothesis would predict that under these conditions, the reduction in failures on depolarization will be markedly less than with a pairing protocol using minimal stimulation of afferent axons. (Failure of the prediction would not, however, invalidate the hypothesis, since spill-over from the collateral terminals might still be sufficient to activate NMDA receptor-mediated responses.)

4.6 Are NMDA and AMPA receptor-mediated components of the evoked response potentiated equally?

If the question posed here can be answered in the affirmative, we would not be much the wiser, at least with regard to the subject of this chapter. However, a presynaptic model would be hard pressed to account for the expression of LTP if only the AMPA receptor-mediated response was potentiated, since an increase in transmitter release should be equally available to both receptor subtypes. Therefore, the announcement by two laboratories in 1988 that exactly this was the case (Kauer et al., 1988; Muller and Lynch, 1988) gave the presynaptic–postsynaptic seesaw an enthusiastic push in the postsynaptic direction. The assertion that the NMDA receptor-mediated component could not be potentiated was disputed by Bashir *et al.* (1991), and several groups have now confirmed that the NMDA response is potentiable (Asztely *et al.*, 1992; Clark and Collingridge, 1995; Kullmann, 1996; O'Connor *et al.*, 1995; Xie *et al.*, 1992). There remains disagreement as to whether the magnitude of LTP of the NMDA component is as great as it is for the AMPA component and, to that extent, therefore, there remains some support for the idea that there is a component of LTP which reflects a differential modification of receptor function.

4.7 Paired-pulse facilitation and depression

Two opposing presynaptic mechanisms control the size of the response evoked by the second of a pair of stimuli delivered to hippocampal synapses when the interstimulus interval is less than 1–2 sec. The first is paired-pulse depression (PPD), which comes into play only if the first stimulus results in the release of transmitter, and which is thought to reflect depletion of the docked or immediately releasable pool of vesicles (Debanne *et al.*, 1996; Thies, 1965). There is an enormous disparity in estimates of how long recovery from this process takes: Stevens and Wang (1995), using the technique of minimal stimulation in area CA1 of acute hippocampal slices, concluded that depression persisted for only 20 msec; in contrast, at single CA3–CA1 connections (with multiple release sites) in organotypic cultures, recovery can take as long as 5 sec (Debanne *et al.*, 1996). The second mechanism is paired-pulse facilitation (PPF), which is usually referred to simply as facilitation. In contrast to PPD, facilitation occurs whether or not the first stimulus is successful in releasing transmitter. The generally agreed explanation for the effect is the residual calcium hypothesis of Katz and Miledi (1968). The increase in calcium concentration at the release zone following the invasion of the terminal and the opening of voltage-dependent calcium channels by the first action potential decays with a time course of a few hundred milliseconds; the calcium flux due to the second action potential adds to this residual calcium, resulting in a higher than normal calcium concentration and a corresponding increase in the probability of release. Direct evidence for the residual calcium hypothesis at Schaffer–collateral terminals has been presented by Wu and Saggau (1994), who measured presynaptic calcium transients and found a linear relationship between facilitation and residual calcium concentration. If, at a given release site, the first response results in release of transmitter, the amplitude of the second response will thus be determined by the interaction of depression and facilitation, with depression being the dominant factor. [Note that in field potential recordings, where activity in many synapses is sampled, the amplitude of the second response will be determined by the distribution of release probabilities, with facilitation or depression reflecting a low or high mean p_r respectively. This point is well illustrated by the medial and lateral branches of the perforant path; the medial component exhibits PPD at all interstimulus intervals, indicating a high mean p_r, while, in the lateral branch, pronounced facilitation is present at short intervals, sugggesting a low mean p_r (McNaughton, 1980).]

What can these two presynaptic mechanisms tell us about the locus of LTP? The question was first posed by McNaughton (1982) in an influential paper examining facilitation in the lateral perforant path of the anesthetized rat. McNaughton found that neither the time course nor the amplitude of facilitation were changed when measured more than a few minutes after the induction of LTP. However, immediately following the tetanus, facilitation was replaced by PPD which gradually gave way to facilitation again, with a time course reflecting that of post-tetanic potentiation. At the neuromuscular junction, post-tetanic potentiation can be accounted for by an increase in the p_r (Augustine *et al.*, 1987; Delaney *et al.*, 1989). Calcium measurements in hippocampal synapses have revealed transient post-tetanic increases in residual calcium (Regehr and Tank, 1991; Wu and Saggau, 1994), strongly suggesting that post-tetanic potentiation in these synapses is also mediated presynaptically. Thus, if facilitation and PPD are mediated presynaptically at hippocampal synapses (see below), the reduction in facilitation during post-tetanic potentiation can be explained by an increase in mean p_r in the population of tetanized terminals, and a consequent switch from facilitation to depression.

Since he observed no such reduction in facilitation during LTP, McNaughton (1982) argued that the expression of LTP cannot be explained by changes in the probability of release, though he did not exclude the possibility of other presynaptic changes (e.g. in the number of release sites, or in the amount of transmitter released per quantum).

The paired-pulse test has become a standard method for analyzing the locus of activity-dependent changes in synaptic efficacy, in the hippocampus and elsewhere in the cortex. In reviewing the hippocampal literature, we shall consider the following issues. First, what is the evidence that facilitation is itself a presynaptic phenomenon in the hippocampus? Second, is there a consensus that facilitation is not affected by LTP? Third, how solid is the inference that a lack of change in facilitation is incompatible with a change in the probability of release?

4.7.1 Are paired-pulse depression and facilitation in the hippocampus presynaptically mediated?

The evidence strongly suggests that this is the case for facilitation in the hippocampus. The magnitude of facilitation is directly proportional to the level of residual calcium in Schaffer-collateral terminals in area CA1 (Wu and Saggau, 1994), and quantal analysis also indicates a presynaptic locus [increase in m, decrease in failures, and a decrease in the coefficient of variation (Foster and McNaughton, 1991)]. Moreover, at putative single synapses in area CA1 of the hippocampal slice, the probability of failure to the second of two stimuli is reduced by the same extent whether or not the first stimulus of the pair produced a response (at time intervals which in this system were sufficiently long such that PPD was no longer a factor; see Stevens and Wang, 1995). It is hard to see how facilitation can be ascribed to postsynaptic changes if it is independent of whether or not postsynaptic responses have been recruited. Direct evidence that PPD is presynaptic has not been obtained; unlike facilitation, it could, in principle, be ascribed to a postsynaptic process such as desensitization, since it only occurs in cases where transmission is successful (Stevens and Wang, 1995). The experiments of Debanne et al. (1996), in which impaled CA3 cells in organotypic cultures were stimulated, and whole-cell recordings made from a target CA1 cell, throw light on this issue. In organotypic cultures, there is extensive proliferation of CA3 axonal arbors, and a target CA1 cell receives many inputs from a single CA3 projection cell (Debanne et al., 1995). When the initial response to the first of a pair of stimuli was large, implying release of transmitter from several afferent terminals, PPD was observed; when the initial response was small, PPF resulted. PPD could be explained either in terms of a presynaptic model, in which release leads to a transient period of depression as vesicle redocking takes place, or by a postsynaptic model, involving, for instance, desensitization of the receptor by transmitter released by the first impulse. However, in organotypic cultures, Debanne et al. (1996) found that PPD is not affected by desensitization blockers; they concluded, therefore, that, as at the neuromuscular junction (Betz, 1970; Thiels et al. 1994), PPD at CA3–CA1 synapses is due to a decrease in the probability of transmitter release.

4.7.2 Most, but not all, reports conclude that facilitation does not interact with LTP in area CA1, but does interact with LTP in mossy fibers

Most experiments have confirmed McNaughton's original findings that facilitation is not affected by LTP (Gustafsson et al., 1988; Muller and Lynch, 1989; Zalutsky and Nicoll, 1990). However, Christie and Abraham (1994) have disputed McNaughton's observations in the lateral perforant path and, in another field recording study, Schulz et al. (1994) found both increases and decreases in facilitation during LTP in area CA1. The changes were inversely correlated with the initial magnitude of facilitation – a large initial value was associated with a reduction in facilitation following the induction of LTP, and a small initial value with an increase. An increase in the number of release sites could explain this pattern of results, providing the recruited sites had a different mean probability of release from the original population – higher or lower, respectively, depending on whether facilitation was depressed or enhanced (an increase in n, with no change in p_r would produce no change in PPF). Other models are possible, although an LTP-associated increase in p_r alone could not explain the experiments in which facilitation was enhanced. Changes in facilitation have also been described in an intracellular study in CA1 (Kuhnt and Voronin, 1994). However, another account, based on whole-cell recording from CA1 neurons, reported no change in facilitation during LTP (Manabe et al., 1993), although a change in CV^2 was noted, a combination of results which is compatible with an increase in the number of release sites (with no change in p_r), or with a postsynaptic mechanism with a nonuniform change in postsynaptic sensitivity (see Faber and Korn, 1992). A striking difference in the way facilitation is affected by LTP at mossy fiber and commissural-associational inputs to CA3 pyramidal cells had been described earlier by Zalutsky and Nicoll (1990). Mossy fiber LTP, which is non-NMDA dependent, was associated with a reduction in PPF, while there was no change in facilitation following the induction of NMDA receptor-dependent LTP in the commissural–associational pathway. Other evidence also suggests that mossy fiber LTP differs significantly in its expression as well as in its induction mechanisms from those responsible for NMDA receptor-dependent LTP (Nicoll and Malenka, 1995; Staubli *et al.*, 1990; Zalutsky and Nicoll, 1990). By a narrow margin, then, the weight of evidence supports the conclusion that facilitation is unchanged in NMDA receptor-dependent LTP, but the matter is unlikely to be settled definitively until experiments between pairs of connected cells have been reported. There do not appear to have been any comparable studies of paired-pulse depression in LTP.

4.7.3 How secure is the conclusion that a lack of change in facilitation is incompatible with a change in the probability of release?

In field potential studies, in which large numbers of afferent fibers are stimulated, the possibility cannot be excluded that different subsets of synapses are contributing to post-tetanic potentiation and LTP. While we do not know if all excitatory synapses in the hippocampus exhibit LTP, there is no reason to believe that post-tetanic potentiation and facilitation are not properties shared by all synapses in the hippocampus, if not the rest of the neural universe. However, even assuming, as minimal stimulation experiments suggest (Stevens and Wang, 1995), that facilitation and LTP occur at the same synapses, it will not necessarily be the case that the two effects are equally well represented. Both LTP and facilitation are activity-dependent, but LTP has additional requirements imposed by cooperativity and the necessity to activate the NMDA receptor. Following a tetanus, most synapses will be subject to post-tetanic potentiation, since even low p_r synapses will have been activated during the tetanus, while LTP will only be induced in a subset of these synapses. Thus, the interaction of facilitation with post-tetanic potentiation will always be greater than its interaction with LTP. If the proportion of synapses in which LTP is induced is small, it is possible that facilitation – which samples the whole population – will not be decreased substantially during LTP. However, this confounding element, though not eliminated, is less serious when recording from single cells with minimal stimulation (Stevens and Wang, 1995). Again, dual impalement experiments are likely to provide a definitive answer. Meanwhile, we shall assume, for the purposes of this discussion, that facilitation and LTP can occur at the same synapses.

The probability that a quantum of transmitter will be released is a power function of the concentration of intraterminal Ca^{2+} (Augustine *et al.*, 1987; Katz and Miledi, 1970). We may think of the process of synaptic transmission as being divided into three components: the initial activity-dependent Ca^{2+} transient; the residual level of cytosolic Ca^{2+} in the terminal; and the sensitivity to Ca^{2+} of the processes of docking, priming and fusion, which determine whether or not release will take place (Wu and Saggau, 1994). A change in any one of these will lead to a change in the probability of release. The first two possibilities have been ruled out convincingly by the calcium measurements of Wu and Saggau (1994), who found no change in calcium transients or in residual calcium during LTP in the Schaffer–commissural pathway in area CA1. There is some evidence, however, that a persistent change in the sensitivity of the release process to Ca^{2+} occurs in LTP. Lynch and Bliss (1986) found that the sensitivity to Ca^{2+} of depolarization-induced glutamate release in the dentate gyrus was increased during LTP, and a prolonged increase in the effects of Ca^{2+} on miniature excitatory postsynaptic potential (EPSP) frequency has been reported in hippocampal cultures during potentiation induced by brief exposure to glutamate (Malgaroli et al., 1992). We are thus led to the conclusion that the presynaptic mechanisms underlying LTP are likely to be different from those responsible for post-tetanic potentiation and facilitation. The data discussed above suggest that the increase in release probability in both post-tetanic potentiation and facilitation is brought about by a transient increase in residual $[Ca^{2+}]$, while in LTP it is due to an increase in the sensitivity to Ca^{2+}. Nevertheless, a reduction in facilitation would still be expected to occur if an increase in p_r contributes to the expression of LTP. Consider a population of n terminals with an average release probability p_r. The first stimulus of the pair will activate $n.p_r$ terminals; because of PPD, none of these $n.p_r$ terminals will respond to the second stimulus, while facilitation will ensure that most if not all the $n-n.p_r$ terminals that failed to respond to the first stimulus will respond to the second transmitter (Stevens and Wang, 1995). The PPF ratio, f, is then approximately:

$$f = (n - n.p_r)/n.p_r = (1 - p_r)/p_r$$

(Where the assumptions underlying this relationship hold, we can estimate p_r from f:

$$p_r = 1/(1 + f)$$

and if f is greater than 1 (facilitation rather than depression) then $p_r < 0.5$). The rate at which facilitation changes with p_r is given by:

$$df/dp_r = -1/p_r^2$$

which is always negative; thus, according to this analysis, if LTP is associated with an increase in p_r, it should also be accompanied by a decrease in facilitation which is inversely proportional to p_r^2.

4.8 Retrograde messengers

The retrograde messenger hypothesis was proposed originally by Bliss *et al.* (1986) to reconcile two apparently incompatible findings: a postsynaptic locus for the induction of NMDA receptor-dependent LTP and a persistent increase in glutamate efflux indicating a presynaptic component to the expression of LTP.

Evidence for the existence of retrograde messengers has waxed and waned in the years since, and an assessment of the relevant literature will not be undertaken here (see Fazeli, 1992, for review). Suffice it to say that evidence for a particular candidate messenger will necessarily provide evidence for an increase in transmitter release and hence for a presynaptic component to the expression of LTP (see, for example, Lynch et al., 1989a; O'Dell et al., 1991).

4.9 Conclusions

What answer can be given to the question we set out to consider at the beginning of this chapter? The evidence that both presynaptic and postsynaptic changes occur in LTP is compelling, but proof that these changes are causally related to the maintenance of LTP is for the most part missing. Thus, an increase in transmitter release does not necessarily mean that maintenance of LTP is wholly or partly presynaptic – an increase in the concentration of transmitter in synaptic vesicles, for instance, would result in an increase in transmitter release, but if the normal level of release saturates the subsynaptic receptor population, the increase in concentration would not translate into an increased response. Conversely, a change in quantal size (classically, a postsynaptic mechanism) would be predicted if more glutamate were packed into each vesicle (a presynaptic effect), and an increase in the response to applied receptor agonist could be the result of a non-specific action of the agonist on presynaptic terminals. These arguments and counter arguments have been so well rehearsed over the past 5 years, that it may be wondered how and if the debate is ever going to be settled. In fact, the situation is even more complex than we have suggested, since much of the evidence relating to the question of the locus of LTP has been derived from *in vitro* experiments in which potentiation is rarely followed for more than an hour or two. At longer intervals, it is probable that morphological changes become the dominant mechanism for potentiation and, if that is the case, both presynaptic and postsynaptic mechanisms of growth will have their roles to play.

References

Abraham WC, Dragunow M, Tate WP. (1991) The role of immediate early genes in the stabilization of long-term potentiation. *Mol. Neurobiol.* 5: 297–314.

Ambros-Ingerson J, Larson J, Peng X, Lynch G. (1991) LTP changes the wave-form of synaptic responses. *Synapse* 9: 314–316.

Aniksztejn L, Roisin MP, Amsellem R, Ben-Ari Y. (1989) Long-term potentiation is not associated with a sustained enhanced release of endogenous excitatory amino acids. *Neuroscience* 28: 387–392.

Asztely F, Wigström H, Gustafsson B. (1992) The relative contribution of NMDA receptor channels in the expression of long-term potentiation in the hippocampal CA1 region. *Eur. J. Neurosci.* 4: 681–690.

Augustine GJ, Charlton MP, Smith SJ. (1987) Calcium action in synaptic transmitter release. *Annu. Rev. Neurosci.* 10: 633–693.

Bashir ZI, Alford S, Davies SN, Randall AD, Collingridge GL. (1991) Long-term potentiation of NMDA receptor-mediated synaptic transmission in the hippocampus. *Nature* 349: 156–158.

Bekkers JM, Stevens CF. (1990) Presynaptic mechanisms for long-term potentiation in the hippocampus. *Nature* **346:** 724–729.

Betz WJ. (1970) Depression of transmitter release at the neuromuscular junction of the frog. *J. Physiol. (Lond.)* **206:** 629–644.

Bliss TVP, Collingridge GL. (1993) A synaptic model of memory – long-term potentiation in the hippocampus. *Nature* **361:** 31–39.

Bliss TVP, Lømo T. (1973) Long-lasting potentiation of synaptic transmission in the dentate area of the anaesthetized rabbit following stimulation of the perforant path. *J. Physiol.* **232:** 331–356.

Bliss TVP, Douglas RM, Errington ML, Lynch MA. (1986) Correlation between long-term potentiation and release of endogenous amino acids from dentate gyrus of anaesthetized rats. *J. Physiol.* **377:** 391–408.

Bliss TVP, Errington ML, Laroche S, Lynch MA. (1987) Increase in K^+-stimulated, Ca^{2+}-dependent release of [^{3}H]glutamate from rate dentate gyrus three days after induction of long-term potentiation. *Neurosci. Lett.* **83:** 107–112.

Bolshakov V, Siegelbaum S. (1994) Postsynaptic induction and presynaptic expression of hippocampal long-term depression. *Science* **264:** 1148–1151.

Christie BR, Abraham WC. (1994) Differential regulation of paired-pulsed plasticity following LTP in the dentate gyrus. *NeuroReport* **5:** 385–388.

Clark KA, Collingridge GL. (1995) Synaptic potentiation of dual-component excitatory postsynaptic currents in the rat hippocampus. *J. Physiol. (Lond.)* **482:** 39–52.

Clark KA, Collingridge, GL. (1996) Evidence that postsynaptic changes are involved in the expression of LTP and LTD of NMDA receptor-dependant EPSCs in area CA1 of the hippocampus. *J. Physiol. (Paris)* Abstract in press.

Cole AJ, Saffen DW, Baraban JM, Worley PF. (1989) Rapid increase of an immediate early gene messenger RNA in hippocampal neurons by synaptic NMDA receptor activation. *Nature* **340:** 474–476.

Collingridge GL, Davies CH, Bashir ZI, Alford S, Bortolotto ZA, Harvey J, Frengnelli BG. (1992) Roles of glutamate receptors in LTP in the CA1 region of the hippocampus. In: *Neuroreceptors, Ion Channels and the Brain* (eds N Kawai, T Nakajima, E Barnard). Elsevier, Amsterdam, pp. 153–159.

Corrie JET, Desantis A, Katayama Y, Khodakah K, Messenger JB, Ogden DC, Trentham DR. (1993) Postsynaptic activation at the squid giant synapse by photolytic release of L-glutamate from a caged L-glutamate. *J. Physiol. (Lond.)* **465:** 1–8.

Davies SN, Lester RAJ, Reymann KG, Collingridge GL. (1989) Temporally distinct pre-synaptic and post-synaptic mechanisms maintain long-term potentiation. *Nature* **338:** 500–503.

Debanne D, Guérineau N, Gähwiler BH, Thompson SM. (1995) Physiology and pharmacology of unitary synaptic connections between pairs of cells in areas CA3 and CA1 of rat hippocampal slice cultures. *J. Neurophysiol.* **73:** 1282–1294.

Debanne D, Guérineau N, Gähwiler BH, Thompson SM. (1996) Paired-pulse facilitation and depression at unitary synapses in rat hippocampus: quantal fluctuation affects subsequent release. *J. Physiol. (Lond.)* **491:** 163–176.

Delaney KR, Zucker RS, Tank DW. (1989) Calcium in motor nerve terminals associated with post-tetanic potentiation. *J. Neurosci.* **9:** 3558–3567.

Dolphin AC, Errington ML, Bliss TVP. (1982) Long-term potentiation of the perforant path *in vivo* is associated with increased glutamate release. *Nature* **297:** 496–498.

Dragunow M, Abraham WC, Goulding M, Mason SE, Robertson HA, Faull R. (1989) Long-term potentiation and the induction of c-fos messenger RNA and proteins in the dentate gyrus of unanesthetized rats. *Neurosci. Lett.* **101:** 274–280.

Edwards FA, Konnerth A, Sakmann B, Takahashi T. (1989) A thin slice preparation for patch clamp recordings from neurons of the mammalian central nervous system. *Eur. J. Physiol.* **414:** 600–612.

Erdemli G, Kullman, D. (1996) Long-term potentiation of NMDA receptor-mediated signals in CA1 pyramidal cells of guinea pig hippocampal slices: evidence for an increase in transmitter release probability. *J. Physiol. (Lond.)* **494:** 81P.

Errington ML, Lynch MA, Bliss TVP. (1987) Long-term potentiation in the dentate gyrus: induction and increased glutamate release are blocked by D(-)aminophosphonovalerate. *Neuroscience* **20:** 279–284.

Faber DS, Korn H. (1992) Applicability of the coefficient of variation method for analyzing synaptic plasticity. *Biophys. J.* **60**: 1288–1294.

Fazeli MS. (1992) Synaptic plasticity: on the trail of the retrograde messenger. *Trends Neurosci.* **15**: 115–117.

Feasey KJ, Lynch MA, Bliss TVP. (1986) Long-term potentiation is associated with an increase in calcium-dependent, potassium-stimulated release of [^{14}C]glutamate from hippocampal slices: an *ex vivo* study in the rat. *Brain Res.* **364**: 39–44.

Foster TC, McNaughton BL. (1991) Long-term enhancement of CA1 synaptic transmission is due to increased quantal size, not quantal content. *Hippocampus* **1**: 79–91.

Friedlander MJ, Sayer RJ, Redman SJ. (1990) Evaluation of long-term potentiation of small compound and unitary epsps at the hippocampal CA3–CA1 synapse. *J. Neurosci.* **10**: 814–825.

Galley PT, Errington ML, Bliss TVP. (1993) Sustained increase in basal and stimulus-dependent glutamate efflux during LTP in dentate gyrus: real-time *in vivo* measurements using a dialysis electrode. *Soc. Neurosci. Abstr.* **19**: 904.

Ghijsen W, Besselsen E, Geukers V, Kamphuis W, Lopes Da Silva FH. (1992) Enhancement of endogenous release of glutamate and gamma-aminobutyric-acid from hippocampus CA1 slices after *in vivo* long-term potentiation. *J. Neurochem.* **59**: 482–486.

Gianotti C, Nunzi MG, Gispen WH, Corradetti R. (1992) Phosphorylation of the presynaptic protein B-50 (GAP-43) is increased during electrically induced long-term potentiation. *Neuron* **8**: 843–848.

Gustafsson B, Huang YY, Wigström H. (1988) Phorbol ester-induced synaptic potentiation differs from long-term potentiation in the guinea-pig hippocampus *in vitro*. *Neurosci. Lett.* **85**: 77–81.

Hessler NA, Shirke AM, Malinow R. (1993) The probability of transmitter release at a mammalian central synapse. *Nature* **366**: 569–572.

Hume RI, Role LW, Fischbach GD. (1983) Acetylcholine release from growth cones detected with patches of acetylcholine receptor-rich membranes. *Nature* **305**: 632–634.

Isaac JTR, Nicoll RA, Malenka RC. (1995) Evidence for silent synapses – implications for the expression of LTP. *Neuron* **15**: 427–434.

Isaacson JS, Nicoll RA. (1991) Aniracetam reduces glutamate receptor desensitization and slows the decay of fast excitatory synaptic currents in the hippocampus. *Proc. Natl Acad. Sci. USA* **88**: 10936–10940.

Katz B, Miledi R. (1968) The role of calcium in neuromuscular facilitation. *J. Physiol. (Lond.)* **195**: 481–492.

Katz B, Miledi R. (1970) Further study of the role of calcium in synaptic transmission. *J. Physiol. (Lond.)* **207**: 789–801.

Katz LC, Dalva MB. (1994) Scanning laser photostimulation: a new approach for analyzing brain circuits. *J. Neurosci. Methods* **54**: 205–218.

Kauer JA, Malenka RC, Nicoll RA. (1988) A persistent postsynaptic modification mediates long-term potentiation in the hippocampus. *Neuron* **1**: 911–917.

Korn H, Faber DS. (1991) Quantal analysis and synaptic efficacy in the CNS. *Trends Neurosci.* **14**: 439–445.

Kuhnt U, Voronin LL. (1994) Interaction between paired-pulse facilitation and long-term potentiation in area CA1 of guinea-pig hippocampal slices: application of quantal analysis. *Neuroscience* **62**: 391–397.

Kullman DM. (1996) Long-term potentiation of NMDA and AMPA receptor-mediated signals in CA1 pyramidal cells – qualitatively similar but quantitatively different. *J. Physiol. (Lond.)* **491P**: 137–138.

Kullman DM, Erdemli G, Asztely F. (1996) Differential long-term potentiation of AMPA and NMDA receptor-mediated signals in the hippocampus can be explained by presynaptic expression and glutamate spill-over. *Neuron*, in press.

Liao DZ, Hessler NA, Malinow R. (1995) Activation of postsynaptically silent synapses during pairing-induced LTP in CA1 region of hippocampal slice. *Nature* **375**: 400–404.

Linden DJ, Wong KL, Sheu F, Routtenberg A. (1988) NMDA receptor blockade prevents the increase in protein kinase C substrate (protein F1) phosphorylation produced by long-term potentiation. *Brain Res.* **458**: 142–146.

Lynch GS, Gribkoff VK, Deadwyler SA. (1976) Long-term potentiation is accompanied by a reduction in dendritic responsiveness to glutamic acid. *Nature* **263**: 151–153.

Lynch MA, Bliss TVP. (1986) On the mechanism of enhanced release of [¹⁴C]glutamate in hippocampal long-term potentiation. *Brain Res.* **369**: 405–408.

Lynch MA, Errington ML, Bliss TVP. (1989a) Nordihydroguaiaretic acid blocks the synaptic component of long-term potentiation and the associated increases in release of glutamate and arachidonate – an *in vivo* study in the dentate gyrus of the rat. *Neuroscience* **30**: 693–701.

Lynch MA, Errington ML, Bliss TVP. (1989b) The increase in [³H]glutamate release associated with long-term potentiation in the dentate gyrus is blocked by commissural stimulation. *Neurosci. Lett.* **103**: 191–196.

Lynch MA, Voss KL, Rodriguez J, Bliss TVP. (1994) Increase in synaptic vesicle proteins accompanies long-term potentiation in the dentate gyrus. *Neuroscience* **60**: 1–5.

Maeda T, Shimoshige Y, Mizukami K, Shimohama Y, Kaneko S, Akaike A, Satoh M. (1995a) Patch sensor detection of glutamate release by a single electrical shock. *Neuron* **15**: 253–257.

Maeda T, Shimoshige Y, Mizukami K, Kaneko S, Akaike A, Satoh M. (1995b) Increased excitatory amino acid release detected with an outside-out patch electrode during mossy fiber LTP in rat hippocampal slices. *Soc. Neurosci. Abstr.* **21**: 603.

Magleby KL, Stevens CF. (1972) A quantitative description of end-plate currents. *J. Physiol. (Lond.)* **223**: 173–197.

Malgaroli A, Tsien RW. (1992) Glutamate-induced long-term potentiation of the frequency of miniature synaptic currents in cultured hippocampal neurons. *Nature* **357**: 134–139.

Malgaroli A, Malinow R, Schulman H, Tsien RW. (1992) Persistent signaling and changes in presynaptic function in long-term potentiation. *Ciba Found. Symp.* **164**: 176–196.

Malgaroli A, Ting AE, Wendland B, Bergamascht A, Villa A, Tsien RW, Scheller RH. (1995) Presynaptic component of long-term potentiation visualized at individual hippocampal synapses. *Science* **268**: 1624–1628.

Malinow R. (1991) Transmission between pairs of hippocampal slice neurons – quantal levels, oscillations, and LTP. *Science* **252**: 722–724.

Malinow R, Tsien R. (1990) Presynaptic changes revealed by whole cell recordings of long-term potentiation in rat hippocampal slices. *Nature* **346**: 177–180.

Manabe T, Nicoll RA. (1994) Long-term potentiation – evidence against an increase in transmitter release probability in the CA1 region of the hippocampus. *Science* **265**: 1888–1892.

Manabe T, Renner P, Nicoll RA. (1992) Postsynaptic contribution to long-term potentiation revealed by the analysis of miniature synaptic currents. *Nature* **355**: 50–55.

Manabe T, Wyllie DJA, Perkel DJ, Nicoll RA. (1993) Modulation of synaptic transmission and long-term potentiation: effects on paired-pulse facilitation and EPSC variance in the CA1 region of the hippocampus. *J. Neurophysiol.* **70**: 1451–1459.

McNaughton BL. (1980) Evidence for two physiologically distinct perforant pathways to the fascia dentata. *Brain Res.* **199**: 1–19.

McNaughton BL. (1982) Long-term synaptic enhancement and short-term potentiation in rat fascia dentata act through different mechanisms. *J. Physiol.* **324**: 249–262.

Meberg PJ, Barnes CA, McNaughton BL, Routtenberg A. (1993) Protein kinase C and F1/GAP-43 gene expression in hippocampus inversely related to synaptic enhancement lasting 3 days. *Proc. Natl Acad. Sci. USA* **90**: 12050–12054.

Muller D, Lynch G. (1988) Long-term potentiation differentially affects two components of synaptic responses in hippocampus. *Proc. Natl Acad. Sci. USA* **85**: 9346–9350.

Muller D, Lynch G. (1989) Evidence that changes in presynaptic calcium currents are not responsible for long-term potentiation in hippocampus. *Brain Res.* **479**: 290–299.

Nicoll RA, Malenka RC. (1995) Contrasting properties of two forms of long-term potentiation in the hippocampus. *Nature* **377**: 115–118.

Nicolls D, Attwell D. (1991) The release and uptake of excitatory amino acids. *Trends Pharmacol.* **11**: 462–468.

O'Connor JJ, Wu J, Rowan MJ, Anwyl R. (1995a) Potentiation of N-methyl-D-aspartate receptor-mediated currents detected using the excised patch technique in the hippocampal dentate gyrus.

Neuroscience **69**: 363–369.

O'Connor J, Rowan MJ, Anwl R. (1995b) Tetanically induced LTP involves a similar increase in the AMPA and NMDA receptor components of the excitatory postsynaptic current: investigations of the involvement of mGlu receptors. *J. Neurosci.* **15**: 2013–2020.

O'Dell T, Hawkins RD, Kandel ER, Arancio O. (1991) Tests of the roles of two diffusible substances in long-term potentiation: evidence for nitric oxide as a possible early retrograde messenger. *Proc. Natl Acad. Sci. USA* **88**: 11285–11289.

Oliet SHR, Malenka RC, Nicoll RA. (1996) Bidirectional control of quantal size by synaptic activity in the hippocampus. *Science* **271**: 1294–1297.

Ramakers GMJ, De Graan PNE, Urban IJA, Kraay D, Tang T, Pasinelli P, Oestreicher AB, Gispen WH. (1995) Temporal differences in the phosphorylation state of presynaptic and postsynaptic protein-kinase-C substrates B50/GAP-43 and neurogranin during long-term potentiation. *J. Biol. Chem.* **270**: 13892–13898.

Raymond LA, Blackstone CD, Huganir RL. (1993) Phosphorylation of amino acid neurotransmitter receptors in synaptic plasticity. *Trends Neurosci.* **16**: 147–153.

Redman S. (1990) Quantal analysis of synaptic potentials in neurons of the central nervous system. *Physiol. Rev.* **70**: 165–198.

Regehr WG, Tank DW. (1991) The maintenance of LTP at hippocampal mossy fiber synapses is independent of sustained presynaptic calcium. *Neuron* **7**: 451–459.

Rosenblum K, Richter-Levin G, Dudai Y. (1996) Long-term potentiation increases tyrosine phosphorylation of the *N*-methyl-D-aspartate receptor for subunit 2B in the rat dentate gyrus *in vivo*. *Proc. Natl Acad. Sci. USA*, in press.

Rosenmund C, Clements JD, Westbrook GL. (1993) Nonuniform probability of glutamate release at a hippocampal synapse. *Science* **262**: 754–757.

Rostas JAP, Brent VA, Voss K, Errington ML, Bliss TVP, Gurd JW. (1996) Enhanced tyrosine phosphorylation of the 2B subunit of the *N*-methyl-D-aspartate receptor in long-term potentiation. *Proc. Natl Acad. Sci. USA*, in press.

Routtenberg A, Lovinger D, Steward O. (1985) Selective increase in phosphorylation of a 47 kDa protein (F1) directly related to long-term potentiation. *Behav. Neural Biol.* **43**: 3–11.

Schrama LH, Degraan P, Zwiers H, Gispen WH. (1986) Comparison of a 52-kDa phosphoprotein from synaptic plasma-membranes related to long-term potentiation and the major coated vesicle phosphoprotein. *J. Neurochem.* **47**: 1843–1848.

Schulz PE, Cook EP, Johnston D. (1994) Changes in paired-pulse facilitation suggest presynaptic involvement in long-term potentiation. *J. Neurosci.* **14**: 5325–5337.

Skrede KK, Malthe-Sørenssen D. (1981) Increased resting and evoked release of transmitter following repetitive electrical tetanization in hippocampus: a biochemical correlate to long-lasting synaptic potentiation. *Brain Res.* **208**: 436–441.

Smirnova T, Laroche S, Errington ML, Hicks AA, Bliss TVP, Mallet J. (1993) Transsynaptic expression of a presynaptic glutamate-receptor during hippocampal long-term potentiation. *Science* **262**: 433–436.

Soderling TR, Tan S-E, McGlade-McCulloch E, Yamamoto H, Fukunaga K. (1994) Excitatory interactions between glutamate receptors and protein-kinases. *J. Neurobiol.* **25**: 304–311.

Staubli U, Larson J, Lynch G. (1990) Mossy fiber potentiation and long-term potentiation involve different expression mechanisms. *Synapse* **5**: 333–335.

Stevens CF, Wang YY. (1994) Changes in reliability of synaptic function as a mechanism for plasticity. *Nature* **371**: 704–707.

Stevens CF, Wang YY. (1995) Facilitation and depression at single central synapses. *Neuron* **14**: 795–802.

Thiels E, Barrionuevo G, Berger T. (1994) Excitatory stimulation during postsynaptic inhibition induces long-term depression in hippocampus *in vivo*. *J. Neurophysiol.* **72**: 3009–3116.

Thies RE. (1965) Neuromuscular depression and the apparent depletion of transmitter in mammalian muscle. *J. Neurophysiol.* **28**: 427–442.

Thomas KL, Laroche S, Errington ML, Bliss TVP, Hunt SP. (1994) Spatial and temporal changes in signal transduction pathways during LTP. *Neuron* **13**: 737–745.

Voronin LL. (1983) Long-term potentiation in the hippocampus. *Neuroscience* **10**: 1051–1069.

Wall, PD. (1977) The presence of ineffective synapses and the circumstances which unmask them. *Phil. Trans. R. Soc. (B)* **278:** 361–372.

Wang YY, Salter MW. (1994) Regulation of NMDA receptors by tyrosine kinases and phosphatases. *Nature* **369:** 233–235.

Weisskopf MG, Nicoll RA. (1995) Presynaptic changes during mossy fiber LTP revealed by NMDA receptor-mediated synaptic responses. *Nature* **376:** 256–259.

Wisden W, Errington ML, Williams S, Dunnet SB, Waters C, Hitchcock D, Evan G, Bliss TVP, Hunt SP. (1990) Differential expression of immediate early genes in the hippocampus and spinal cord. *Neuron* **4:** 603–614.

Worley PF, Bhat RV, Baraban JM, Erickson CA, McNaughton BL, Barnes CA. (1993) Thresholds for synaptic activation of transcription factors in hippocampus: correlation with long-term enhancement. *J. Neurosci.* **13:** 4776–4786.

Wu L, Saggau P. (1994) Presynaptic calcium is increased during normal synaptic transmission and paired-pulse facilitation, but not in long-term potentiation in area CA1 of hippocampus. *J. Neurosci.* **14:** 645–654.

Xie XP, Berger TW, Barrionuevo G. (1992) Isolated NMDA receptor-mediated synaptic responses express both LTP and LTD. *J. Neurophysiol.* **67:** 1009–1013.

Yuste R, Denk W. (1995) Dendritic spines as a basic functional units of neuronal integration. *Nature* **375:** 682–684.

Zalutsky RA, Nicoll RA. (1990) Comparison of two forms of long-term potentiation in single hippocampal neurons. *Science* **248:** 1619–1624.

5

Post-translational mechanisms which could underlie the postsynaptic expression of LTP and LTD

Timothy A. Benke, Iris Bresink, Valerie J. Collett, Andrew J. Doherty, Jeremy M. Henley and Graham L. Collingridge

5.1 Introduction

Most of the synapses in the vertebrate brain which exhibit long-term potentiation (LTP) and/or long-term depression (LTD) utilize glutamate as their neurotransmitter. At these synapses, the glutamate receptors both mediate basal synaptic transmission and are involved in the induction of the change in synaptic efficiency (see Chapter 2). The induction involves a transient alteration in the frequency of activation of glutamate receptors; for example, in the hippocampus, transient high frequency stimulation (e.g. 100 Hz, 1 sec) leads to LTP and prolonged 'low' frequency stimulation (e.g. 1 Hz, 15 min) leads to LTD. The induction of both phenomena involves second messenger systems, for example the activation of protein kinases and phosphatases are implicated in the induction of hippocampal LTP and LTD, respectively (see Chapter 3). What happens downstream of these initial events is less clear. There continues to be a vigorous debate as to whether LTP at a given synapse is expressed by pre- and/or postsynaptic modifications (see Chapter 4). This is particularly true for the synapse between Schaffer collateral–commissural fibers and CA1 pyramidal neurons in the hippocampus, where most LTP studies are carried out. The locus of expression of LTD is less contentious, perhaps because it has not yet been studied so extensively. Most LTD studies have focused on the parallel synapse on cerebellar Purkinje cells (see Chapter 11) and various pathways within

---

Cortical Plasticity, edited by M.S. Fazeli and G.L. Collingridge.
© 1996 BIOS Scientific Publishers Ltd, Oxford.

the neocortex (see Chapter 10), though recently attention has turned to homosynaptic LTD in the hippocampus. This phenomenon was first described as a means of reversing LTP, where it is referred to as depotentiation.

It is not the purpose of this chapter to enter further into the debate concerning the locus of expression. It seems most likely that both presynaptic and postsynaptic changes will be important for the expression of LTP and LTD. The extent to which one or other is involved will probably depend upon the synapse in question and may range from purely presynaptic to purely postsynaptic. Other critical factors may include the developmental stage of the animal, the method of induction and the time after induction. This chapter will focus on some of the possible mechanisms that could underlie the postsynaptic component of expression of LTP and LTD. The evidence for, and possible mechanisms of, postsynaptic changes associated with LTP in area CA1 is then considered, since most information is available for this synaptic modification. This chapter therefore complements Chapter 4, which concerns the possible presynaptic mechanisms. This discussion will be limited to post-translational modifications; the possible involvement of transcriptional regulation and other nuclear events for longer term modifications are the subject of Chapter 6.

There are several broad categories of mechanisms that could underlie the postsynaptic expression of LTP and LTD. However, since both processes are input specific (i.e. nonconditioned inputs on to the same neuron are not altered) it is most likely that the critical changes are confined to the synaptic spine. The possibility that changes in the electrical properties of spines, perhaps associated with alterations in spine morphology, are involved has been considered. Structural changes are likely to be involved, particularly at later stages of the plastic processes, and are considered in Chapters 4 and 6. However, the most likely targets for post-translational modification are the glutamate receptors themselves. For example, glutamate receptors expressed on the plasma membrane of the synapse may be modified so as to alter their affinity for L-glutamate or their availability to respond to glutamate. Alternatively, the trafficking rates to and from the plasma membrane might be altered so as to affect the total number of glutamate receptors expressed at the postsynaptic membrane of the synapse. This chapter starts with a summary of the subtypes, molecular structure and properties of vertebrate glutamate receptors.

5.2 Glutamate receptors

5.2.1 Subtypes of glutamate receptor

The glutamate receptor subtypes were discovered pharmacologically. Using selective, and in some cases highly specific, agonists and antagonists, four types of glutamate receptor were identified and were named after prototypic agonists as NMDA (*N*-methyl-D-aspartate), kainate, quisqualate and L-AP4 [(*S*)-2-amino-4-phosphonobutanoate] receptors (Watkins and Evans, 1981). This classification has been refined following the realization that glutamate also couples to G-protein-linked receptors, termed metabotropic glutamate receptors (mGluRs). Quisqualate receptors have been renamed as AMPA (α-amino-3-hydroxy-5-methyl-4-isoxazolepropionate) receptors, since quisqualate also activates certain subtypes of mGluRs, and L-AP4 receptors have been incorporated within the mGluR family. There has been considerable confusion over the distinction between AMPA and kainate receptors, mainly because kainate is also a potent agonist at AMPA receptors.

The cloning of genes encoding glutamate receptor subunits has confirmed the pharmacological classification of the glutamate receptor subtypes and has provided considerable additional information about their structures and diversity.

5.2.2 Molecular identity of glutamate receptors

AMPA receptors are heteromeric assemblies of subunits (probably pentamers) encoded by four separate genes (*GluR1*, *GluR2*, *GluR3* and *GluR4*; Figure 5.1). Different combinations of these subunits confer different properties; in particular the absence of a GluR2 subunit confers significant Ca^{2+} permeability to AMPA receptors. The diversity is enhanced further by alternative splicing. Each of the four genes can encode one of two splice variants termed 'flip' and 'flop' which form channels with different conductance properties. An additional complexity is provided by RNA editing; the *GluR2* gene codes for RNA which contains a glutamine (Q) codon (CAG) within the putative second 'transmembrane domain' but this is converted into an arginine (R) codon (CGG) by the action of an RNA adenosine deaminase. The edited and unedited forms display different properties; most notably, the presence of the edited GluR2 subunit within a heteromeric complex greatly limits the calcium permeability of the receptor. Normally, the GluR2 subunit is expressed in its fully edited form. [See Bettler and Mulle (1995) for further details and a full bibliography on the molecular biology of AMPA receptors.] The possibility that changes in the relative level of expression of GluR subunits, in their alternative splicing, or level of editing may contribute to the expression of LTP at later stages is discussed in Chapter 6.

Kainate receptors are probably heteromeric assemblies of GluR5, GluR6, GluR7, KA-1 and KA-2, in unknown combinations. Since kainate receptors have not (yet) been implicated in LTP these will not be considered further.

NMDA receptors are heteromeric assemblies of subunits encoded by five separate genes (*Figures 5.2* and *5.3*). NR1 is essential for function and requires co-expression with at least one NR2 subunit to confer the properties of native NMDA receptors. The four NR2 subunits NR2A, NR2B, NR2C and NR2D influence the fundamental properties of the NMDA receptor, such as the voltage dependence of Mg^{2+} block, the Ca^{2+} permeability, regulation by phosphorylation and channel kinetics. NR1 exists in eight alternatively spliced variants and a truncated short form; thus the potential diversity of native NMDA receptors is enormous. [See Mori and Mishina (1995) for further details and a full bibliography on the molecular biology of NMDA receptors.]

mGluRs are a family of at least eight G-protein-linked receptors. Each gene encodes a distinct receptor (mGluR1–mGluR8) with different properties. These

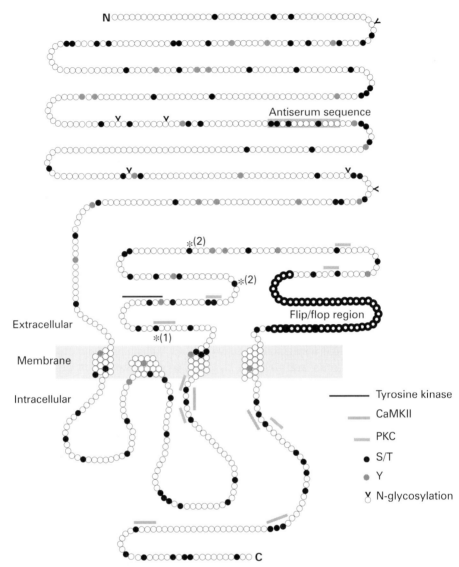

Figure 5.1. Schematic representation of the proposed structure of GluR1$_{flop}$. The locations of a number of potential kinase consensus sequences are indicated. Consensus sequences located on the N-terminal domain are not shown. The consensus sequences used were [RK]-x(2,3)-[DE]-x(2,3)-Y (tyrosine kinase), R-x-x-[S,T] (CaMKII), and [S,T]-x[R,K] (PKC). Locations of individual serine and/or threonine (black circles) and tyrosine (orange) residues are also shown. Those marked by * are likely targets for phosphorylation (1, Yakel *et al.*, 1995; 2, Nakazawa *et al.*, 1995). The peptide product from the alternatively spliced exon giving rise to flip and flop isoforms of GluR1 is indicated by heavy circles, and the location of the peptide sequence used for the preparation of polyclonal antisera against the N terminus is also indicated (Richmond *et al.*, 1996).

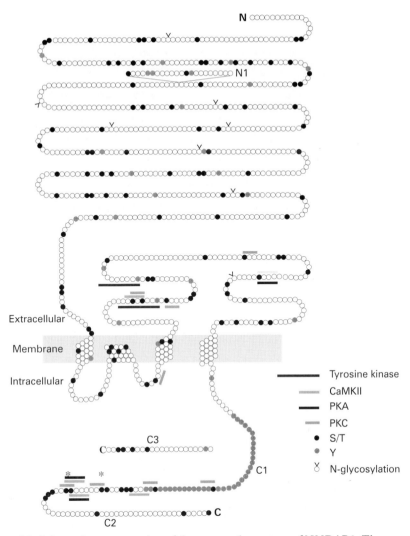

Figure 5.2. Schematic representation of the proposed structure of NMDAR1. The locations of a number of potential kinase consensus sequences are indicated. Consensus sequences are as those in *Figure 5.1* with the addition of PKA ([R,K]-x-x-[S,T]). Individual residues are indicated as in *Figure 5.1* and those likely to be phosphorylated indicated by *. The locations of peptide regions derived from alternatively spliced exons are indicated as N1 (21 amino acid sequence which may be inserted in the N terminus), C1 (37 amino acid sequence which may be deleted from the C terminus) and C2 [C-terminal tail may be replaced by an unrelated 21 amino acid sequence (C3)]. The splice variant nomenclature used here is that of Hollmann *et al.* (1993); a suffix of 1–4 is given for different combinations of used exons in the C-terminal tail. A suffix of '1' indicates a receptor containing C1 and C2 peptides, '2' indicates a C1 deletion, '3' indicates the use of C3 instead of C2 and '4' indicates both a C1 deletion and C2 replacement with C3. In addition, variants without the N1 insert have a suffix 'a' while those with the N1 insert have a suffix 'b'. This gives eight possible splice variants; NMDAR1-1a/b, -2a/b, -3a/b and -4a/b. The receptor subunit shown here is NMDAR1-1a.

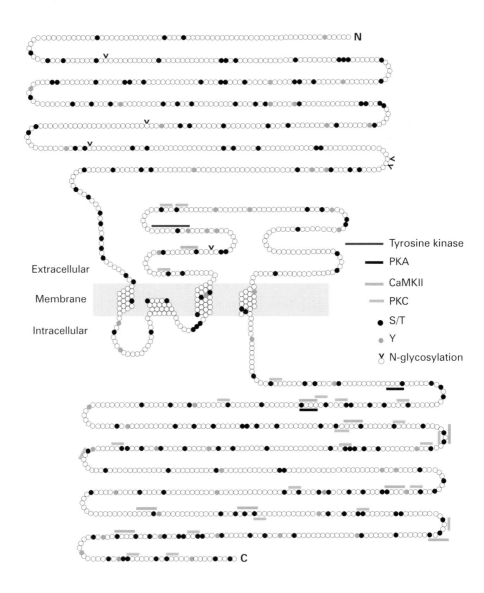

Figure 5.3. Schematic representation of the proposed structure of NMDAR2A. The locations of a number of potential kinase consensus sequences are indicated. Consensus sequences are as those in *Figure 5.2*. Individual residues are indicated as in *Figure 5.1*. Unlike the NMDAR1 subunit, the three other NMDAR2 receptor subunits (NMDAR2B, C and D) are encoded by separate genes rather than being generated by splice variation.

receptors are important for the induction of LTP (Chapter 2). However, mGluRs do not contribute directly to the synaptic potentials which are modified during LTP and so will not be considered further in this chapter.

5.2.3 Structure of AMPA and NMDA receptors

To be able to determine the nature of any direct modification of glutamate receptors, it is helpful to consider the topology of the receptor once it is incorporated within the plasma membrane. In this way, for example, the more likely sites of phosphorylation can be established. There is, however, controversy regarding the topology of both AMPA and NMDA receptor subunits. Although it is generally agreed that the N-terminal region is extracellular and the C-terminal region is intracellular and therefore there is an odd number of transmembrane-spanning regions, it is still unresolved as to where these spanning regions are located within the polypeptide. Popular current models for the topology of AMPA and NMDA receptor subunits are shown in *Figures 5.1–5.3*.

5.2.4 Phosphorylation of AMPA receptors

Basal phosphorylation. It has been shown that phosphorylation may be needed to prevent run-down of AMPA receptor-mediated currents (Wang *et al.*, 1991). The GluR1 subunit in cultured cortical neurons is basally phosphorylated on serine residues within a single tryptic phosphopeptide (Blackstone *et al.*, 1994). It has also been shown that recombinant GluR1 (flop) expressed in HEK (human embryonic kidney) 293 cells is basally phosphorylated on serine residues on the same peptide (Blackstone *et al.*, 1994; Moss *et al.*, 1993).

Protein kinase A. It has been found that AMPA receptors (activated by kainate) in cultured hippocampal neurons can be functionally regulated by protein kinase A (PKA) (Greengard *et al.*, 1991; Wang *et al.*, 1991). Thus, extracellular application of either forskolin or the permeable cAMP analog Sp-cAMPS or intracellular application of PKA enhanced kainate-induced currents approximately 1.5- to 2-fold; single channel analysis showed that the potentiation was due to an increase in both channel open time and the frequency of opening. A competitive inhibitor of PKA (Rp-cAMPS) (Wang *et al.*, 1991) or a peptide inhibitor of the catalytic subunit of PKA [PKI(5–24)] (Rosenmund *et al.*, 1994) depressed kainate-induced currents, in the absence of exogenous PKA, suggesting that AMPA receptors are regulated by endogenous PKA. Kainate currents, and AMPA receptor-mediated synaptic currents, were also depressed by an inhibitor of PKA-anchoring proteins (Ht31), suggesting that specific localization of PKA is required for it to be able to modulate AMPA receptor function (Rosenmund *et al.*, 1994). Activation of PKA also enhanced kainate-induced currents in oocytes expressing recombinant GluR1 and GluR3 (Keller *et al.*, 1992), and forskolin increased phosphorylation of GluR1 in cultured cortical neurons and of recombinant GluR1 in HEK 293 cells (Blackstone *et al.*, 1994); but see Moss *et al.* (1993). In contrast to these positive findings, it has been shown that PKA is unable to phosphorylate GluR1 in hippocampal postsynaptic densities (McGlade-McCulloch *et al.*, 1993). The reason for this discrepancy is not known. Also it should be noted that GluR1–4 lack consensus PKA sites and so the PKA-mediated phosphorylation is either at an atypical site or on a regulatory protein. Finally, PKA is able to enhance kainate responses of cells expressing recombinant GluR6, probably via phosphorylation of Ser684 (Raymond *et al.*, 1993; Wang *et al.*, 1993) but, as discussed in Section 4.2.2, GluR6 is a kainate rather than an AMPA receptor subunit.

Protein kinase C. There is evidence that PKC can regulate AMPA receptors. In particular, the proteolytically activated form of PKC (PKM) enhances the activation of hippocampal AMPA receptors by high concentrations of AMPA or kainate (i.e. above the EC_{50} value) but depresses their activation when low agonist concentrations are used (Wang *et al.* 1994). PKC also phosphorylates GluR1 in hippocampal postsynaptic densities (McGlade-McCulloch *et al.*, 1993). Furthermore, GluR1 in cultured hippocampal (Tan *et al.*, 1994) and cortical (Blackstone *et al.*, 1994) neurons can be phosphorylated by the action of phorbol esters. However, modulation by PKC is not invariably seen. For example, PKC did not affect AMPA- or kainate-induced currents in trigeminal neurons (Chen and Huang, 1992) and recombinant GluR1 or GluR1/GluR2 combinations expressed in HEK 293 cells showed little or no phosphorylation following treatment with a phorbol ester (Moss *et al.*, 1993).

It has been shown that treatment of cerebellar slices with AMPA leads to transient phosphorylation of GluR2 as detected using an antibody raised against a peptide containing phosphorylated Ser696 (Nakazawa *et al.*, 1995). This site, in the synthetic peptide, can also be phosphorylated by protein kinase G. Ser662 was also found to be a candidate phosphorylation site. Both peptide sequences are common to GluR1– GluR4 and so these sites may be of physiological relevance in the regulation of all four AMPA receptor subunits by PKC.

In oocytes, injected with total brain or subunit-specific mRNAs, variable effects on kainate-induced currents have been reported following treatment with phorbol esters (e.g. Moran and Dascal, 1989). It is well established that AMPA receptor-mediated synaptic transmission is enhanced by activation of PKC following the extracellular application of phorbol esters; however, this effect probably involves, at least in part, presynaptic alterations leading to enhanced glutamate release. Nevertheless, enhancement of AMPA receptor-mediated synaptic responses has been observed following the intracellular injection of PKC in hippocampal slices (Hu *et al.*, 1987) and PKM in hippocampal cultures (Wang *et al.*, 1994a).

CaMKII. It has been shown that calcium/calmodulin-dependent protein kinase II (CaMKII) strongly phosphorylates GluR1 in hippocampal postsynaptic densities and potentiates approximately threefold the activation of AMPA receptors by kainate in cultured hippocampal neurons (McGlade-McCulloch *et al.*, 1993). Furthermore, in cultured hippocampal neurons glutamate was able to cause phosphorylation of GluR1 via activation of NMDA receptors and CaMKII (Tan *et al.*, 1994). An important finding was that a point mutation of Ser627 to an alanine residue abolished CaMKII regulation of GluR1 expressed in oocytes (Yakel *et al.*,

1995). It was also found that a synthetic peptide corresponding to the residues 620–638 in GluR1 is phosphorylated by CaMKII but not by PKC or PKA. A constitutively active form of CaMKII injected into CA1 pyramidal neurons enhanced both AMPA receptor-mediated synaptic transmission and responses to exogenously applied AMPA (Lledo et al., 1995).

Tyrosine kinases. Co-expression in HEK 293 cells of GluR1 (or GluR1/GluR2) and the protein tyrosine kinase v-src leads to tyrosine phosphorylation of GluR1 (Moss et al., 1993). However, immunoprecipitation with anti-phosphotyrosine antibodies failed to provide any evidence that GluR1–4 subunits are normally tyrosine phosphorylated (Lau and Huganir, 1995).

Phosphatases. AMPA receptors in cultured hippocampal neurons can be regulated, directly or indirectly, by endogenous phosphatases, since the phosphatase inhibitors okadaic acid (Wang et al., 1991) and microcystin-LR (McGlade-McCulloch et al., 1993) can enhance kainate-induced currents. The direction of the modulation with okadaic acid, like that of PKM, was dependent on the agonist concentration, with depression of currents induced by low kainate concentrations (Wang et al., 1994a).

5.2.5 Phosphorylation of NMDA receptors

Basal phosphorylation. It has been shown that phosphorylation may be needed to prevent run-down of NMDA receptor-mediated currents (MacDonald et al., 1989); but see Rosenmund and Westbrook (1993). It has also been shown that recombinant NR1 is basally phosphorylated on serine residues (Tingley et al., 1993), whilst recombinant NR2A and NR2B are basally phosphorylated on tyrosine residues (Lau and Huganir, 1995).

PKA. Activation of PKA is able to prevent glycine-independent desensitization of NMDA receptor-mediated currents caused by dephosphorylation and is required for recovery from this desensitization (Raman et al., 1996). This suggests that tonically active PKA phosphorylates NMDA receptors. See under *Phosphatases* for more details.

PKC. Activation of PKC by phorbol esters potentiates NMDA-induced currents in oocytes (Kelso et al., 1992; Urushihara et al., 1992). In recombinant systems, this regulation is specific for NR1/NR2A and NR1/NR2B heteromeric receptors (Mori et al., 1993; Wagner and Leonard, 1996). Native NMDA receptors in trigeminal neurons are positively regulated by PKC (Chen and Huang, 1992); it was found that PKC increased the probability of channel opening and reduced the level of Mg^{2+} blockade. However, it is unlikely that the latter mechanism can be generalized to other systems; firstly, an effect on Mg^{2+} block accounted for a negligible amount of the phorbol ester potentiation of recombinant receptors (Wagner and Leonard, 1995) and, secondly, phorbol esters do not invariably potentiate native NMDA responses. For example, PKC activators have been found to either depress (Markham and Segal, 1992) or have no effect on NMDA responses in hippocampal neurons (Harvey and Collingridge, 1993).

Recombinant NR1 is multiply phosphorylated, largely on serine residues, by treatment with phorbol esters. Most of the sites are contained within a single alternatively spliced exon (amino acids 864–900; serines at 889, 890, 896, 897) in the C-terminal domain (Tingley et al., 1993). However, phosphorylation of these sites does not underlie the phorbol ester potentiation of NMDA responses, since potentiation is still observed in a mutant NR1 lacking serines and threonines within the C-terminal domain (Yamakura et al., 1993). Furthermore, analysis of splice variants of recombinant NR1 showed a greater potentiation by activators of PKC of subunits which contained the N-terminal insert but lacked the C-terminal insert (Durand et al., 1993). Indeed, functional analysis of heteromeric channels containing chimeric NR2 receptors has shown that it is the C-terminal domain of the NR2 subunit which is primarily responsible for phorbol potentiation of NR1/NR2 receptors (Mori and Mishina, 1995).

CaMKII. There is also evidence that NMDA receptors are regulated by CaMKII. In substantia gelatinosa neurons, injection of CaMKII enhances both AMPA and NMDA currents (Kolaj et al., 1994). Also, it has been found that recombinant NR2B can be phosphorylated by CaMKII (Moon et al., 1995).

Tyrosine kinases. There is strong evidence that NMDA receptors are modulated by tyrosine phosphorylation. Thus, the inhibitors of protein tyrosine kinases (PTKs), genistein and lavendustin A, were found to depress whilst the PTK pp60c-src augmented native NMDA receptor-mediated currents (Wang and Salter, 1994). Recombinant NMDA receptors are similarly regulated by tyrosine phosphorylation (Chen and Leonard, 1996; Köhr and Seeburg, 1996). This modulation could be mediated directly via tyrosine phosphorylation of NMDA receptor subunits, since native NR2A and NR2B subunits, but not NR1, are basally tyrosine phosphorylated (Lau and Huganir, 1995; Moon et al., 1994). However, this tyrosine phoshorylation of NR2B may be maximal since only NR2A can be tyrosine phosphorylated further (six- to eightfold) (Lau and Huganir, 1995). It is likely that the PTK fyn is responsible for this modulation since fyn is highly enriched within the postsynaptic density and phosphorylates both native NR2A and NR2B (Suzuki and Okumura-Noji, 1995). Fyn phosphorylation of NR2 subunits may be involved in setting the threshold for the induction of LTP since fyn$^{-/-}$ mutants show impaired LTP (Grant et al., 1992). Recently, it has been shown that src and fyn kinases increase glutamate-mediated currents specifically at NR1-NR2A channels, probably via phosphorylation within the C-terminal region (Köhr and Seeburg, 1996).

Phosphatases. There is evidence that NMDA receptors are regulated by endogenous serine/threonine protein phosphatases (PPs). Firstly, both calyculin A and okadaic acid, applied at concentrations that inhibit PP1 and PP2A (but not PP2B or PP2C) enhance, whilst addition of PP1 or PP2A to the internal membrane depress, NMDA-induced activity in cultured hippocampal neurons (Wang et al., 1994b). There is also evidence that PP2B (calcineurin) regulates NMDA receptor activity, since the selective antagonist FK506 prolonged, whilst calcineurin itself applied to the cytoplasmic surface of the membrane inhibited, NMDA channel activity in acutely dissociated dentate granule neurons (Lieberman and Mody, 1994). This Ca^{2+}-dependent change in NMDA receptor properties by calcineurin is thought to involve an increase in glycine-insensitive desensitization (Tong and Jahr, 1994). Sufficient Ca^{2+} can enter during the synaptic activation of NMDA receptors to produce a short-lived (i.e. a few seconds) calcineurin-dependent desensitization of subsequent NMDA receptor-mediated synaptic responses (Tong et al., 1995). Recovery from desensitization is due to phosphorylation mediated by basally active PKA (Raman et al., 1996). There is also evidence that endogenous protein tyrosine phosphatases (PTPs) regulate NMDA receptors, since the PTP inhibitor sodium orthovanadate potentiated NMDA currents up to twofold (Wang and Salter, 1994).

5.2.6 Other post-translational modifications

Most studies have focused on phosphorylation of AMPA and NMDA receptors. There are, however, other post-translational modifications that may affect receptor function. For example, palmitoylation, the covalent attachment of palmitic acid on to cysteine residues, is a post-translational modification which might be involved in protein targeting and receptor activity and its modulation. Palmitoylation has been demonstrated for mGluR4 (Alaluf et al., 1995) and for GluR6 (Pickering et al., 1995) but not, as yet, for any of the AMPA or NMDA receptor subunits.

5.2.7 Regulation by modulatory proteins

NMDA receptors, but not AMPA receptors, are influenced by the state of actin polymerization (Rosenmund and Westbrook, 1993). Thus, in cultured hippocampal neurons, phalloidin, which prevents actin de-polymerization, prevented run-down of NMDA currents seen in the presence of Ca^{2+}. Conversely, cytochalasins, which enhance actin-ATP hydrolysis, induced run-down of NMDA currents recorded in the presence of an ATP-regenerating system, which otherwise prevented Ca^{2+}-dependent run-down. It was suggested that actin filaments maintain a soluble regulatory protein in a position where it prevents NMDA receptor channel run-down. A potential candidate for such a protein is calmodulin (CaM), which recently was shown to interact with NR1 but not NR2 subunits (Ehlers et al., 1996). CaM binding to NR1 was Ca^{2+} dependent and resulted in the inactivation of both homomeric NR1 complexes and heteromeric NR1–NR2 complexes. A high affinity CaM binding site was found to be located within the C1 exon cassette in the C terminus (*Figure 5.2*). It was suggested that this could provide a mechanism for the regulation of NMDA receptor association with the cytoskeleton, through competitive binding of CaM and cytoskeletal elements to the C1 region of the NR1 subunit.

An AMPA receptor-associated modulatory protein (GluR-MP) was first proposed from radiation inactivation experiments done in rat cortex (Honoré and Nielsen, 1985). Using this technique, linear decay curves for proteins are usually obtained, but curvilinear inactivation curves were obtained for [^{3}H]AMPA binding. This biphasic curve was explained by the inactivation at lower doses of radiation of an allosteric inhibitory protein associated with the smaller ligand-binding subunit. Similar radiation inactivation effects have been seen in chick telencephalon and *Xenopus* brain for [^{3}H]AMPA but not [^{3}H]kainate binding (Henley *et al.*, 1992). Taken together, these results suggest the presence of an inhibitory modulatory protein associated with AMPA receptors, but the mechanism of interaction remains unclear.

The two-hybrid *in vivo* genetic approach in yeast has identified the key role of the 95-kDa postsynaptic density protein (PSD-95) family of proteins in anchoring, and perhaps modulating the function of, NMDA receptors (Kornau *et al.*, 1995). PSD-95 is a multidomain protein with three PDZ repeats, an SH3 domain and an extended homology to yeast membrane-associated guanylate kinases. A seven-residue (*t*SXV) region at the cytoplasmic C-terminal domains of the NR2 subunits and certain NR1 alternative splice variants is responsible for the interaction with PSD-95. Intriguingly, Shaker-type K$^+$ channels have also been found to interact with PSD-95 (Kornau *et al.*, 1995), and from sequence homology comparisons it is predicted that various other receptors and ion channels interact with PSD-95 as well (see Kornau *et al.*, 1995 for details). The K$^+$ channel–PSD-95 interaction was found to mediate channel clustering in transfected COS cells.

Protein kinase A anchor proteins (e.g. AKA79) are a group of proteins which function in the regulation of type II PKA by restricting its subcellular distribution to defined microenvironments containing specific kinase target proteins (Coghlan *et al.*, 1995). In the two-hybrid system using AKA79 as a target protein, interactions have been identified with both the β isoform of calcineurin subunit A (CaN) and the regulatory subunit of type II PKA (RII). A ternary complex of AKA79, CaN and RII was found localized specifically in the neurites of cultured hippocampal cells. Since the PKA–AKA79 interaction is known to be necessary for the proper modulation of glutamate receptor channels, it was proposed that the ternary complex is involved in the regulation of synaptic transmission.

5.2.8 Cellular localization and trafficking of AMPA and NMDA receptors

In addition to altering the properties of receptors that are already at their postsynaptic sites, synaptic efficiency could be modified by alterations in the number of receptors at a synapse. For example, a 'silent' population, perhaps in the form of a cluster, of receptors could be moved to or from the synapse from extrasynaptic sites. These extrasynaptic receptors might already be incorporated within the plasma membrane but at sites away from the influence of synaptically released glutamate. Alternatively, they may be present intracellularly, in the form of a receptor reserve. At present, little is known about the cell biology of glutamate receptors and so these issues are largely a matter of speculation.

Localization. Several antibodies have been raised which recognize, often specifically, glutamate receptor subunits and have been used to localize these subunits within the brain. Most, however, have been raised against intracellular epitopes and so can only be used after cells have been permeabilized to enable antibody access.

Antibodies raised against an extracellular epitope of GluR1 (*Figure 5.1*) have been used to map the distribution of this AMPA receptor subunit in fixed tissue (Molnár *et al.*, 1993) and on living cultured neurons (Richmond *et al.*, 1996). These studies provide evidence that GluR1 is localized at synaptic and extra-synaptic sites within the plasma membrane and are also present within dendritic spines. Relevant to the silent synapse hypothesis, some spines of living neurons lacked detectable GluR1 on their membrane surface whilst they contained GluR1 within them (as determined after fixation and permeabilization). Since, in living neurons, neighboring spines were labeled positively for GluR1, it is unlikely that the failure of some spines to be labeled was due to a lack of sensitivity or access of the antibody to receptors.

NMDA receptors have been localized on living cultured hippocampal neurons using a fluorescent derivative of a toxin, conantokin-G, which blocks NMDA receptors (Benke *et al.*, 1993). Considerable co-localization with a synaptic marker was evident. By using fluorescent photobleach recovery, it was determined that the majority of labeled receptors were immobilized on the membrane. Approximately 25% of the receptors were, however, mobile and moved at rates of 7×10^{-10} cm^2 sec^{-1}. This corresponds to movement from the periphery to the center of a spine head in about 1 sec.

There is interest in how NMDA receptors may be targeted to discrete regions of the cell. It has been shown that alternative splicing and phosphorylation regulates the subcellular distribution of NR1 in QT6 quail fibroblasts (Ehlers *et al.*, 1995). It was found that the presence of the C1 exon cassette was necessary for the subunit to form discrete receptor-rich domains, which were found to be localized at or near the plasma membrane. Phosphorylation of two serines within the C1 cassette (Ser889, Ser890) disrupted the domains in a reversible manner. However, for cell surface expression of NR1, it is likely that NR2 subunits are required (McIlhinney *et al.*, 1996).

5.3 LTP in the CA1 region of the hippocampus

The possible involvement of modification of glutamate receptors in synaptic plasticity is considered with respect to LTP in the CA1 region of the hippocampus, since there is considerably more information available for this synapse than for any other.

5.3.1 The identity of the glutamate receptor subtypes which mediate modifiable responses

As discussed in Chapter 2, it is established that both AMPA and NMDA receptor-mediated components of synaptic transmission can express LTP. Early studies suggested a specific (Kauer *et al.*, 1988) or preferential (Muller and Lynch, 1988) enhancement of the AMPA receptor-mediated response; however, it is likely that the extent of LTP of the NMDA receptor-mediated component was underestimated in these studies. Many subsequent studies have reported substantial LTP of the NMDA receptor-mediated synaptic response (see Chapter 2) and, indeed, two recent reports suggest that the magnitude of potentiation of the two components may be similar (Clark and Collingridge, 1995; O'Connor et al., 1995). Theoretically, this could be because LTP is expressed purely by a presynaptic change (see Chapter 4), but the most popular view is that both AMPA and NMDA receptor-mediated synaptic responses are modified by postsynaptic mechanisms.

5.3.2 Modification of AMPA receptors

Manipulation of a variety of signaling systems by the intracellular injection of inhibitors has been shown to interfere with the generation of LTP of AMPA receptor-mediated synaptic transmission (see Chapter 3). Of course, it is not possible to tell from such experiments whether the molecules targeted are normally mediating their effects by directly modulating AMPA receptors. They could be part of a cascade which ultimately modifies AMPA receptors or some other aspect of the synapse; indeed the final modification could even be presynaptic if a retrograde messenger is involved (Chapter 4).

There is, however, direct evidence for a postsynaptic modification. LTP is associated with an increase in sensitivity of neurons to exogenously applied AMPA receptor ligands. In these studies, the increase in sensitivity was not immediate; it started within a few minutes but developed gradually over a period of approximately 1 h and then remained stable (Davies et al., 1989) (*Figure 2.3* in Chapter 2). This time course paralleled the conversion of short-term potentiation (STP) to LTP, suggesting that whilst changes in AMPA receptor function are important for LTP, STP is expressed differently (presumably presynaptically). It is generally found that inhibitors of various protein kinases prevent the induction of LTP but spare STP (see Chapter 3). Considered together, these results suggest that the increase in AMPA receptor sensitivity might be associated specifically with kinase-dependent processes. Consistent with this possibility, the protein kinase inhibitor K-252b does not affect STP but completely inhibits LTP and the increase in AMPA receptor function (Reymann et al., 1990). Furthermore, activation of mGluRs can induce LTP, without STP, via a kinase-dependent mechanism, and this effect is exactly paralleled by an increase in AMPA receptor sensitivity (Bortolotto and Collingridge, 1995). Finally, inhibition of mGluRs can inhibit LTP but not STP (see Chapter 2) and prevents the increase in AMPA receptor sensitivity (Seergueva et al., 1993). Again it is not possible to determine whether the modification is due to direct phosphorylation of AMPA receptors or of some intermediate process. In summary, the kinase-dependent component of synaptic potentiation is paralleled in time and magnitude by an increase in postsynaptic sensitivity to AMPA.

More recently, evidence has been presented that LTP may involve the recruitment of AMPA receptor clusters to functionally silent synapses (Isaac et al., 1995;

Kullmann, 1994; Liao et al., 1995). This effect has a much more rapid time course than the increase in sensitivity detected by exogenous application of AMPA receptor ligands. The mechanism that underlies this modification has not been explored. Furthermore, other recent findings cannot be simply reconciled with the silent synapse hypothesis (Bolshakov and Siegelbaum, 1995; Stevens and Wang, 1994; see Kullmann and Siegelbaum, 1995).

An understanding of what determines the kinetics of AMPA receptor-mediated excitatory postsynaptic currents (EPSCs) can help to address the question of what changes postsynaptically. It is found generally that LTP does not result in an alteration of the time course of EPSCs [e.g. Isaacson and Nicoll (1991) but see Ambros-Ingerson et al. (1993)]. Hence, once the factors which determine this are established they can be excluded as the major modifiable elements in LTP. Unfortunately this simple fact has been difficult to establish for the Schaffer collateral–commissural pathway. The time constant of decay of AMPA receptor-mediated EPSCs is slower than would be expected on the basis of what is known about the behavior of dendritic AMPA-gated channels (Spruston et al., 1995). This is probably due to electrotonic filtering of the synaptic response (Spruston et al., 1994). However, studies using cultured cells have suggested that AMPA receptor-mediated EPSC decay is governed by both desensitization and deactivation (e.g. Clements et al., 1992).

The EPSC is sensitive to agents which facilitate AMPA receptors, such as the nootropic drug aniracetam (Isaacson and Nicoll, 1991; Ito et al., 1990; Tang et al., 1991) and cyclothiazide (Rammes et al., 1996). On the basis of studies with aniracetam, it was suggested that the time course of AMPA receptor-mediated EPSCs is determined by desensitization. However, recent studies using cyclothiazide favor deactivation of AMPA receptors as the primary determinant of EPSC decay (Rammes et al., 1996). Thus, cyclothiazide prolonged EPSC decay to a similar extent and at concentrations which, in studies on cultured neurons, it prolonged the decay of the current following removal of AMPA. In contrast, desensitization was affected much more markedly and at lower doses of cyclothiazide in both preparations. Furthermore, 2,3-benzodiazepines, such as GYKI 52466, antagonized the effects of cyclothiazide on deactivation and EPSC decay but did not antagonize its effects on desensitization. Since the deactivation of AMPA receptors is determined by agonist affinity, it follows, since LTP does not involve changes in AMPA receptor-mediated EPSC decay, that LTP is unlikely to involve an alteration in AMPA receptor affinity.

5.3.3 Modification of NMDA receptors

Less is known concerning the mechanisms involved in the expression of LTP of NMDA receptor-mediated synaptic transmission. However, if it is assumed that LTP of AMPA receptor-mediated synaptic transmission in area CA1 is mediated postsynaptically, rather than presynaptically, then there must also be postsynaptic modifications of NMDA receptor-mediated synaptic transmission to account for LTP of this component. Results from sensitivity experiments of the sort performed with AMPA have not been reported; a technical problem being that application of NMDA can inhibit the generation of LTP (Izumi et al., 1992).

The time course of NMDA receptor-mediated EPSCs in hippocampal cultures is determined by the kinetics of NMDA receptor-gated channels (Lester *et al.*, 1990). The decay time course is determined mainly by the unbinding rate, and to a lesser extent by desensitization (Lester and Jahr, 1992). It is likely that these parameters apply also to NMDA receptor-mediated synaptic transmission in slices. The fact, therefore, that LTP of NMDA receptor-mediated synaptic transmission does not involve an alteration in the time course of NMDA receptor-mediated EPSCs (Bashir *et al.*, 1991) suggests that LTP does not involve a change in NMDA receptor affinity.

5.6 Concluding remarks

Cellular and molecular studies are starting to identify possible mechanisms by which the functioning of AMPA and NMDA receptors may be altered. Considerable work will be required, however, to establish which of these mechanisms are actually responsible for, or contribute to, the changes in synaptic strength which underlie LTP and LTD in the vertebrate brain.

References

Alaluf S, Mulvihill ER, McIlhinney RAJ. (1995) The metabotropic glutamate receptor mGluR4, but not mGluR1α, is palmitoylated when expressed in BHK cells. *J. Neurochem.* **64**: 1548–1555.

Ambros-Ingerson J, Xiao P, Larson J, Lynch G. (1993) Waveform analysis suggests that LTP alters the kinetics of synaptic receptor channels. *Brain Res.* **620**: 237–244.

Bashir ZI, Alford S, Davies SN, Randall AD, Collingridge GL. (1991) Long-term potentiation of NMDA receptor-mediated synaptic transmission in the hippocampus. *Nature* **349**: 156–158.

Benke TA, Jones OT, Collingridge GL, Angelides KJ. (1993) N-methyl-D-aspartate receptors are clustered and immobilized on dendrites of living cortical neurons. *Proc. Natl Acad. Sci. USA* **90**: 7819–7823.

Bettler B, Mulle C. (1995) Review: neurotransmitter receptors II. AMPA and kainate receptors. *Neuropharmacology* **34**: 123–139.

Blackstone C, Murphy TH, Moss SJ, Baraban JM, Huganir RL. (1994) Cyclic AMP and synaptic activity-dependent phosphorylation of AMPA-preferring glutamate receptors. *J. Neurosci.* **14**: 7585–7593.

Bolshakov VY, Siegelbaum SA. (1995) Regulation of hippocampal transmitter release during development and long-term potentiation. *Science* **269**: 1730–1734.

Bortolotto ZA, Collingridge GL. (1995) On the mechanism of long-term potentiation induced by (1S,3R)-1-aminocyclopentane-1,3-dicarboxylic acid (ACPD) in rat hippocampal slices. *Neuropharmacology* **34**: 1003–1014.

Chen L, Huang L-YM. (1992) Protein kinase C reduces Mg^{2+} block of NMDA-receptor channels as a mechanism of modulation. *Nature* **356**: 521–523.

Chen S-J, Leonard JP. (1996) Protein tyrosine kinase-mediated potentiation of currents from cloned NMDA receptors. *J. Neurochem.* **67**: 194–200.

Clark KA, Collingridge GL. (1995) Synaptic potentiation of dual-component excitatory postsynaptic currents in the rat hippocampus. *J. Physiol. (Lond.)* **482**: 39–52.

Clements JD, Lester RAJ, Tong G, Jahr CE, Westbrook GI. (1992) The time course of glutamate in the synaptic cleft. *Science* **258**: 1498–1501.

Coghlan VM, Perrino BA, Howard M, Langeberg LK, Hicks JB, Gallatin WM, Scott JD. (1995) Association of protein kinase A and protein phosphatase 2B with a common anchoring protein. *Science* **267**: 108–111.

Davies SN, Lester RAJ, Reymann KG, Collingridge GL. (1989) Temporally distinct pre- and postsynaptic mechanisms maintain long-term potentiation. *Nature* **338**: 500–503.

Durand GM, Bennett MVL, Zukin RS. (1993) Splice variants of the *N*-methyl-D-aspartate receptor NR1 identify domains involved in regulation by polyamines and protein kinase C. *Proc. Natl Acad. Sci. USA* **90**: 6731–6735.

Ehlers MD, Tingley WG, Huganir RL. (1995) Regulated subcellular distribution of the NR1 subunit of the NMDA receptor. *Science* **269**: 1734–1737.

Ehlers MD, Zhang S, Bernhardt JP, Huganir RL. (1996) Inactivation of NMDA receptors by direct interaction of calmodulin with the NR1 subunit. *Cell* **84**: 745–755.

Grant SGN, O'Dell TJ, Karl KA, Stein PL, Soriano P, Kandel ER. (1992) Impaired long-term potentiation, spatial learning, and hippocampal development in *fyn* mutant mice. *Science* **258**: 1903–1910.

Greengard P, Jen J, Nairn AC, Stevens CF. (1991) Enhancement of the glutamate response by cAMP-dependent protein kinase in hippocampal neurones. *Science* **253**: 1135–1138.

Harvey J, Collingridge GL. (1993) Signal transduction pathways involved in the acute potentiation of NMDA responses by 1S,3R-ACPD in rat hippocampal slices. *Br. J. Pharmacol.* **109**: 1085–1090.

Henley JM, Nielsen M, Barnard EA. (1992) Characterisation of an allosteric modulatory protein associated with α-[^{3}H]amino-3-hydroxy-5-methylisoxazolepropionate binding sites in chick telencephalon: effects of high-energy radiation and detergent solubilisation. *J. Neurochem.* **58**: 2030–2036.

Hollmann M, Boulter J, Maron C, Beasley L, Sullivan J, Pecht G, Heinemann S. (1993) Zinc potentiates agonist-induced currents at certain splice variants of the NMDA receptor. *Neuron* **10**: 942–954

Honoré T, Nielsen M. (1985) Complex structure of quisqualate-sensitive glutamate receptors in rat cortex. *Neurosci. Lett.* **54**: 27–32.

Hu G, Hvalby O, Walaas SI, Albert KA, Skjeflo P, Andersen P, Greengard P. (1987) Protein kinase C injection into hippocampal pyramidal cells elicits features of long-term potentiation. *Nature* **328**: 426–429.

Isaac JTR, Nicoll RA, Malenka RC. (1995) Evidence for silent synapses: implications for the expression of LTP. *Neuron* **15**: 427–434.

Isaacson JS, Nicoll RA. (1991) Aniracetam reduces glutamate receptor desensitization and slows the decay of fast excitatory synaptic currents in the hippocampus. *Proc. Natl Acad. Sci. USA* **88**: 10936–10940.

Ito I, Tanabe S, Kohda A, Sugiyama H. (1990) Allosteric potentiation of quisqualate receptors by a nootropic drug aniracetam. *J. Physiol. (Lond.)* **424**: 533–543.

Izumi Y, Clifford DB, Zorumski CF. (1992) Low concentrations of *N*-methyl-D-aspartate inhibit the induction of long-term potentiation in rat hippocampal slices. *Neurosci. Lett.* **137**: 245–248.

Kauer JA, Malenka RC, Nicoll RA. (1988) A persistent postsynaptic modification mediates long-term potentiation in the hippocampus. *Neuron* **1**: 911–917.

Keller BU, Hollmann M, Heinemann S, Konnerth A. (1992) Calcium influx through subunits GluR1/GluR3 of kainate/AMPA receptor channels is regulated by cAMP dependent protein kinase. *EMBO J.* **11**: 891–896.

Kelso SR, Nelson TE, Leonard JP. (1992) Protein kinase C-mediated enhancement of NMDA currents by metabotropic glutamate receptors in *Xenopus* oocytes. *J. Physiol. (Lond.)* **449**: 705–718.

Köhr G, Seeburg PH. (1996) Subtype-specific regulation of recombinant NMDA receptor-channels by protein tyrosine kinases of the src family. *J. Physiol.* **492**: 445–452.

Kolaj M, Cerne R, Cheng G, Brickey DA, Randic M. (1994) Alpha-subunit of calcium–calmodulin dependent protein kinase enhances excitatory amino-acid and synaptic responses of rat spinal dorsal horn neurones. *J. Neurophysiol.* **72**: 2525–2531.

Kornau H-C, Schenker LT, Kennedy MB, Seeburg PH. (1995) Domain interaction between NMDA receptor subunits and the postsynaptic density protein PSD-95. *Science* **269**: 1737–1739.

Kullmann DM. (1994) Amplitude fluctuations of dual-component EPSCs in hippocampal pyramidal cells: implications for long-term potentiation. *Neuron* **12**: 1111–1120.

Kullmann DM, Siegelbaum SA. (1995) The site of expression of NMDA receptor-dependent LTP: new fuel for an old fire. *Neuron* **15**: 997–1002.

Lau L-F, Huganir RL. (1995) Differential tyrosine phosphorylation of N-methyl-D-aspartate receptor subunits. *J. Biol. Chem.* **270**: 20036–20041.

Lester RAJ, Jahr CE. (1992) NMDA channel behavior depends on agonist affinity. *J. Neurosci.* **12**: 635–643.

Lester RAJ, Clements JD, Westbrook GI, Jahr CE. (1990) Channel kinetics determine the time course of NMDA receptor-mediated synaptic currents. *Nature* **346**: 565–567.

Liao D, Hessler NA, Malinow R. (1995) Activation of postsynaptically silent synapses during pairing-induced LTP in CA1 region of hippocampal slice. *Nature* **375**: 400–404.

Lieberman DN, Mody I. (1994) Regulation of NMDA channel function by endogenous Ca^{2+}-dependent phosphatase. *Nature* **369**: 235–239.

Lledo PM, Hjelmstad GO, Mukherji S, Soderling TR, Malenka RC, Nicoll RA. (1995) Calcium calmodulin-dependent kinase II and long-term potentiation enhance synaptic transmission by the same mechanism. *Proc. Natl Acad. Sci. USA* **92**: 11175–11179.

MacDonald JF, Mody I, Salter MW. (1989) Regulation of N-methyl-D-aspartate receptors revealed by intracellular dialysis of murine neurones in culture. *J. Physiol. (Lond.)* **414**: 17–34.

Markram H, Segal M. (1992) Activation of protein kinase C suppresses responses to NMDA in rat CA1 hippocampal neurones. *J. Physiol. (Lond.)* **457**: 491–501.

McGlade-McCulloh E, Yamamoto H, Tan S-E, Brickey DA, Soderling TR. (1993) Phosphorylation and regulation of glutamate receptors by calcium/calmodulin-dependent protein kinase II. *Nature* **362**: 640–642.

McIlhinney RAJ, Molnár E, Atack JR, Whiting PJ. (1996) Cell surface expression of the human N-methyl-D-aspartate receptor subunit 1a requires the co-expression of the NR2A subunit in transfected cells. *Neuroscience* **70**: 989–997.

Molnár E, Baude A, Richmond SA, Patel PB, Somogyi P, McIlhinney RAJ. (1993) Biochemical and immunocytochemical characterization of antipeptide antibodies to a cloned GluR1 glutamate receptor subunit: cellular and subcellular distribution in the rat forebrain. *Neuroscience* **53**: 307–326.

Moon IS, Apperson ML, Kennedy MB. (1994) The major tyrosine-phosphorylated protein in the postsynaptic density fraction is N-methyl-D-aspartate receptor subunit 2B. *Proc. Natl Acad. Sci. USA* **91**: 3954–3958.

Moon IS, Jin DH, Ko BH. (1995) Phosphorylation at N-methyl-D-aspartate receptor subunit 2B by Ca^{2+} calmodulin-dependent protein-kinase-II. *Molecules and Cells* **5**: 475–480.

Moran O, Dascal N. (1989) Protein kinase C modulates neurotransmitter responses in *Xenopus* oocytes injected with rat brain RNA. *Mol. Brain Res.* **5**: 193–202.

Mori H, Mishina M. (1995) Structure and function of the NMDA receptor channel. *Neuropharmacology* **34**: 1219–1237.

Mori H, Yamakura T, Masaki H, Mishina M. (1993) Involvement of the carboxyl-terminal region in modulation by TPA of the NMDA receptor channel. *NeuroReport* **4**: 519–522.

Moss SJ, Blackstone CD, Huganir RL. (1993) Phosphorylation of recombinant non-NMDA glutamate receptors on serine and tyrosine residues. *Neurochem. Res.* **18**: 105–110.

Muller D, Lynch G. (1988) Long-term potentiation differentially affects two components of synaptic responses in hippocampus. *Proc. Natl Acad. Sci. USA* **85**: 9346–9350.

Nakazawa K, Tadakuma T, Nokihara K, Ito M. (1995) Antibody specific for phosphorylated AMPA-type glutamate receptors at GluR2 Ser-696. *Neurosci. Res.* **24**: 75–86.

O'Connor JJ, Rowan MJ, Anwyl R. (1995) Tetanically induced LTP involves a similar increase in the AMPA and NMDA receptor components of the excitatory postsynaptic current: investigations of the involvement of mGlu receptors. *J. Neurosci.* **15**: 2013–2020.

Pickering DS, Taverna FA, Salter MW, Hampson DR. (1995) Palmitoylation of the GluR6 kainate receptor. *Proc. Natl Acad. Sci. USA* **92**: 12090–12094.

Raman IM, Tong G, Jahr CE. (1996) β-adrenergic regulation of synaptic NMDA receptors by cAMP-dependent protein kinase. *Neuron* **16**: 415–421.

Rammes G, Swandulla D, Collingridge GL, Hartmann S, Parsons CG. (1996) Interactions of 2,3-benzodiazepines and cyclothiazide at AMPA receptors: patch clamp recordings in cultured neurones and area CA1 in hippocampal slices. *Br. J. Pharmacol.* **117**: 1209–1221.

Raymond LA, Blackstone CD, Huganir RL. (1993) Phosphorylation and modulation of recombinant GluR6 glutamate receptors by cAMP-dependent protein kinase. *Nature* 361: 637–641.

Reymann KG, Davies SN, Matthies H, Kase H, Collingridge GL. (1990) Activation of a K-252b-sensitive protein kinase is necessary for a post-synaptic phase of long-term potentiation in area CA1 of rat hippocampus. *Eur. J. Neurosci.* 2: 481–486.

Richmond SA, Irving AJ, Molnár E, McIlhinney RAJ, Michelangeli F, Henley JM, Collingridge GL. (1996) Localisation of the glutamate receptor subunit GluR1 on the surface of living and within cultured hippocampal neurones. *Neuroscience* (in press).

Rosenmund C, Westbrook GI. (1993) Calcium-induced actin depolymerization reduces NMDA channel activity. *Neuron* 10: 805–814.

Rosenmund C, Carr DW, Bergeson SE, Nilaver G, Scott JD, Westbrook GI. (1994) Anchoring of protein kinase A is required for modulation of AMPA/kainate receptors on hippocampal neurons. *Nature* 368: 853–856.

Sergueeva OA, Fedorov NB, Reymann KG. (1993) An antagonist of glutamate metabotropic receptors, (RS)-α-methyl-4-carboxyphenylglycine, prevents the LTP-related increase in postsynaptic AMPA sensitivity in hippocampal slices. *Neuropharmacology* 32: 933–935.

Spruston N, Jaffe DB, Johnston D. (1994) Dendritic attenuation of synaptic potentials and currents: the role of passive membrane properties. *Trends Neurosci.* 17: 161–166.

Spruston N, Jonas P, Sakmann B. (1995) Dendritic glutamate receptor channels in rat hippocampal CA3 and CA1 pyramidal neurons. *J. Physiol. (Lond.)* 482: 325–352.

Stevens CF, Wang Y. (1994) Changes in reliability of synaptic function as a mechanism for plasticity. *Nature* 371: 704–707.

Suzuki T, Okamura-Noji K. (1995) NMDA receptor subunits ε1 (NR2A) and ε2 (NR2B) are substrates for Fyn in the postsynaptic density fraction isolated from the rat brain. *Biochem. Biophys. Res. Commun.* 216: 582–588.

Tan S-E, Wenthold RJ, Soderling TR. (1994) Phosphorylation of AMPA-type glutamate receptors by calcium/calmodulin-dependent protein kinase II and protein kinase C in cultured hippocampal neurons. *J. Neurosci.* 14: 1123–1129.

Tang CM, Shi QY, Katchman A, Lynch G. (1991) Modulation of the time course of fast EPSCs and glutamate channel kinetics by aniracetam. *Science* 254: 288–290.

Tingley WG, Roche KW, Thompson AK, Huganir RL. (1993) Regulation of NMDA receptor phosphorylation by alternative splicing of the C-terminal domain. *Nature* 364: 70–73.

Tong G, Jahr CE. (1994) Regulation of glycine-insensitive desensitization of the NMDA receptor in outside-out patches. *J. Neurophysiol.* 72: 754–761.

Tong G, Shepherd D, Jahr CE. (1995) Synaptic desensitization of NMDA receptors by calcineurin. *Science* 267: 1510–1512.

Urushihara H, Tohda M, Nomura Y. (1992) Selective potentiation of *N*-methyl-D-aspartate-induced current by protein kinase C in *Xenopus* oocytes injected with rat brain RNA. *J. Biol. Chem.* 267: 11697–11700.

Wagner DA, Leonard JP. (1996) Effect of protein kinase-C activation on the Mg^{2+}-sensitivity of cloned NMDA receptors. *Neuropharmacology* 35: 29–36.

Wang L-Y, Salter MW, MacDonald JF. (1991) Regulation of kainate receptors by cAMP-dependent protein kinases and phosphatases. *Science* 253: 1–3.

Wang L-Y, Taverna FA, Huang X-P, MacDonald JF, Hampson DR. (1993) Phosphorylation and modulation of a kainate receptor (GluR6) by cAMP-dependent protein kinase. *Science* 259: 1173–1175.

Wang L-Y, Dudek EM, Browning MD, MacDonald JF. (1994a) Modulation of AMPA/kainate receptors in cultured murine hippocampal neurones by protein kinase C. *J. Physiol. (Lond.)* 475: 431–437.

Wang L-Y, Orser BA, Brautigan DL, MacDonald JF. (1994b) Regulation of NMDA receptors in cultured hippocampal neurons by protein phosphatases 1 and 2A. *Nature* 369: 230–232.

Wang YT, Salter MW. (1994) Regulation of NMDA receptors by tyrosine kinases and phosphatases. *Nature* 369: 233–235.

Watkins JC, Evans RH. (1981) Excitatory amino acid transmitters. *Annu. Rev. Pharmacol. Toxicol.* 21: 165–204.

Yakel JL, Vissavajjhala P, Derkach VA, Brickey DA, Soderling TR. (1995) Identification of a Ca^{2+}/calmodulin-dependent protein kinase II regulatory phosphorylation site in non-N-methyl-D-aspartate glutamate receptors. *Proc. Natl Acad. Sci. USA* **92**: 1376–1380.

Yamakura T, Mori H, Shimoji K, Mishina M. (1993) Phosphorylation of the carboxy-terminal domain of the ζ1 subunit is not responsible for potentiation by TPA of the NMDA receptor channel. *Biochem. Biophys. Res. Commun.* **196**: 1537–1544.

6

Postsynaptic gene expression and long-term potentiation in the hippocampus

K.L. Thomas and S.P. Hunt

6.1 Introduction

It is widely believed that long-term, but not short-term, storage of information in the nervous system of both vertebrates and invertebrates requires protein synthesis and gene expression (Davis and Squire, 1984; Montarolo *et al.*, 1986). However, we know remarkably little about any genes involved in memory formation. Hippocampal long-term potentiation (LTP; Bliss and Collingridge, 1993) possesses many of the physiological characteristics necessary for long-term storage of information and is still, 30 years after its discovery, regarded as the most attractive model for studying memory processes in mammals.

LTP is a persistent, activity-dependent form of synaptic modification induced by brief high frequency stimulation of inputs to hippocampal neurons. The mechanisms for the induction and maintenance of LTP have been described in detail in other chapters in this book. Mechanistically, LTP can be divided into several categories. LTP1 or E-LTP (early LTP) which decays in a few hours after induction. This form of LTP is independent of new protein synthesis. The rise in postsynaptic calcium via N-methyl-D-aspartate (NMDA) receptors after LTP-inducing stimuli is believed to result in the activation of serine/threonine kinases in four parallel, but possibly interdependent, intracellular phosphorylation signaling pathways, leading to the phosphorylation and alteration of the functional characteristics of target proteins which may include glutamate receptors (see below). As such, LTP1 is maintained by post-translational modifications of existing proteins. LTP2 and LTP3 [or (L)ate-LTP] are recruited by increasing the

---

Cortical Plasticity, edited by M.S. Fazeli and G.L. Collingridge.
© 1996 BIOS Scientific Publishers Ltd, Oxford.

number of tetanic trains and are longer lasting forms, persisting for days and weeks respectively when induced in freely moving rats. LTP2 and LTP3 are both dependent on new protein synthesis (Frey *et al.*, 1989; Krug *et al.* 1984; Otani and Abraham, 1989; Otani *et al.* 1989; Stanton and Sarvey, 1984), but new gene transcription is critical for LTP3 (Nguyen *et al.*, 1994). The requirement for new mRNA and protein synthesis for LTP2/3 occurs during a relatively brief time window around the time of induction, suggesting that there is an initiation of a gene induction program. Since activation of serine/threonine kinases leads to the modulation of transcription factor expression within the nucleus (Hunter and Karin, 1992), it is likely that their increased activity during LTP induction would alter the expression of late onset genes which support L-LTP. Much work has focused on identifying which gene candidates are involved, in particular the immediate-early genes (see below).

There are at least two possible approaches to the search for genes that modulate or maintain the capacity for neuronal plasticity. Primarily, we have monitored the activity of previously described genes that had been shown to be or were suspected of being involved in plasticity, measuring changes in levels of mRNA after LTP with *in situ* hybridization. This technique not only allows for the quantification of changes in the expression of individual genes but also for alternative splice variants of the gene. There are enormous numbers of possible candidate genes and particular weight was given to those genes that coded for glutamate receptor subunits, serine/threonine kinases which mediate many of the nuclear and other intracellular effects of glutamate receptor activation, transcription factors and growth factors. We suspected that these groups of genes would be involved in any long-term changes in postsynaptic neuronal sensitivity if rapid post-translational events, such as receptor phosphorylation or enzyme activation, had at later time points to be backed up by synthesis of new protein. The second approach which is currently underway in a number of laboratories, including our own, is to identify, using recently introduced molecular screening techniques, the collection of genes, including novel genes, whose expression is associated with distinct phases of LTP. This suggests, of course, that alterations to known proteins may be insufficient to account for long-term changes in synaptic plasticity and that new mechanisms need to be discovered. Novel genes are indeed being identified, although full characterizations have not yet been published. Surveying the patterns of gene expression that follow LTP induction will help us in an attempt to uncover molecular mechanisms that may be common to the establishment and maintenance of synaptic change in LTP and to learning and memory. It seems almost certain that a full description of the molecular biology of long-term information storage is going to require both of the above approaches to the identification of candidate genes.

In this chapter, we will describe the transcriptional events in the postsynaptic neuron that follow the induction of NMDA receptor-dependent LTP within the hippocampus. The first changes to be measured lie within a group of genes called immediate-early genes (IEGs), so called because of their very rapid expression following stimulation. This will be followed by a description of the changes that occur in certain second messenger pathways and in the expression of glutamate receptor genes. Finally, we will review those alterations in gene expression which may be closely related to structural changes that are thought to maintain the expression of LTP. Such remodeling of the synapse may require the directed synthesis of growth factors, cell adhesion molecules and many other proteins, particularly those related to the cytoskeleton.

6.2 Immediate-early genes

IEGs are classically defined as those genes whose transcription is activated rapidly and transiently within minutes by extracellular stimuli and whose induction is not dependent on protein synthesis (Morgan and Curran, 1991). IEGs encode regulatory proteins, such as transcription factors that control the expression of late response genes. Transcription factors themselves are proteins which can bind to specific short DNA sequences in the regulatory regions of genes and regulate the expression of the gene. Most transcription factors contain specific regions which modulate their ability to bind DNA and other regions which mediate their ability to activate transcription by interacting with other factors once bound. Families of transcription factors are described by specific structural DNA binding motifs (Latchman, 1991).

6.2.1. Zinc finger transcription factors

The zinc finger transcription factors all contain one or more repeats of a 30 amino acid motif, Tyr/Phe-X-Cys-X-Cys-X_{24}-Cys-X_3-Phe-X_5-Leu-X_2-His-X_{34}-His-X_5, where X is a variable amino acid. The two pairs of cysteine and histidine residues coordinate a single zinc atom resulting in a finger-like structure. It is the basic residues, phenylalanine and leucine which bind to 5'-GCGGGGGCG-3' DNA sequences.

The zinc finger transcription factor NGFI-A (also known as zif268, Egr1, Krox 24 and T1S8; Milbrandt, 1987) was one of the first IEGs identified that was induced dramatically and transiently in postsynaptic neurons of the hippocampus by synaptic stimuli in association with LTP (Cole et al., 1989; Wisden et al., 1990). *In situ* hybridization techniques revealed that following unilateral induction of LTP *in vivo* at the perforant path–dentate granule cell synapse there was an increase in mRNA levels for NGFI-A appearing in dentate granule cells as early as 15 min and lasting up to 3 h after stimulation (*Figure 6.1a*). The unilateral increase in mRNA for NGFI-A was mirrored by a later increase in protein. Since, blockade of the NMDA receptor prevented the unilateral induction of LTP (*Figure 6.1b*) and expression of NGFI-A in the dentate gyrus, and the fact that the stimulus intensity/frequency required to increase NGFI-A mRNA was similar to that needed to induce LTP strongly suggested that the two phenomena were regulated by similar mechanisms. Furthermore, stimulation of inhibitory commissural fibers originating in the hilus of the opposite dentate gyrus and terminating on ipsilateral dentate granule cells

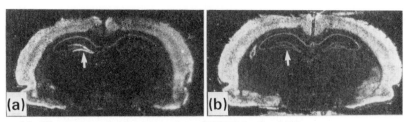

Figure 6.1. The increase in the expression of NGFI-A after LTP.
(a) NGFI-A mRNA increases in the dentate gyrus (arrow) 30 min after LTP. (b) This effect was blocked in the presence of the NMDA receptor antagonist APV.

given before perforant path stimulation prevented LTP induction and NGFI-A expression. However, commissural stimulation alone was able to induce the expression of NGFI-A, which indicates that expression of this transcription factor may not always be associated with stimulation protocols that induce long-term plastic changes. The persistence of LTP in the dentate gyrus has been shown to be highly correlated with the initial magnitude of NGFI-A expression in the granule cells (Richardson *et al.*, 1992). How NGFI-A might contribute to LTP is unclear. The consensus binding sequence for this gene is found on a number of synaptic vesicle-associated protein genes, and there is not only evidence that expression of synapsin I can be regulated *in vitro* by this transcription factor (Thiel *et al.*, 1994) but the levels of synapsin I (and synaptotagmin and synaptophysin) protein increase in synaptosomal preparations 3 h after LTP induction in the perforant path (Lynch *et al.*, 1994). Although in this case synapsin I levels were essentially elevated in the presynaptic compartment, this does not rule out NGFI-A mediated increases in synapsin I expression in the postsynaptic cell, with LTP and consequential effects on neurotransmitter release from the axon terminals of the postsynaptic cell.

Recently, another zinc finger transcription factor, Egr3, only homologous to NGFI-A within the zinc finger DNA, binding motif, has been reported to be regulated by synaptic activity in the brain (Yamagata *et al.*, 1994). Originally isolated as a differentially expressed gene from seizure-stimulated hippocampus, Erg3 is expressed at low levels and with a similar distribution to NGFI-A in the normal rat brain. Following unilateral high frequency stimulation of the perforant path *in vivo*, Erg3 mRNA was strongly induced in the dentate granule cells ipsilateral to stimulation. However, the induction of Erg3 was more robust than that for NGFI-A, lasting for at least 4 h; but, like NGFI-A expression, it was blocked by antagonists of the NMDA receptor.

6.2.2. Basic region-leucine zipper family

Transcription factors of the basic leucine zipper (bZIP) family include the fos and jun families of IEGs. This group of IEGs has attracted considerable interest as they are thought to function in neuronal transcriptional programs induced by such diverse stimulation as glutamate application to cultured cells or peripheral noxious stimulation and seizure induction in the whole animal (Morgan and

Curran, 1991). Fos family members include c-fos, fra1, fra2 and fosB which can form heterodimeric transcription complexes with members of the jun family, c-jun, junB and junD, via their leucine zipper domains. The resulting heterodimers bind, via their adjacent basic regions, with high affinity to the AP1 consensus DNA sequence. *In vitro*, the efficacy of binding of the heterodimers to the AP1 site is determined by the fos–jun combination.

LTP induced in freely moving rats at the perforant path–dentate granule cell synapse has been associated with reliable, rapid and transient increases in the expression of the AP1 binding transcription factors including *fos*-related genes (Dragunow *et al.*, 1989; Jeffery *et al.*, 1990; Nikolaev *et al.*, 1991) and members of the jun family (Demmer *et al.*, 1993). However, LTP induced at the same site but in anesthetized rats was less reliably correlated with increases in the expression of junB, c-jun and c-fos (Cole *et al.*, 1989; Wisden *et al.*, 1990). This suggested that anesthesia may be attenuating the response of these IEGs following LTP. Indeed, the NMDA receptor-dependent induction of c-jun-, junB-, junD- and fos-related proteins by a given LTP-inducing protocol was prevented by pentobarbital, although LTP was not blocked (Demmer *et al.*, 1993). To address this discrepancy between the pattern of IEG induction between anesthetized and unanesthetized LTP preparations, two independent groups tested the effect on IEG expression of increasing the number of stimulus trains given to the perforant path in unanesthetized rats (Abraham *et al.*, 1993; Worley *et al.*, 1993). Both studies revealed that fos–jun expression was observed with the higher but not lower stimulation regimes, whereas NGFI-A expression was seen associated with all LTP-inducing stimulations. Furthermore, increasing the number of stimulus trains given was highly correlated with the persistence, measured in days, of LTP. This suggests that induction of the fos–jun IEGs was seen only with those LTP-inducing paradigms which produce the longest form of LTP, LTP3. In support of this idea, pentobarbital significantly decreased the persistence of LTP (Jeffery *et al.* 1990).

The cAMP response element-binding protein (CREB) is a member of the bZIP family of transcription factors which activate the transcription of genes that possess cAMP response elements (CREs) in their regulatory region (Gallin and Greenberg, 1994). CREB does not conform strictly to the classical definition of an IEG as it is constitutively expressed in neurons. However, it exists in a latent form and transcriptional activity is conferred upon it by post-translational modifications in response to extracellular events. CREB has a nuclear localization and is bound to DNA either as a homodimer, or as a heterodimer with other members of the bZIP family. A critical step in the activation of CREB-mediated transcription is the phosphorylation at Ser133 in response to rises in the intracellular concentration of calcium or cAMP. Phosphorylation at Ser133 results in the recruitment of a secondary factor, CREB binding protein and the formation of an active transcription complex (Arias *et al.*, 1994; Kwok *et al.*, 1994). Protein kinase A (PKA) and some calcium/calmodulin-dependent protein kinases (CaMKII) have been shown to phosphorylate Ser133 directly *in vitro* (Dash *et al.*, 1991).

Mice with a targeted mutation of the CREB gene (so-called 'knockout' mutation) cannot sustain LTP induced *in vitro* at the Schaffer collateral–CA1 synapse in the hippocampus for more than 90 min, although other measures of synaptic function appeared normal (Bourtchuladze et al., 1994). Studies in invertebrates have shown that direct genetic manipulation of CREB-related genes produces both a loss and gain of function in neuronal plasticity and learning. In *Aplysia*, long-term facilitation is an extensively studied form of neuronal plasticity between the sensory and motor neurons which normally participate in the tail-withdrawal reflex in the whole animal and which in isolated culture is induced by the repeated or sustained application of serotonin (Montarolo et al., 1986). It was shown that microinjection of CRE sequences into the nucleus of the sensory neuron blocks the serotonin-induced increase in synaptic strength (Dash et al., 1990). An inhibitory form of CREB called ApCREB2 was isolated from an *Aplysia* central nervous system library and shown to be a substrate for serine/threonine protein kinases and expressed in sensory neurons (Bartsch et al., 1995). Following injection of anti-ApCREB2 antibodies into the sensory neurons, long-term facilitation could be elicited by a single application of serotonin, a manipulation which normally produces only short-term facilitation. Furthermore, in *Drosophila*, long-term memory of Pavlovian olfactory learning was not only disrupted by the heat-induced expression of the transgene dCREB2-b, another inhibitory form of CREB (Yin et al., 1994) but *subthreshold* training paradigms which normally did not produce long-term olfactory memory could now do so after the heat-induced expression of hs-CREB2-a, an activator CREB isoform (Yin et al., 1995).

We have observed that, under resting conditions, CREB protein is seen in the nucleus of virtually every neuron in the brain, while the levels of the phosphorylated, active form are very low. Our preliminary evidence suggests that phosphorylation of CREB is not related specifically to LTP but simply to afferent activation. Although increases in the levels of the Ser133 phosphorylated form of this receptor were ipsilaterally increased in the postsynaptic dentate granule cells 30 min after a unilateral LTP-inducing stimulation of the perforant path in anesthetized rats, 90 min later increases were also seen in the contralateral dentate gyrus which had only received low frequency test stimulations (*Figure 6.2*). Diesseroth et al. (1996), working with hippocampal neurons *in vitro*, were able to show that postsynaptic CREB phosphorylation at Ser133 was evoked rapidly both by high frequency stimulations that evoked LTP and by low frequency stimulations that induce long-term depression (LTD), but stimulations of an intermediate frequency were ineffective. CREB can, however, also be phosphorylated at Ser142, which results in an inhibition of CREB's ability to activate transcription. We therefore need to know whether CREB can be differentially phosphorylated by distinct types of afferent stimulation and how this alters CREB transcriptional activity.

A role for another serine/threonine kinase-regulated transcription factor, the CCAAT enhancer-binding protein (C/EBP), in neuronal plasticity was reported in *Aplysia*. The long-term synaptic enhancement between siphon sensory neurons and gill motor neurons produced by repeated or prolonged applications of serotonin was associated with transient increases in mRNA for the *Aplysia* homolog of C/EBP, ApC/EBP (Alberini et al., 1994). Under normal conditions, ApC/EBP expression is absent, but it peaks 2 h after serotonin exposure. Furthermore, its

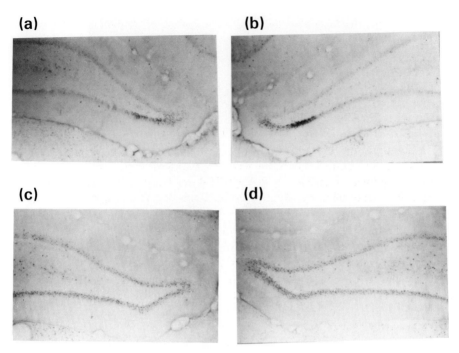

Figure 6.2. The bilateral increase in the Ser133 phosphorylated form of CREB after unilateral LTP. Phospho-CREB immunoreactivity was localized to the nuclei in the medial region of the ventral blade of the dentate gyrus after LTP induction on the potentiated (a) and contralateral nonpotentiated (b) sides. There are apparently more positive nuclei in the potentiated dentate gyrus. There was almost no phospho-CREB immunoreactivity in the granule cells of naive rats (data not shown). Two hours after LTP induction, immunoreactive nuclei were observed scattered throughout the potentiated (c) and nonpotentiated (d) dentate gyrus.

expression was attenuated completely by specific antibodies to ApC/EBP and short antisense DNA oligonucleotide sequences that contain the DNA-binding element with which ApC/EBP associates. Since short-term facilitation produced by a single application of serotonin was not affected by ApC/EBP antibody or antisense treatment, the authors proposed that ApC/EBP expression was linked to the consolidation of the increased synaptic strength between the sensory and motor neurons.

Recently, we have looked at the expression of C/EBPβ, the mammalian homolog of ApC/EBP, in granule cells after LTP induction at the perforant path–dentate gyrus synapses in freely moving rats. Like *Aplysia* ApC/EBP, in situ hybridization revealed that C/EBPβ mRNA levels in the brain are virtually absent; however, 2 h after LTP induction there was a dramatic increase in its expression in the dentate granule cells (*Figure 6.3*). Similar increases were seen in the expression of junD. The increased expression of C/EBPβ after LTP induction was transient, as the hybridization signal was no longer present by 24 h. Furthermore, unlike the case for active CREB protein, the increase in C/EBPβ

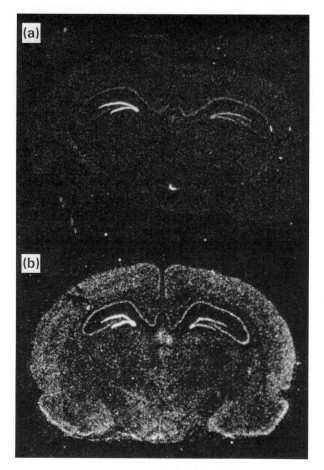

Figure 6.3. The increase in the expression of C/EBPβ and junD after LTP. (a) C/EBPβ and (b) junD mRNA increases in the dentate gyrus 2 h after unilateral LTP induction were confined to the potentiated hemisphere.

was confined only to those postsynaptic neurons which were potentiated. Since C/EBPβ is capable of forming heterodimers with CREB and junD, and transcription events associated with both are calcium regulated (for references see Gallin and Greenberg, 1995), it is possible that CREB as a C/EBPβ–CREB or junD–CREB heteromer can mediate the transcriptional events linked specifically with LTP.

6.3 Secondary messenger pathways: serine/threonine kinases

Inhibitor studies have shown that members of the serine/threonine superfamily of kinases, particularly the CaMKII, protein kinase C (PKC) and PKA families, play an important role in the induction and maintenance of LTP (Frey *et al.*,

1993; Huang and Kandel, 1994; Klann et al., 1993; Lovinger et al., 1987; Malenka et al., 1989; Malinow et al., 1988, 1989; Matthies and Reymann, 1993; Pettit et al., 1994; Reymann et al., 1988; Sacktor et al., 1993; Wang and Feng, 1992). Studies in which transgenes have been introduced into mice to disrupt the expression of specific proteins have shown the importance of particular isoforms of these kinases, principally the α isoform of CaMKII (Silva et al., 1992), the γ isoform of PKC (Abeliovich et al., 1993) and the RIβ and CIβ subunits of PKA (Brandon et al., 1995; Huang et al., 1995: Qi et al., 1996), in the induction of LTP. These data, and the fact that the expression of LTP is dependent on new protein synthesis and transcription, suggest that changes in the expression of kinases that act to modify pre-existing protein targets could have profound effects on the long-term responsiveness of postsynaptic cells. For example, as a consequence of serine/threonine kinase-directed phosphorylation, the conductance through ionotropic glutamate receptors increases (Greengard et al., 1991; McGlade McCulloh et al., 1993; Tan et al., 1994; Tingley et al., 1993; Wang et al., 1994). Indeed, LTP is associated with enhanced NMDA as well as non-NMDA receptor-induced currents (Aszetly et al., 1992; Bashir et al., 1991). Also, cytoskeletal components such as MAP2 and tau are substrates for CaMKII, PKC and the serine/threonine kinase, extracellular signal-regulated kinase [ERK; also known as mitogen activated protein (MAP) kinase] and the phosphorylation state of these structural components determines the stability of the cytoskeletal network (Drewes et al., 1992; Goedert et al., 1991; Hoshi et al., 1988; Matus, 1988; Olmstead, 1986). As such, the induction and activation of the serine/threonine kinases following LTP may facilitate the morphological remodeling phase after LTP (see Edwards, 1995, for references). In addition to modifying pre-existing proteins to alter their functional characteristics, these kinases may alter the expression of as yet unidentified proteins during the later stages of LTP via their direct action on transcription factors (Hunter and Karin, 1992).

Our working hypothesis was that particular proteins crucial for LTP could be identified by alterations in their expression at the mRNA level. This hypothesis has been confirmed substantially by correlation with the results emerging from gene deletion and pharmacological studies.

Using *in situ* hybridization to estimate mRNA levels in granule cells of the dentate gyrus at between 2 and 48 h after unilateral LTP induction in freely moving rats allowed us to investigate the role of individual isoforms of serine/threonine kinases in LTP (Thomas et al., 1994a). We studied changes in the postsynaptic expression of calcium-dependent CaMKIIs (α and β isoforms), calcium-dependent PKCs (α, βI, βII and γ isoforms) and calcium-independent PKCs (δ, ε and ζ isoforms). In addition, we investigated changes in the expression of two other families of kinases that rely on phosphorylation by other kinases rather than calcium for their activation: the ERKs (ERK-1, ERK-2 and ERK-3) and Rafs (Raf-1, Raf-A and Raf-B). These latter kinases are thought to be members of the same signaling pathway, Raf functioning upstream of ERK (Kyriakis et al., 1992). We found changes in the expression of only one isoform of each of these four serine/threonine kinase families after LTP, suggesting a specific role for these isoforms in neuronal plasticity.

6.3.1 CaMKII

mRNA for the α isoform of CaMKII is found not only in cell bodies of the granule cells of the dentate gyrus and pyramidal cells of the hippocampus, but also in their associated dendritic fields (Burgin *et al.*, 1990). Two hours after unilateral LTP induction, transient increases in α CaMKII expression in the cell bodies of the ipsilateral dentate granule cells were seen in comparison with the contralateral granule cells (*Figure 6.4*). However, within the dendritic field, there was subsequent and sustained elevation in mRNA levels for this gene appearing 24 and 48 h after induction in the proximal and distal fields respectively. The expression of β CaMKII did not change following LTP induction. A recent *in vitro* study in hippocampal slices showed that LTP was associated with increases in dendritic α CaMKII mRNA levels (Roberts *et al.*, 1995), confirming our *in vivo* results.

Since mRNA for the α isoform of CaMKII is localized to the dendritic fields as well as to cell bodies of central neurons and is associated with the postsynaptic densities at the base of dendritic spines and polyribosomes (Burgin *et al.*, 1990), the dendritic synthesis of α CaMKII protein could be predicted to change in response to local stimulation. The increase in the expression of α CaMKII mRNA in the cell bodies and in the dendritic fields following LTP supports this suggestion. Furthermore, calcium-dependent autophosphorylation of CaMKII yields a constitutively active calcium-independent and relatively stable form of the enzyme *in vitro* (Davis *et al.*, 1987; Miller and Kennedy, 1986). Indeed, high frequency stimulation of CA1 afferents resulted in a long-lasting increase in calcium-independent activity of this enzyme (Fukunaga *et al.*, 1993). Therefore, phosphorylated α CaMKII itself may serve as a mechanism for maintaining increased synaptic strength in the absence of calcium transients. Our study also implies that the increase in α CaMKII mRNA at 2 h in the soma of the granule cells but not at later time points, and its subsequent enhancement in the proximal then proximal and distal dendritic fields, 24 and 48 h after LTP, reflects the dendritic transport of mRNA for α CaMKII from the nucleus to the site of stimulation in response to a brief tetanic stimulus. The temporal increases in α CaMKII expression in the dendritic field follows the predicted rate of mRNA transport into the dendrites (Davis *et al.*, 1989).

6.3.2 PKC

The proposed mechanisms for the enhanced PKC activation during LTP maintenance include a persistent translocation of PKC to the plasma membrane which results in an enduring activation of the enzyme (Klann *et al.*, 1993), a long-lasting increase in a constitutively active proteolytic cleavage product (PKM) of PKC in the cytosol (Sacktor *et al.*, 1993) and a tenacious cytosolic PKC activity due to its autophosphorylation or phosphorylation by other kinases. Persistently activated

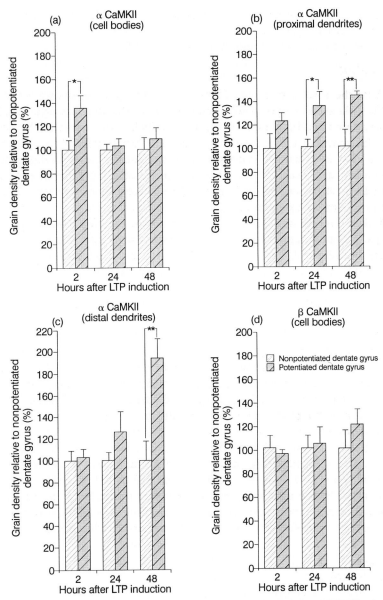

Figure 6.4. The specific increase in the expression of α CaMKII after LTP. The density of silver grains in dentate granule cells in the potentiated hemisphere (darkly shaded bars) are expressed as a percentage of the density of silver grains counted in the nonstimulated hemisphere (lightly shaded bars) as the mean ± SEM. *Post hoc* analysis (Dunn's test) revealed that α CaMKII mRNA expression was increased over the cell bodies at 2 h post-induction, while it was increased over the proximal dendritic field at 24 and 48 h and increased over the distal dendritic field 48 h following induction (*$p<0.05$ and **$p<0.01$). (d) There was no effect of tetanic stimulation on βCaMKII mRNA levels in the dentate granule cell bodies.

PKC is likely to be postsynaptically located since injection of selective peptide inhibitors of PKC into the postsynaptic neuron can reverse the maintenance of LTP (Malenka *et al.*, 1989; Malinow *et al.*, 1988, 1989).

The calcium-dependent α, βI, βII and γ isoforms of PKC (Nishizuka, 1988) have the widest expression distributions in the rat brain, and all are found in the hippocampus. Within the hippocampus, these PKC isoforms show a higher relative distribution in the CA1, CA2 and CA3 pyramidal cells of Ammon's horn than in the granule cells of the dentate gyrus, with the exception of the mRNAs for the I and II splice variants of β PKC, which appear to be absent from CA2 (*Figure 6.5*). Of the calcium-independent isoforms, ε PKC appears to be more highly expressed in the dentate gyrus and CA3 than CA2/1, while ζ PKC expression is equally distributed in the hippocampus. We were unable to detect the expression of δ PKC in any cells of the hippocampus.

There was a transient induction only in the expression of γ PKC in the granule cells of the dentate gyrus after LTP induction (*Figure 6.6*). There was also an apparent increase in ζ PKC expression following LTP induction. This result is interesting in view of the demonstration of a persistent activation of ζ PKC in the CA1 region following LTP induction in hippocampal slices (Sacktor *et al.*, 1993). However, the differences in ζ PKC mRNA levels between the potentiated and nonpotentiated dentate granule cells at individual time points were not statistically significant due to the large standard errors produced by the low density of grains generated by the ζ PKC-specific probe.

The lack of an apparent change in the expression of *calcium-independent* PKCs (δ, ε and ζ isoforms) after LTP was unexpected. Constitutively active catalytic domains of PKC isoforms (PKM) are generated by proteolytic cleavage (Inoue *et al.*, 1977; Pontremoli *et al.*, 1986) and a recent study in hippocampal slices shows that formation of such a PKC ζ PKM fragment occurs in the maintenance phase of LTP in the CA1 (Sacktor *et al.*, 1993). A similar cleavage of *calcium-dependent* γ PKC to yield a persistently active γ PKM may occur in our system. It is interesting to note that γ PKC, the only isoform to increase rapidly 2 h following LTP, is a brain-specific isoform with expression reaching a maximum in the brain around the time that the hippocampus matures (Nishizuka, 1988). This suggests that γ PKC may play a special role in synaptic plasticity in the adult brain.

6.3.3 ERKs

ERK-1 and ERK-2 isoforms show differential expression patterns in the adult rat brain, as previously described (Thomas and Hunt, 1993). In the hippocampus, ERK-1 expression is high in the dentate gyrus and CA3 regions, weaker in the CA2 region and almost absent in the CA1 region, while ERK-2 expression is high in all neuronal components of the hippocampus. ERK-3 mRNA is expressed at its highest levels early in development in the rat central nervous system and, in adulthood, hindbrain expression is greater than that in the forebrain (Boulton *et al.*, 1991). Consistent with these observations, we found no specific ERK-3 expression in the hippocampus (data not shown). Twenty-four hours after LTP induction, ERK-2

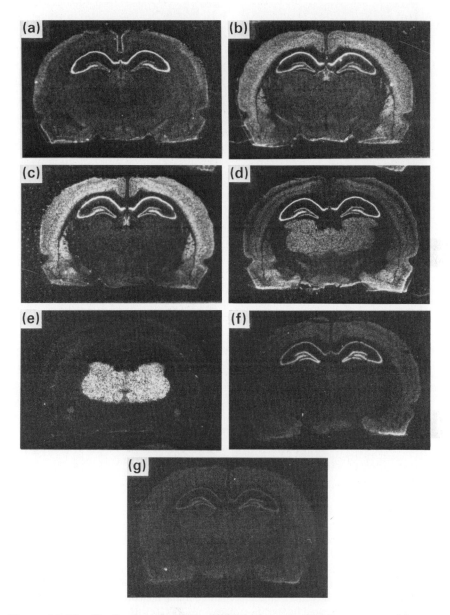

Figure 6.5. The distribution of mRNA of PKC isoforms in the hippocampus of the normal rat. Autoradiographs from X-ray film show the differential expression of α PKC (a), βI PKC (b), βII PKC (c), γ PKC (d), δ PKC (e), ε PKC (f) and ζ PKC (g) in coronal sections through the brain. All PKC isoforms were expressed in the granule cells of the dentate gyrus and pyramidal cells of the CA regions of the hippocampus, with the exception of δ PKC.

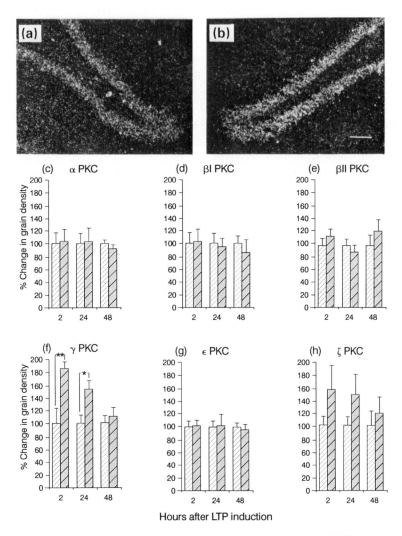

Figure 6.6. The increase in γ PKC mRNA in dentate granule cells after LTP. Dark field photomicrographs showing silver grains generated by a ^{35}S-labeled probe specific for γ PKC associated with dentate granule cells in the nonpotentiated hemisphere (a) and potentiated hemisphere (b) 2 h after unilateral LTP induction. The density of silver grains in dentate granule cells in the potentiated hemisphere (darkly shaded bars) were expressed as a percentage of the density of silver grains counted in the nonstimulated hemisphere (lightly shaded bars) generated by specific probes for α PKC (c), βI PKC (d), βII PKC (e), γ PKC (f), ε PKC (g) and ζ PKC (h) mRNAs. *Post hoc* analysis (Dunn's test) showed that γ PKC mRNA was increased in the dentate granule cell bodies 2 and 24 h after LTP induction (*$p < 0.05$ and **$p < 0.01$). Tetanic stimulation did not effect constitutive α, βI, βII, or ε PKC expression. In addition, there was no effect of LTP on ζ PKC expression following induction and no differences in ζ PKC mRNA levels between the potentiated and nonpotentiated dentate granule cells at individual time points. Neither was the expression of δ PKC induced following LTP (data not shown). Scale bar 500 μm.

expression was increased in the potentiated granule cells (*Figure 6.7*). Forty-eight hours after LTP, the expression had returned to normal levels. There were no alterations in the mRNA levels for ERK-1 at any time after LTP induction. ERK-3 expression was not induced by LTP.

Raf-1 and Raf-B mRNA expression was observed in the granule cells of the dentate gyrus and the pyramidal cells of CA1, CA2 and CA3, while no signal for Raf-A could be detected in the hippocampus (*Figure 6.8*). Following the induction of LTP, an increase in the expression of brain-specific Raf-B (Storm *et al.*, 1990) was seen at 24 h, which returned to normal by 48 h (*Figure 6.8*). Raf-1 expression was not changed and Raf-A expression was not induced in the dentate gyrus following LTP.

Rafs and ERKs are calcium-independent serine/threonine kinases that rely on phosphorylation for their activation (Nishida and Gotoh, 1993). In addition, these kinases function within the same signal transduction pathway. Although an up-regulation of Raf protein has been reported (Mihaly *et al.*, 1990), it is not known whether there is an increase in activity of particular isoforms of ERK or Raf after LTP. In addition, it is unclear how they might be activated. It has been

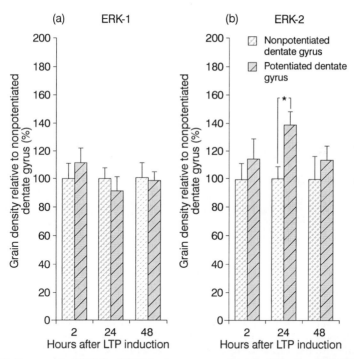

Figure 6.7. Increase in ERK-2 mRNA in dentate granule cells 24 h after LTP. The density of silver grains in dentate granule cells in the potentiated hemisphere (darkly shaded bars) expressed as a percentage of the density of silver grains counted in the nonstimulated hemisphere (lightly shaded bars) show that there was no effect of tetanic stimulation on mRNA levels for ERK-1 (a) but a significant increase in ERK-2 mRNA levels (*$p<0.05$, Dunn's test) were seen 24 h after LTP induction (b). The expression of ERK-3 was not induced following LTP (data not shown).

CORTICAL PLASTICITY

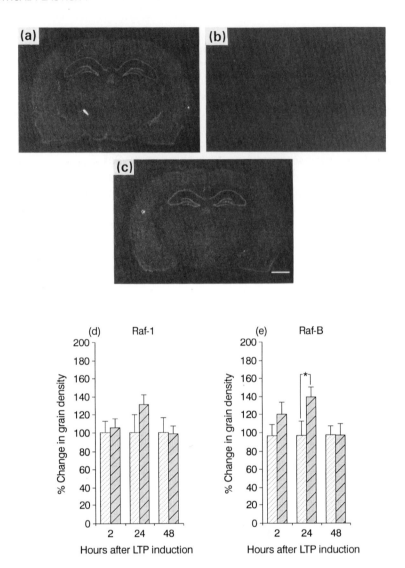

Figure 6.8. An increase in Raf-B mRNA in dentate granule cells 24 h after LTP. The distribution of mRNA of Raf isoforms in the hippocampus of the normal rat. Autoradiographs from X-ray film show the expression of Raf-1 (a), Raf-A (b) and Raf-B (c) in coronal sections through the brain. There is no Raf-A in adult rat brain. Scale bar 2000 μm. The density of silver grains in dentate granule cells in the potentiated hemisphere (darkly shaded bars) expressed as a percentage of the density of silver grains counted in the nonstimulated hemisphere (lightly shaded bars) show that there was no effect of tetanic stimulation on mRNA levels for Raf-1 (d) but a significant increase in Raf-B mRNA levels (e) were seen in the 24 h after induction of LTP. *Post hoc* analysis showed that mRNA for Raf-B was significantly increased 24 h post-tetanic stimulation (*$p<0.05$, Dunn's test). The expression of Raf-A was not induced following LTP.

suggested that *in vitro* the calcium-dependent serine/threonine CaMKII and ERK mediate independent transduction pathways to the nucleus (Bading *et al.*, 1993) and, as such, the ERK/Raf pathway may be activated after LTP by activation of cell surface tyrosine kinase receptors by growth factors. However, there is increasing evidence that activation of PKC (Kolch *et al.*, 1993; Wood *et al.*, 1992) or G-protein-linked surface receptors (Daub *et al.*, 1996; Wan *et al.*, 1996) can in turn increase ERK activity. This may be mediated by the novel protein tyrosine kinase, PYK2 (Lev *et al.*, 1995). Therefore, during LTP, initial direct activation of α CaMKII and γ PKC by calcium entering via the NMDA receptor or, in the case of γ PKC, indirectly by metabotropic glutamate receptor (mGluR) stimulation (see below) may lead to PYK2-directed activation of the calcium-independent intracellular signaling pathway. Inhibitor studies (O'Dell *et al.*, 1991) and gene deletion experiments (Grant *et al.*, 1992) have indicated that LTP induction is dependent on tyrosine kinase activation, particularly *fyn*. The extracellular or intracellularly mediated activation of the ERK-2/Raf-B pathway clearly indicates a mechanism by which post-translational and transcriptional events associated with LTP can outlast the initial intracellular calcium transient.

6.3.4 PKA

Activation of PKA is critical for the late phase of LTP or L-LTP (Frey *et al.*, 1993; Huang and Kandel, 1994; Matthies and Reymann, 1993). PKA is a complex molecule (Taylor *et al.*, 1992). To date, two catalyic subunit isoforms, Cα and Cβ (three splice variants), and four regulatory isoforms, RIα, RIβ, RIIα and RIIβ of PKA have been described. In its inactive state, two catalytic and two regulatory subunits are covalently associated within the cytoplasm of the cell. However, in the presence of cAMP, the regulatory subunits dissociate and allow the active catalytic subunits to translocate to the nucleus where they can act on a variety of substrates including the transcription factor CREB. A recent report shows that mice in which genes coding for the βI isoforms of the catalytic and regulatory PKA subunits were deleted were unable to support the NMDA-independent and presynaptically maintained LTP at the synapses of mossy fibers and CA3 pyramidal cells of the hippocampus (Huang *et al.*, 1995). However, differences in NMDA receptor-dependent LTP between the two mutants were observed. CβI mutants showed an inability to maintain LTP at the Schaffer collateral–CA1 synapse in the long-term (Qi *et al.*, 1996), which reflects the role of PKA in the late phases of LTP. However, the RIβ mutant mice showed perfectly normal LTP at this hippocampal synapse (Brandon *et al.*, 1995). The studies can be reconciled, in part, since the RIβ mutant mice show a compensatory increase in RIα. Furthermore, these transgenic studies suggest that NMDA receptor-dependent and -independent LTP at different sites in the hippocampus rely on distinct subunits of PKA. It is not known whether LTP induction at any of the hippocampal synapses is associated with changes in the expression of individual PKA subunits. Presumably, an increase in catalytic or a decrease in regulatory subunits would result in a potential increase in PKA activity and PKA-directed gene expression. Indeed, studies in *Aplysia* show a precedent for this idea. Applications of serotonin which induce the long-term facilitation of responses between co-cultured sensory and motor neurons not only induce a translocation of the PKA subunits to the nucleus of sensory neurons (Bacskai *et al.*, 1993) and activation of gene transcription (Kaang *et al.*, 1993), but by 24 h after serotonin application the protein levels of the PKA regulatory subunit are significantly reduced (Bergold *et al.*, 1992).

6.4 Signal transduction pathways: protein phosphatases

Lisman and Goldring (1988) proposed that the expression of the two main forms of hippocampal synaptic plasticity, LTP and LTD (for LTD review, see Bear and Abraham, 1996) depended on the differential activation and the opposing actions of protein kinases and protein phosphatases on their substrates. There is accumulating evidence to show that the expression of hippocampal LTD in young rats is dependent on the activation of the wide substrate specificity protein phosphatases PP1, PP2A and PP2B (Mulkey *et al.*, 1993, 1994). Therefore, LTP may be associated with a decrease in phosphatase activity and gene expression. A recent report showed that although E-LTP in the adult rat was dependent on postsynaptic cAMP and presumably PKA activation, cAMP application alone was not sufficient to enhance synaptic efficacy (Blitzer *et al.*, 1995). Moreover, the study demonstrated that when protein phosphatase inhibitors were injected postsynaptically, E-LTP was no longer dependent on PKA activation. This suggested that, under normal circumstances, the expression of LTP is prevented by the constitutive activity of these protein phosphatases but that, during LTP induction, cAMP-mediated PKA activation leads to a relief of this constraint. Indeed, PKA has been reported to negatively regulate PP1 (Shinolikar and Nairn, 1991). At present, there is no evidence to show that expression of these phosphatase genes is altered in the long term after LTP.

An increase in the expression of a novel protein phosphatase has been seen with increased synaptic activity (Qian *et al.*, 1994). A differential molecular screen of a hippocampal cDNA library and one from animals that had undergone metrazole-induced seizure has shown that the expression of one of the brain activity-dependent genes isolated (BAD), BAD2, is also increased after LTP. BAD2 expression was rapidly but transiently increased in potentiated cells within 1 h of LTP induction. The gene encoded for a dual-purpose threonine- and tyrosine-protein phosphatase which specifically dephosphorylates ERK and as such suggests that the increased expression of BAD2 with LTP may provide a negative feedback control.

6.5 Glutamate receptors

There are three major families of glutamate receptor genes, and many of the family members are differentially spliced to produce large numbers of potentially functional receptor isoforms (Hollmann and Heinemann, 1994). The AMPA

(α-amino-3-hydroxy-5-methyl-4-isoxazole propionate) and NMDA ionotropic receptors and the mGluRs have all been shown to have a role in the induction and potential roles in the maintenance of LTP (see Chapter 5).

6.5.1 AMPA receptors

Early reports suggest that increases in the number of AMPA receptors could underlie the increase in synaptic efficacy. These reports indicated increases in the density of [^{3}H]-AMPA binding in several regions of the hippocampus within 1 h of unilateral LTP induction in the perforant path–dentate gyrus synapses (Maren et al., 1993; Tocco et al., 1992). However, the observed increases were bilateral, which may call into question the specificity of changes with LTP. Other reports showed that the affinity of agonists for the AMPA receptor increases after LTP (Shahi and Baudry, 1992), suggesting that not only do the numbers of AMPA receptors increase but also their kinetic properties alter with LTP. One mechanism which may underlie the change in the kinetics of these receptors would involve post-translational modifications (see Chapter 5). Another mechanism that has been proposed recently to be capable of supporting the increased AMPA receptor responses associated with LTP is the functional uncovering of active AMPA receptors which, prior to induction, were 'silent' (Issac et al., 1995; Liao et al., 1995). The rapid acquisition of AMPA responses at single CA1 synapses in hippocampal slices again suggests that post-translational modifications are paramount, at least during the early phases of LTP. It would be expected that in the long-term these post-translational modifications would be superseded by transcriptional events which would support the same functional objective.

The four AMPA receptor subunits, GluR-1–4, are derived from four separate genes and are thought to associate to form heteromeric channels similar to the nicotinic acetylcholine receptor. This family of glutamate receptors shows a very great functional diversity. The subunits can form functional homomers as well as heteromers, each with different pharmacologies and channel properties when (co-)expressed in oocytes or transfected cells. Furthermore, each subunit exists in two forms, 'flip' and 'flop', generated by the alternative splicing of a C-terminal exon. 'Flip' forms of the receptor are more efficient than 'flop' forms in their response to glutamate, and it had been suggested that a limited substitution of the 'flip' for the 'flop' forms on hippocampal neurons could account for the change in synaptic efficacy that characterizes LTP. However, while there is some indication that this may occur in pathological brain states (Kamphuis et al., 1992), there currently is no evidence to suggest that any change in AMPA receptor gene expression occurs during LTP.

Finally, a mechanism known as RNA editing, by which a single codon in a pre-mRNA is altered by site-selective adenosine deamination, creates yet another level of AMPA receptor diversity (Melcher et al., 1996; Sommer et al., 1991). Editing at a codon for the channel determinant (Q/R) in GluR-2 controls the calcium permeability of the AMPA receptor. In theory, an increase in the AMPA receptor subunits which mediate calcium inflow may result in long-lasting changes in synaptic efficacy associated with LTP.

6.5.2 NMDA receptors

The NMDA receptor is an oligomeric receptor, composed of NR1 subunits, which form the lining of the ion channel itself, and modulatory NMDA receptor 2 (NR2) subunits (A–D), in an unknown stoichiometry (Hollmann and Heinemann, 1994). Co-expression studies *in vitro* and the differential expression of individual subunits in the brain suggest that the subunit composition of the NMDA receptor dictates its functional characteristics. Subsequent to the cloning of NR1 by Nakanishi in 1991 (Moriyoshi *et al.*, 1991), further studies have identified at least eight isoforms of NR1 (Sugihara *et al.*, 1992). These isoforms are generated from one gene by a combination of the alternative splicing of two exons which code for amino acids in the C-terminal domain of the receptor protein and the insertion of an amino acid cassette towards the N terminus. The product of one of the C-terminal exons (exon 21) contains several potential sites for phosphorylation by PKC, based on the characterized consensus sequence for PKC (Kennelly and Kreb, 1991). *In vitro* studies have indicated that NR1 can be phosphorylated by PKC to potentiate receptor function (Durand *et al.*, 1992; Kutsawada *et al.*, 1992; Urushihara *et al.*, 1992). Moreover, the augmentation of the responsiveness of the NMDA receptor by PKC occurred by the phosphorylation of site(s) in the C terminus of NR1 coded for by exon 21 (Tingley *et al.*, 1993).

Since NMDA- as well as non-NMDA-mediated currents appear to be enhanced in the maintenance phase of LTP, the phosphorylation of both NMDA and non-NMDA glutamate receptors may underlie this process in LTP. Indeed, support for this hypothesis was shown when direct injection of a combination of PKC inhibitors into the postsynaptic neuron, which may prevent phosphorylation of glutamate receptors, blocked the maintenance phase of LTP (Wang and Feng, 1992). The question arose as to whether there was an increased expression of exon 21-coded variants of NMDA receptor in the maintenance phase of LTP (Thomas *et al.*, 1994b).

Both species of NR1 subunits are expressed in the granule cells of the dentate gyrus and pyramidal cells of the CA1–3 region. NR1 subunits that do not contain a PKC phosphorylation site are highly expressed in the CA1–3 region and moderately expressed in the dentate gyrus. NR1 species that do contain the PKC phosphorylation site are highly expressed in both these cell layers of the hippocampus.

Over the 48 h after LTP induction, there was no change in the expression of NR1 splice variants of the NMDA receptor that do not contain a phosphorylation site for PKC coded for by exon 21. However, 48 h after LTP induction, there was a 50% increase in splice variants that contain mRNA for the PKC phosphorylation consensus site and coded for by exon 21 (*Figure 6.9*). This result, together with the previous data suggesting a specific increase in the γ PKC isoform and the deleterious effect of γ PKC gene deletion on LTP, suggests that the enhanced responsiveness of postsynaptic NMDA receptors in the maintenance phase of LTP may possibly be achieved initially by the increased activity of PKC towards the NR1, and then by the increased expression of modifiable NR1 at later time points.

These observations demonstrated that the maintenance phase of LTP was associated with an increase in the expression of NR1 subunits that contain a consensus

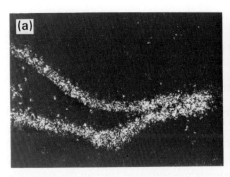 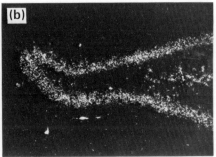

Figure 6.9. Increase in the expression of NR1 subunits of the NMDA receptor which contain a PKC phosphorylation site in dentate granule cells 48 h after LTP induction. The dark field micrograph (a) shows the silver grains (white) generated by the radiolabeled probe specific for mRNA of NR1 subunits of the NMDA receptor which encode the exon 21 PKC phosphorylation site over granule cells of the potentiated dentate gyrus 48 h after high frequency stimulation, while that of (b) shows the hybridization signal; over the contralateral, nonstimulated dentate gyrus of the same rat.

site for PKC-mediated phosphorylation. It is not known at present whether the elevated levels of mRNA coding for these NR1 subunits result in a net increase in the modifiable protein or whether the increase in their expression is a homeostatic response to the specific activation and subsequent breakdown of the PKC-sensitive NR1 subunits of the NMDA receptor. An increase in the expression of the PKC-sensitive but not PKC-insensitive NR1 subunits may indicate a change in the stoichiometry of the heteromeric NMDA receptor after LTP. In all cases, the change in the expression of the PKC-sensitive subunits may indicate a change in the phosphorylation and functional states of the NMDA receptor during LTP.

The NR2 subunits, particularly NR2A and NR2B, can also be phosphorylated, but, in contrast to NR1 subunits, by tyrosine kinases (Lau and Huganir, 1995; Moon et al., 1994). NR2B subunits are highly enriched in the postsynaptic density (PSD) and are the most prominently tyrosine phosphorylated proteins of the PSD (Moon et al., 1994). These authors suggested that tyrosine phosphorylation of NR2B may permit the interaction of NMDA receptors with other proteins of the PSD and create an assembly point for signal transduction complexes. A change in the phosphorylation state of NR2B following LTP would have important ramifications for the functional state of the postsynaptic cell. While mice that lack functional NR2A subunits show a reduced LTP at CA1 synapses (Sakimura et al., 1995), NR2B-deficient mice do not live more than 2 weeks although they cannot support juvenile LTD as do their NR2B-intact littermates (Kutsuwada et al., 1996). Thus, both NR2A and NR2B may be important for neuronal plasticity. However, preliminary data from our laboratory identifies an increase in the expression of NR2B but not other NR2 subunits in dentate granule cells after perforant path LTP (data not shown). The increase was transient but like that for PKC-sensitive NR1 subunits, was evident 48 h after induction. The increased expression of NR1 subunits that can undergo phosphorylation many hours after

LTP induction would provide a mechanism by which not only NMDA receptor responsiveness to released glutamate, but also its expression, is enhanced and this signifies that the efficacy of downstream processes are also improved.

6.5.3 Metabotropic glutamate receptors

Currently, eight genes of the G-protein-linked mGluRs have been cloned and characterized (Pin and Boekaert, 1995; Pin and Duvoisin, 1995). The mGluRs can be classified into three groups based on their signal transduction mechanisms and their pharmacology. The mGluR1 and mGluR5 isoforms are G-protein-linked receptors coupled to phospholipase C, which when activated initiate the release of diacylglycerol (DAG) and inositol trisphosphate (IP_3) from membrane bound phospholipids. DAG can directly activate PKC, while IP_3 can indirectly activate the calcium-dependent isoforms of PKC after releasing calcium from intracellular stores. Increased activity of this class of mGluRs or an enhancement of their sensitivity to released glutamate via a PKC-mediated process may contribute to the synaptic mechanisms enabling the maintenance of LTP (Asztely *et al.*, 1992; O'Conner *et al.*, 1995). Moreover, genetic studies show that mutant mice which are deficient for mGluR1 show diminished LTP (Aiba *et al.*, 1994; Conquet *et al.*, 1994). Since increases in the expression of γ PKC (Thomas *et al.*, 1994a) and PKC-sensitive NR1s (Thomas *et al.*, 1994b) accompany LTP, it is possible that LTP may also be associated with increased mGluR expression.

Within the adult hippocampus, mGluR1a, mGluR1b and mGluR1c all showed the highest levels of expression in the dentate granule cells, while there was no apparent expression in the CA1 pyramidal cells (data not shown). mGluR1a and mGluR1b expression was seen at moderate levels in CA3 pyramidal cells. There was no apparent signal for mGluR1c mRNA in CA3 pyramidal neurons. In contrast, the highest levels of mGluR5a and mGluR5b expression in the hippocampus were observed in the CA1 and CA3 pyramidal cells. mGluR5a showed a relatively low level of expression in the dentate granule cells, while mGluR5b showed a moderate expression in these cells.

Our preliminary observations suggested that there were no changes in the expression of mGluR1a and mGluR1b in the dentate granule cells 2–96 h following the induction of unilateral LTP in the perforant path–dentate granule cell synapses in freely moving rats when compared with the levels of expression in the contralateral nonpotentiated dentate granule cell. However, there was a slow onset but persistent increase in mGluR1c expression in the dentate granule cells (*Figure 6.10*).

Ultrastructural studies have shown that unlike mGluR1a which is localized on postsynaptic elements at the periphery of postsynaptic densities (Baude *et al.*, 1993; Martin *et al.*, 1992), the shorter mGluR1b and mGluR1c are also found on nerve terminals, at least in the rat striatum (Fotuhi *et al.*, 1993). Moreover, we have seen mGluR1-like immunoreactivity in the mossy fibers (the axons of the dentate granule cells), and their terminals as well as in the granule cell soma and their associated dendritic field (data not shown). Since phospholipase C-coupled

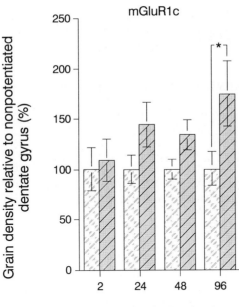

Figure 6.10. The selective increase in the expression of mGluR1c in dentate granule cells of freely moving rats after LTP induction. A slow onset but persistent increase in the expression of mGluR1c ($F = 5.2$ $^{df}1,14$ $p<0.05$, repeated measures ANOVA) was seen in dentate granule cells that were the postsynaptic targets of perforant path fibers that had recieved a high frequency tetanic stimulation (darkly shaded bars) relative to the contralateral side that did not receive this stimulation (lightly shaded bars). By 96 h after LTP, this increase was significantly above normal expression (* $p<0.05$, Student's t-test). Results are presented as the means ± SEM.

mGluRs potentiate glutamate release from synaptosomes and at glutamatergic synapses (Herrero *et al.*, 1992; McBain *et al.*, 1994), presynaptic mGluR1c present on mossy fiber terminals may function to enhance glutamate release at this site after LTP induction.

Changes in the postsynaptic expression of proteins which are then targeted to the presynaptic terminals of these neurons during LTP would have important implications for the long-lasting increases in synaptic strength in hippocampal circuits. Rapid and transient increases in the expression of GR33 were observed in dentate granule cells after LTP-inducing stimulations were applied to the perforant path (Smirnova *et al.*, 1993a). GR33 protein was identical to the presynaptic protein syntaxin 1B which functions in the docking and priming of vesicles for calcium-dependent neurotransmitter release (Sudhof, 1995), but in a recombinant form was shown to have similar pharmacological (but distinct electrophysiological) properties to NMDA receptors when expressed in oocytes (Smirnova *et al.*, 1993b). The increase in GR33 would have profound consequences for the release of neurotransmitter at the axon terminals of granule cells, and as such may support the NMDA-independent, transynaptically mediated LTP at the mossy fiber–CA3 synapses observed after tetanic stimulation of the perforant path (Yeckel and Berger, 1990).

6.6 LTP and growth-related changes in gene expression

There is fairly good evidence that some sort of morphological change accompanies the establishment of LTP (see Edwards, 1995, for references). This is thought to be preceded by an activation of growth factor expression, a breakdown of existing local synaptic circuitry by the action between proteases on adhesion molecules and cytoskeletal components and finally a re-establishment of a new synaptic environment by terminal growth and dendritic remodeling.

6.6.1 Growth factors

Levels of brain-derived neurotrophic factor (BDNF) mRNA are increased in hippocampal slice preparations in CA1 cells (Hughes *et al.*, 1993; Patterson *et al.*, 1992) and *in vivo* in dentate granule cells (Bramham *et al.*, 1996; Castren *et al.*, 1994) following LTP inducing paradigms. The neurotrophins may function in consolidating changes in synaptic strength. However, the *in vivo* studies showed that the increase in BDNF expression in the dentate gyrus was not confined to the potentiated hippocampus but was also seen to the same degree in the contralateral side. Assuming that the contralateral side was not potentiated by the LTP-inducing stimulation to the other side (the authors do not always report this), then it is unlikely that BDNF expression is linked specifically to LTP but to afferent activity *per se*. In one study, mRNA for neurotrophin-3 (NT3) was decreased bilaterally by LTP (Castren *et al.*, 1994), while in another unilateral increases in NT3 and nerve growth factor (NGF) expression were observed (Bramham *et al.*, 1996). The lack of consensus in these studies may simply reflect the stimulation paradigms used. These results do suggest that individual neurotrophic factors play different roles in neuronal plasticity. Indeed, deletion of the BDNF gene in mice resulted in a substantial reduction of LTP in CA1, while other physiological parameters of neuronal transmission appeared normal (Korte *et al.*, 1995). Bramham and colleagues (1996) did report an up-regulation in the BDNF receptor, *trk*B (but not *trk*C), expression which was confined to the postsynaptic cells in the potentiated hemisphere. It is possible that the unilateral change in *trk*B expression confers specificity of action of increased BDNF levels after LTP. The synthesis and release of neurotrophin from the postsynaptic neuron could potentially act at a number of sites and, therefore, have far-reaching effects; for example, on the postsynaptic neuron itself in an autocrine fashion, or on the presynaptic axon terminal facilitating the release of neurotransmitter (Thoenen, 1995).

6.6.2 Proteases

In the hippocampus, tissue levels of mRNAs of the proteases, tissue plasminogen activator (tPA) and brain-specific dipeptidyl peptidase-like protein were rapidly and transiently induced by stimulation of the perforant pathway (Qian et al., 1993). In freely moving rats, mRNA levels for tPA were elevated in dentate gyrus granule cells within 1 h of unilateral LTP induction, returning to basal levels within 24 h. The tPA expression was confined to the dentate granule cells and was attenuated by NMDA receptor blockade. Since tPA release is correlated with morphological differentiation, it was suggested that it played a similar role in mediating any structural changes that would accompany the LTP-associated increase in synaptic efficacy.

The expression of 16C8, a gene originally isolated using a subtractive and differential cloning strategy between cDNA libraries constructed from dentate granule cells from naive and kainic acid-treated rats, was induced in the 6 h following LTP induction (Nedivi et al., 1993). 16C8 encoded a protease inhibitor. As such this protein could function generally in a fashion similar to that of the phosphatase BAD2 on ERK, by forming a negative feedback loop to control for or dampen excessive protease activity in the later phases of LTP after structural changes would have occurred.

6.6.3 Cell adhesion molecules

One possible family of protease targets are the cell adhesion molecules (CAMs). CAMs are a group of proteins that include neural cell adhesion molecule (NCAM), L1, Tag1 and Thy1 belonging to the superfamily of surface glycoproteins which stabilize process–process neuronal connections by homo- and heterophilic interactions in the adult brain (Doherty et al., 1995).

In *Aplysia*, long-term facilitation is associated not only with the formation of new synaptic connections between the sensory and motor neurons, but also with the internalization and subsequent decrease in *Aplysia* NCAMs on the surface of the presynaptic sensory cells (Bailey et al., 1992; Mayford et al., 1992). This down-regulation was thought to destabilize cell–cell interactions which normally prevent growth and lead to the redistribution of membrane proteins to newly forming synapses.

In mammals, the adhesion molecules NCAM and L1 also appear to be crucial to the establishment of hippocampal LTP. Reports show that LTP induction in CA1 neurons in slices was attenuated by the application of NCAM antibodies, recombinant L1 fragments and by treatments that destabilize NCAM–L1 heterophilic interactions (Luthi et al., 1994; Ronn et al., 1995). Increases in the extracellular concentrations of NCAM and amyloid precursor protein, which can function as a protease inhibitor itself, were seen 90 min after LTP induction *in vivo* (Fazeli et al., 1994), indicating that protease-mediated cleavage of CAMs does indeed accompany synaptic plasticity. However, it is not known which of the many isoforms of NCAM are involved in adult plasticity. Individual NCAM isoforms are generated from one gene by the alternative splicing of many of its 26 exons (Barthels et al., 1992; Hemperly et al., 1986). The up-regulation of one particular N-terminal exon, VASE, which may modulate NCAM–NCAM recognition, is correlated with the decrease in polysialic acid glycosylation and synaptic plasticity associated with brain maturation (Seki and Arai, 1991; Theodosis *et al.*, 1991). Viable adult NCAM-deficient mice have been generated (Cremer *et al.*, 1994). These mice showed an almost total loss of protein-bound polysialic acid and displayed deficits in a hippocampal-dependent learning task. Unfortunately, whether these mice could support hippocampal LTP was not tested. In addition, it is possible that LTP could be correlated with an increase in the expression of VASE-containing NCAM.

6.6.3 Cytoskeletal components

Synaptic remodeling would be expected to involve the destabilization and restructuring of the cytoskeleton. An early report suggested that LTP was associated with activation of proteases and cleavage of fodrin, a specific form of the major cytoskeletal protein spectrin, which is localized specifically to the soma and dendrite of neurones (Lynch and Baudry, 1987). More recently, a gene initially isolated by differential cloning between hippocampal cDNA libraries obtained from naive rats and rats in which electroconvulsive seizure was induced demonstrated the very rapid (within 30 min) but transient (absent by 24 h) expression in postsynaptic dentate granule cells after LTP induction (Lyford *et al.*, 1995). This gene encoded for *arc*, an activity-regulated cytoskeletal-associated protein, which shows homology to spectrin and is localized to dendrites. The LTP-associated expression of *arc* could facilitate structural rearrangement in the dendrites.

Indirect evidence has shown that the dendritic MAP2, which cross-links and stabilizes microtubules (Matus, 1988; Olmstead, 1986), may play a role in neuronal plasticity. mRNA for MAP2, like α CaMKII mRNA, is transported into dendrites (Davis *et al.*, 1987), which likewise suggests that the regulation of MAP2 could occur locally and rapidly. Phosphorylation of MAP2 impairs its ability to promote microtubule polymerization and actin binding *in vitro* (Hoshi *et al.*, 1988; Matus, 1988; Olmstead, 1986); *in vivo* this could lead to a destabilization of the cytoskeleton. Recently, it has been demonstrated that a brief application of glutamate to hippocampal slices results initially in an mGluR-dependent increase in MAP2 phosphorylation which is followed by PP2B dephosphorylation to below normal levels (Quinlan and Halpain, 1996). This study may have important consequences for neuronal plasticity. The biphasic profile of MAP2 phosphorylation, first facilitating destabilization and then a restabilization of the local cytoskeleton, would contribute significantly to structural remodeling accompanying LTP. As yet there is no direct evidence for the transcriptional or post-translational control of MAP2 in LTP. However, restricted application of NMDA (or nitric oxide donors) to the molecular layer of the dentate gyrus in slices has been shown to cause a local increase in MAP2 mRNA (Johnstone and Morris, 1994). This increase in MAP2 presumably could lead to a stabilization of postsynaptic structural changes.

6.7 Summary

We have summarized much of the available data on gene expression during LTP, particularly that related to the establishment of NMDA-dependent LTP in the hippocampus. We suggest that spatially and temporally distinct patterns of gene expression may reinforce many of the post-translational changes that occur immediately following the induction of LTP. We describe a series of changes which could reflect the actions of a series of biochemically linked parallel pathways leading from the cell surface to the nucleus, and from there back to the cytoplasm and neuronal membrane. It is not inconceivable that activation of each family of glutamate receptors leads to an independent cascade of intracellular events which then act in concert to produce LTP. Activation of each of these pathways would have consequences both post-translationally on proteins within the cytoplasm and cell membrane and on transcription factor activation within the nucleus leading to prolonged changes in the neuronal machinery. Nevertheless, postsynaptic changes in gene expression could well be adaptive changes to increased presynaptic release of neurotransmitter, preparing the synthetic machinery for an extended period of increased activity. However, it is equally possible that the changes in gene expression described here reflect alterations in the postsynaptic neuron which maintain the increased excitability of the neuron in the absence of enhanced presynaptic activity. It seems likely that both processes are at work during LTP (Chapter 4).

It is unclear how local morphological or physiological changes to a region of the dendritic tree could be targeted by increased expression of a protein following LTP. The mechanism of targeting just part of the dendritic tree and those synapses which are potentiated is indeed totally unknown. Such a mechanism would be expected to involve a local 'sign' indicating the induction of LTP and interrupting the flow of mRNA or protein from the nucleus to consolidate the local changes that are taking place. In this context, considerable weight has been given to the mRNA of α CaMKII which is one of the very few mRNAs that is transported into dendrites. Many questions remain to be answered, notably our almost complete ignorance of what occurs to gene expression presynaptically following LTP.

Acknowledgment

KLT is supported by the Human Frontiers of Science Program.

References

Abeliovich A, Chen C, Goga Y, Silva A, Stevens CF, Tonegawa S. (1993) Modified hippocampal long-term potentiation in PKC γ mutant mice. *Cell* 75: 1253–1262.
Abraham WC, Mason SE, Demmer J, Williams JM, Richardson CL, Tate WP, Lawlor PA, Dragunow M. (1993) Correlations between immediate early gene induction and the persistence of long-term potentiation. *Neuroscience* 56: 717–727.
Aiba A, Chen C, Herrup K, Rosenmund C, Stevens CF, Tonegawa S. (1994) Reduced hippocampal long-term potentiation and context-specific deficit in associative learning in mGluR1 mutant mice. *Cell* 79: 365–375.

Alberini CM, Ghirardi M, Metz R, Kandel ER. (1994) C/EBP is an immediate-early gene required for the consolidation of long-term facilitation in *Aplysia*. *Cell* **76**: 1099–1114.

Arias J, Alberts AS, Brindle P, Claret FX, Smeal T, Karin M, Feramisco J, Montminy M. (1994) Activation of cAMP and mitogen responsive genes relies on a common nuclear factor. *Nature* **370**: 226–229.

Asztely F, Wigstrom H, Gustafsson B. (1992) The relative contribution of NMDA receptor channels in the expression of long-term potentiation in the hippocampal CA1 region. *Eur. J. Neurosci.* **4**: 681–690.

Bacskai BJ, Hochner B, Mahaut-Smith M, Adams SR, Kaang B-K, Kandel ER, Tsien RY. (1993) Spatially resolved dynamics of cAMP and protein kinase subunits in *Aplysia* sensory neurons. *Science* **260**: 222–226.

Bading H, Ginty DM, Greenberg ME. (1993) Regulation of gene expression in hippocampal neurons by distinct calcium signalling pathways. *Science* **260**: 181–186.

Bailey CH, Chen M, Keller F, Kandel ER. (1992) Serotonin-mediated endocytosis of apCAM: an early step of learning related synaptic growth in *Aplysia*. *Science* **256**: 645–649.

Barthels D, Vopper G, Boned A, Cremer H, Wille W. (1992) High degree of NCAM diversity generated by alternative RNA splicing in brain and muscle. *Eur. J. Neurosci.* **4**: 327–337.

Bartsch D, Ghiradi M, Skehel PA, Karl KA, Herder SP, Chen M, Bailey CH, Kandel ER. (1995) *Aplysia* CREB2 represses long-term facilitation: relief of repression converts transient facilitation into long-term functional and structural change. *Cell* **83**: 979–992.

Bashir ZI, Alford S, Davis SN, Randall AD, Collingridge GL. (1991) Long-term potentiation of the NMDA receptor mediated synaptic transmission in the hippocampus. *Nature* **349**: 156–158.

Baude A, Nusser Z, Roberts JDB, Mulvihill E, McIlhinnney RAJ, Somogyi P. (1993) The metabotropic glutamate receptor (mGluR1α) is concentrated at perisynaptic membranes of neuronal subpopulations as detected by immunogold reaction. *Neuron* **11**: 771–787.

Bear MF, Abraham WC. (1996) Long-term depression in the hippocampus. *Annu. Rev. Neurosci.* **19**: 437–462.

Bergold PJ, Beushausen SA, Sacktor T, Cheley S, Bayley H, Schwartz JH. (1992) A regulatory subunit of the cAMP-dependent protein kinase own-regulated in *Aplysia* sensory neurons during long-term sensitization. *Neuron* **8**: 387–397.

Blitzer RD, Wong T, Nouranifar R, Iyengar R, Landau EM. (1995) Postsynaptic cAMP gates early LTP in hippocampal CA1 region. *Neuron* **15**: 1403–1414.

Bliss TVP, Collingridge GL. (1993) A synaptic model of memory; long-term potentiation in the hippocampus. *Nature* **361**: 31–39.

Boulton TG, Nye SH, Robbins DJ, *et al.* (1991) ERKs: a family of protein-serine/threonine kinases that are activated and tyrosine phosphorylated in response to insulin and NGF. *Cell* **65**: 663–675.

Bourtchuladze R, Frenguelli B, Blendy J, Cioffi D, Schutz G, Silva AJ. (1994) Deficient long-term memory in mice with a targeted mutation of the cAMP-responsive element-binding protein. *Cell* **79**: 59–68.

Bramham CR, Southard T, Sarvey JM, Herkenham M, Brady LS. (1996) Unilateral LTP triggers bilateral increases in hippocampal neurotrophin and trk receptor mRNA expression in behaving rats: evidence for interhemispheric communication. *J. Comp. Neurol.* **368**: 371–382.

Brandon EP, Zhou M, Huang YY, Qi M, Gerhold KA, Burton KA, Kandel ER, McKnight GS, Idzera RL. (1995) Hippocampal long-term depression and depotentiation are defective in mice carrying a targeted disruption of the gene encoding the RIβ subunit of cAMP-dependent protein kinase. *Proc. Natl Acad. Sci. USA* **92**: 8851–8855.

Burgin KE, Waxman MN, Rickling S, Westgate SA, Mobley WC, Kelly PT. (1990) *In situ* hybridization histochemistry of Ca^{2+}/calmodulin-dependent protein kinase in developing rat brain. *J. Neurosci.* **10**: 1788–1798.

Castren E, Pitanen M, Sirvio J, Parsadanian A, Lindholm D, Thoenen H, Riekkinen PJ. (1994) The induction of LTP increases BDNF and NGF mRNA but decreases NT-3 mRNA in the dentate gyrus. *NeuroReport* **4**: 895–898.

Cole AJ, Safen DW, Baraban JM, Worley PF. (1989) Rapid increase of an immediate early gene messenger RNA in hippocampal neurons by synaptic NMDA receptor activation. *Nature* **340**: 470–476.

Conquet F, Bashir ZI, Davis CH, et al. (1994) Motor deficit and impairment of synaptic plasticity in mice lacking mGluR1. *Nature* **372**: 237–243.

Cremer H, Lange R, Christoph A, et al. (1994) Inactivation of the N-CAM gene in mice results in size reduction of the olfactory bulb and deficits in spatial learning. *Nature* **367**: 455–459.

Dash PK, Hochner B, Kandel ER. (1990) Injection of the cAMP responsive element into the nucleus of *Aplysia* sensory neurons blocks long-term facilitation. *Nature* **345**: 718–721.

Dash PK, Karl KA, Colicos MA, Prywes R, Kandel ER. (1991) cAMP response element binding protein is activated by Ca^{2+}/calmodulin- as well as cAMP-dependent protein kinase. *Proc. Natl Acad. Sci. USA* **88**: 5061–5065.

Davis HP, Squire LR. (1984) Protein synthesis and memory: a review. *Psychol. Bull.* **96**: 518–559.

Davis L, Banker GA, Steward O. (1987) Selective dendritic transport of RNA in hippocampal neurons in culture. *Nature* **330**: 477–479.

Davis L, Burger B, Banker GA, Steward O. (1989) Dendritic transport: quantitative analysis of the time course of somatodendritic transport of recently synthesised RNA. *J. Neurosci.* **10**: 3056–3068.

Daub H, Weiss FU, Wallasch C, Ullrich A. (1996) Role of transactivation of the EGF receptor in signalling by G-protein-coupled receptors. *Nature* **379**: 557–560.

Demmer J, Dragunow M, Lawlor PA, Mason SE, Leah JD, Abraham WC, Tate WP. (1993) Differential expression of fos and jun immediate early genes after hippocampal long-term potentiation in awake rats. *Mol. Brain Res.* **17**: 279–286.

Diesseroth K, Bito H, Tsien RW. (1996) Signaling from synapse to nucleus: postsynaptic CREB phosphorylation during multiple forms of hippocampal synaptic plasticity. *Neuron* **16**: 89–101.

Doherty P, Fazeli MS, Walsh FS. (1995) The neural cell adhesion molecule and synaptic plasticity. *J. Neurobiol.* **26**: 437–446.

Dragunow M, Abraham WC, Goulding M, Mason SE, Robertson HA, Faull RLM. (1989) Long-term potentiation and the induction of c-fos mRNA and protein in the dentate gyrus of unanaesthetized rats. *Neurosci. Lett.* **101**: 274–280.

Drewes G, Lichtenberg-Kraag B, Doring F, Mandelkow E-M, Biernat J, Goris J, Doree M, Mandelkow E. (1992) Mitogen activated protein (MAP) kinase transforms tau protein into an Alzheimer-like state. *EMBO J.* **11**: 2131–2138.

Durand GM, Bennett MVL, Zukin RS. (1992) Splice variants of the *N*-methyl-D-aspartate receptor NR1 identify domains involved in regulation by polyamine and protein kinase C. *Proc. Natl Acad. Sci. USA* **90**: 6731–6735.

Edwards FA. (1995) LTP – a structural model to explain the inconsistencies. *Trends Neurosci.* **18**: 250–255.

Fazeli MS, Breen K, Errington ML, Bliss TVP. (1994) Increase in extracellular NCAM and amyloid precursor protein in the dentate gyrus of anaesthetized rats. *Neurosci. Lett.* **169**: 77–80.

Fotuhi M, Sharp AH, Glatt CE, Hwang PM, von Krosigk M, Snyder SH, Dawson TM. (1993) Differential localization of phosphoinositide-linked metabotropic glutamate receptor (mGluR1) and the inositol 1,4,5-triphosphate receptor in rat brain. *J. Neurosci.* **13**: 2001–2012.

Frey U, Krug M, Brodemann R, Reymann K, Matties H. (1989) Long-term potentiation induced in dendrites separated from rats CA1 pyramidal somata does not establish a late phase. *Neurosci. Lett.* **97**: 135–139.

Frey U, Huang Y-Y, Kandel ER. (1993) Effects of cAMP stimulate a late stage of LTP in hippocampal CA1 neurons. *Science* **260**: 1661–1664.

Fukunaga K, Stoppini L, Miyamoto E, Muller D. (1993) Long-term potentiation is associated with increased activity of Ca^{2+}/calmodulin-dependent protein kinase II. *J. Biol. Chem.* **268**: 7863–7867.

Gallin WJ, Greenberg ME. (1995) Calcium regulation of gene expression in neurons: the mode of entry matters. *Curr. Opin. Neurobiol.* **5**: 367–374.

Goedert M, Crowther RA, Garner CC. (1991) Molecular characterisation of microtubule-associated proteins tau and MAP2. *Trends Neurosci.* **14**: 193–199.

Gozlan H, Chinestra P, Diabira D, Ben-Ari Y. (1994) NMDA redox site modulates longterm potentiation of NMDA but not AMPA receptors. *Eur. J. Pharmacol.* **262**: R3–R4.

Grant SGN, O'Dell TJ, Karl KA, Stein PL, Soriano P, Kandel ER. (1992) Impaired long-term potentiation, spatial-learning and hippocampal development in *fyn* mutant mice. *Science* **258**: 1903–1910.

Greengard P, Jen J, Nairn AC, Stevens CF. (1991) Enhancement of the glutamate response by cAMP dependent protein kinase in hippocampal neurons. *Science* **253**: 1135–1138.

Hemperly JJ, Murray BA, Edelman GM, Cunningham BA. (1986) Sequence of a cDNA clone encoding the polysialic acid rich, cytoplasmic domains of the neural cell-adhesion molecule N-CAM. *Proc. Natl Acad. Sci. USA* **83**: 3037–3041.

Herrero I, Mirasportugal MT, Sanchezprieto J. (1992) Positive feedback of glutamate exocytosis by metabotropic presynaptic receptor stimulation. *Nature* **360**: 163.

Hollmann M, Heinemann S. (1994) Cloned glutamate receptors. *Annu. Rev. Neurosci.* **17**: 31–108.

Hoshi M, Akiyama T, Shinohara Y, Miyata Y, Ogawar H, Nishida E, Sakai H. (1988) Protein-kinase C-catalyzed phosphorylation of the microtubule-binding domain of the microtubule-associated protein 2 inhibits its ability to induce tubulin polymerization. *Eur. J. Biochem.* **174**: 225–230.

Huang Y-Y, Kandel ER. (1994) Recruitment of long-lasting and protein kinase A-dependent long-term potentiation in the CA1 region of hippocampus requires repeated tetanization. *Learn. Mem.* **1**: 74–82.

Huang Y-Y, Kandel ER, Varshasky L, Brandon EP, Qi M, Idzerda RL, McKnight GS, Bourtchuladze R. (1995) A genetic test of the effects of mutations in PKA on mossy fibre LTP and its relation to spatial, contextual learning. *Cell* **83**: 1211–1222.

Hughes P, Bielharz E, Gluckman P, Drugunow M. (1993) Brain-derived neurotrophic factor is induced as an immediate early gene following N-methyl-D-aspartate receptor activation. *Neuroscience* **57**: 319–328.

Hunter K, Karin M. (1992) The regulation of transcription by phosphorylation. *Cell* **70**: 375–387.

Inoue M, Kashimoto A, Takai Y, Nishizuka Y. (1977) Studies on a catalytic nucleotide-independent protein kinase and its proenzyme in mammalian tissues. *J. Biol. Chem.* **252**: 7610–7616.

Isaac JTR, Nicoll RA, Malenka RC. (1995) Evidence for silent synapses – implications for the expression of LTP. *Neuron* **15**: 427–434.

Jeffery KJ, Abraham WC, Dragunow M, Mason SE. (1990) Induction of fos like immunoreactivity and the maintenance of long-term potentiation in the dentate gyrus of unanaesthetised rats. *Mol. Brain Res.* **8**: 267–274.

Johnstone HM, Morris BJ. (1994) NMDA and nitric-oxide increase microtubule-associated protein-2 gene-expression in hippocampal granule cells. *J. Neurochem.* **63**: 379–382.

Kaang B-K, Kandel ER, Grant SGN. (1993) Activation of cAMP-responsive genes by stimuli that produce long-term facilitation in *Aplysia* sensory neurons, *Neuron* **10**: 427–435.

Kamphuis W, Monyer H, Derijk TC, Da Silva FHL. (1992) Hippocampal kindling increases the expression of glutamate receptor-A flip and receptor-B flip messenger-RNA in dentate granule cells. *Neurosci. Lett.* **148**: 51–54.

Kennelly RJ, Kreb EG. (1991) Consensus sequences as substrate specificity determinants for protein kinases and protein phosphatases. *J. Biol. Chem.* **266**: 15555–15558.

Klann E, Chen SJ, Sweatt JD. (1993) Mechanism of protein kinase C activation during the induction and maintenance of long-term potentiation probed using a selective peptide substrate. *Proc. Natl Acad. Sci. USA* **90**: 8337–8341.

Kolch W, Heidecker G, Kochs G, Hummel R, Vahidi H, Mischak H, Finkenzeller G, Marme D, Rapp UR. (1993) Protein kinase C activates RAF-1 by direct phosphorylation. *Nature* **364**: 249–252.

Korte M, Carroll P, Wolf E, Brem G, Thoenen H, Bonhoeffer T. (1995) Hippocampal long-term potentiation is impaired in mice lacking brain-derived neurotrophic factor. *Proc. Natl Acad. Sci. USA* **92**: 8856–8860.

Krug M, Lossner B, Ott T. (1984) Anisomycin blocks the late phases of long-term potentiation in the dentate gyrus of freely moving rats. *Brain Res. Bull.* **13**: 39–42.

Kutsawada T, Kashiwabi N, Mori H, *et al.* (1992) Molecular diversity of the NMDA receptor. *Nature* **358**: 36–41.

Kutsawada T, Sakimura K, *et al.* (1996) Impairment of suckling response, trigeminal neuronal pattern formation, and hippocampal LTD in NMDA receptor $\epsilon 2$ subunit mutant mice. *Neuron* **16**: 333–344.

Kwok RPS, Lundblad JR, Chrivia JC, Richards JP, Bachinger HP, Brennan RG, Green MR, Goodman RH. (1994) Nuclear protein CBP is a coactivator for the nuclear transcription factor CREB. *Nature* 370: 223–226.

Kyriakis JM, App H, Zhang X-F, Banerjee P, Brautigan DL, Rapp UR, Avruch J. (1992) Raf-1 activates MAP kinase–kinase. *Nature* 358: 417–421.

Latchman DS. (1991) *Eukaryotic Transcription Factors.* Academic Press, London.

Lau L-F, Huganir RL. (1995) Differential tyrosine phosphorylation of N-methyl-D-aspartate receptor subunits. *J. Biol. Chem.* 270: 20036–20041.

Lev S, Moreno H, Martinez R, Canoll P, Peles E, Musacchio JM, Plowman GD, Rudy B, Schlessinger J. (1995) Protein tyrosine kinase PYK2 involved in Ca^{2+}-induced regulation of ion channel and MAP kinase functions. *Nature* 376: 737–745.

Liao DZ, Hessler NA, Malinow R. (1995) Activation of postsynaptically silent synapses during pairing-induced LTP in CA1 region of hippocampal slice. *Nature* 375: 400–404.

Lisman JE, Goldring MA. (1988) Feasibility of long-term storage of graded information by the Ca^{2+}/calmodulin-dependent protein kinase molecules of the postsynaptic density. *Proc. Natl Acad. Sci. USA* 85: 5320–5324.

Lovinger DM, Wong KL, Murakami K, Routtenberg A. (1987) Protein kinase C inhibitors eliminate hippocampal long-term potentiation. *Brain Res.* 436: 177–183.

Luthi A, Laurent JP, Figurov A, Muller D, Schacher M. (1994) Hippocampal long-term potentiation and pleural cell-adhesion molecules L1 and NCAM. *Nature* 372: 777–779.

Lyford GL, Yamagata K, Kaufmann W, *et al.* (1995) Arc, a growth factor, activity-regulated gene, encodes a novel cytoskeleton-associated protein that is enriched in neuronal dendrites. *Neuron* 14: 433–455.

Lynch G, Baudry M. (1987) Brain spectrin, calpain and long-term changes in synaptic efficiency. *Brain Res. Bull.* 18: 809–815.

Lynch MA, Voss KL, Rodriguez J, Bliss TVP. (1994) Increase in synaptic vesicle proteins accompanies long-term potentiation in the dentate gyrus. *Neuroscience* 60: 1–5.

Malenka RC, Kauer JA, Perkel DJ, Mauk MD, Kelly PT, Nicoll RA, Waxham MN. (1989) An essential role for postsynaptic calmodulin and protein kinase activity in long-term potentiation. *Nature* 340: 554–556.

Malinow R, Madison D, Tsien RW. (1988) Persistent protein kinase activity underlying long-term potentiation. *Nature* 335: 820–824.

Malinow R, Schulman H, Tsien RW. (1989) Inhibition of postsynaptic PKC or CaMKII blocks induction but not expression of LTP. *Science* 245: 862–866.

Maren S, Tocco G, Standley S, Baudry M, Thompson RF. (1993) Postsynaptic factors in the expression of long-term potentiation (LTP) – increased glutamate-receptor binding following LTP induction *in vivo*. *Proc. Natl Acad. Sci. USA* 90: 9654–9658.

Martin LJ, Blackstone CD, Huganir RL, Price DL. (1992) Cellular localization of a metabotropic glutamate receptor in rat brain. *Neuron* 9: 259–270.

Masu M, Tanabe Y, Shigemoto R, Nakanishi S. (1991) Sequence, expression of a metabotropic glutamate receptor. *Nature* 349: 760–765.

Matthies H, Reymann KG. (1993) Protein kinase A inhibitors prevent the maintenance of hippocampal long-term potentiation. *NeuroReport* 4: 712–714.

Matus A. (1988) Microtubule-associated proteins: their role in determining neuronal morphology. *Annu. Rev. Neurosci.* 11: 29–44.

Mayford M, Barzilai A, Kellor F, Schacher S, Kandel ER. (1992) Modulation of an NCAM-related adhesion molecule with long-term plasticity in *Aplysia*. *Science* 256: 638–644.

McBain CJ, Dichiara TJ, Kauer JA. (1994) Activation of metabotropic glutamate receptors differentially affects 2 classes of hippocampal interneurons and potentiates excitatory synaptic transmission. *J. Neurosci.* 14: 443

McGlade-McCulloh E, Yamamoto H, Tan S-E, Brickey DA, Soderling TR. (1993) Phosphorylation and regulation of receptors by calcium/calmodulin-dependent protein kinase II. *Nature* 362: 640–642.

Melcher T, Maas S, Herb A, Spengel R, Seeburg PH, Higuchi M. (1996) A mammalian RNA editing enzyme. *Nature* **379**: 460–464.

Mihaly A, Olah Z, Krug M, Kuhnt U, Matthies H, Rapp UR, Joo F. (1990) Transient increase of raf protein kinase like immunoreactivity in the rat dentate gyrus during long-term potentiation. *Neurosci. Lett.* **116**: 45–50.

Milbranndt J. (1987) A nerve growth factor-induced gene encodes a possible transcriptional regulatory factor. *Science* **238**: 797–799.

Miller SG, Kennedy MB. (1986) Regulation of brain type II Ca^{2+}/calmodulin-dependent protein kinase by autophosphorylation: a Ca^{2+}-triggered molecular switch. *Cell* **44**: 861–887.

Montarolo PG, Goelet P, Castellucci VF, Morgan J, Kandel ER. (1986) A critical period for macromolecular synthesis in long-term heterosynaptic facilitation in *Aplysia*. *Science* **234**: 1249–1254.

Moon IS, Apperson ML, Kennedy MB. (1994) The major tyrosine-phosphorylated protein in the postsynaptic density fraction is N-methyl-D-aspartate receptor subunit 2B. *Proc. Natl Acad. Sci. USA* **91**: 3954–3958.

Morgan JI, Curran T. (1991) Stimulus–transcription coupling in the nervous system: involvement of the inducible proto-oncogenes *fos* and *jun*. *Annu. Rev. Neurosci.* **14**: 421–51.

Moriyoshi K, Masu M, Ishii T, Shigemoto R, Mizuno N, Nakanishi S. (1991) Sequence and expression of a metabotropic glutamate receptor. *Nature* **354**: 31–37.

Mulkey RM, Herron CE, Malenka RC. (1993) An essential role for protein phosphatases in hippocampal long-term depression. *Science* **261**: 1051–1055.

Mulkey RM, Endo S, Shenolikar S, Malenka RC. (1994) Involvement of a calcineurin/inhibitor-1 phosphatase cascade in hippocampal long-term depression. *Nature* **369**: 486–488.

Nedivi E, Hevroni D, Naot D. Israeli D, Citri Y. (1993) Numerous candidate plasticity-related genes revealed by differential cDNA cloning. *Nature* **363**: 718–722.

Nguyen PV, Abel T, Kandel ER. (1994) Requirement of a critical period of transcription for induction of a late phase of LTP. *Science* **265**: 1104–1107.

Nikolaev E, Tischmeyer W, Krug M, Matthies H, Kaczmarek L. (1991) c-*fos* protooncogene expression in rat hippocampus and entorhinal cortex following tetanic stimulation of the perforant path. *Brain Res.* **560**: 346–349.

Nishida E, Gotoh Y. (1993) The MAP kinase cascade is essential for diverse signal transduction pathways. *Trends Biochem. Sci.* **18**: 128–131.

Nishizuka Y. (1988) The molecular heterogeneity of protein kinase C and its implications for cellular regulation. *Nature* **334**: 661–665.

O'Dell TJ, Kandel ER, Grant SGN. (1991) Long-term potentiation in the hippocampus is blocked by tyrosine kinase inhibitors. *Nature* **353**: 558–560.

O'Connor JJ, Rowen MJ, Anwyl R. (1995) Tetanically induced LTP involves a similar increase in the AMPA, NMDA receptor components of the excitatory postsynaptic current: investigations of the involvement of mGlu receptors. *J. Neurosci.* **15**: 2013–2020.

Olmstead JB. (1986) Microtubule-associated proteins. *Annu. Rev. Cell Biol.* **2**: 421–457.

Otani S, Abraham WC. (1989) Inhibition of protein synthesis in the dentate gyrus, but not entorhinal cortex blocks the maintenance of long-term potentiation in rats. *Neurosci. Lett.* **106**: 175–180.

Otani S, Marshal CJ, Tate WP, Goddard GV, Abraham WC. (1989) Maintenance of long-term potentiation in rat dentate gyrus requires protein synthesis but not messenger RNA synthesis immediately post-tetanization. *Neuroscience* **28**: 519–526.

Patterson SL, Grover LM, Schwartzkroin PA, Bothwell M. (1992) Neurotrophin expression in rat hippocampal slices: a stimulus paradigm inducing LTP in CA1 evokes increase in BDNF and NT3 mRNAs. *Neuron* **9**: 1081–1088.

Pettit DL, Perlman S, Malinow R. (1994) Potentiated transmission and prevention of further LTP by increased CaMKII activity in postsynaptic hippocampal slice neurons. *Science* **266**: 1881–1885.

Pin JP, Boekaert J. (1995) Get receptive to metabotropic glutamate receptors. *Curr. Opin. Neurobiol.* **5**: 342–349.

Pin JP, Duvoisin R. (1995) The metabotropic glutamate receptors – structure and functions. *Neuropharmacology* **34**: 1–26.

Pontremoli S, Melloni E, Michetti M, Sacco O, Salamino F, Sparatore B, Horecker BL. (1986) Biochemical responses in activated human neutrophils mediated by protein kinase C and a Ca^{2+}-requiring proteinase. *J. Biol. Chem.* **261**: 8309–8313.

Qi M, Zhou M, Skalhegg BS, Brandon EP, Kandel ER, McKnight GS, Idzerda RL. (1996) Impaired hippocampal plasticity in mice lacking the CβI catalytic subunit of cAMP-dependent protein kinase. *Proc. Natl Acad. Sci. USA* **93**: 1571–1576.

Qian Z, Gilbert ME, Colicos MA, Kandel ER, Kuhl D. (1993) Differential screening for activity-dependent genes in the brain reveals that tissue plasminogen activator is an immediate early gene that is induced during seizure, kindling and long-term potentiation. *Nature* **361**: 453–457.

Qian Z, Gilbert M, Kandel ER. (1994) Temporal and spatial regulation of the expression of BAD2, a MAP kinase phosphatase, during seizure, kindling and long-term potentiation. *Learn. Mem.* **1**: 180–188.

Quinlan EM, Halpain S. (1996) Postsynaptic mechanisms for bidirectional control of MAP2 phosphorylation by glutamate receptors. *Neuron* **16**: 357–368.

Reymann KG, Frey U, Jork R, Matthies H. (1988) Inhibitors of calmodulin and protein kinase C block different phases of long-term potentiation. *Brain Res.* **440**: 305–314.

Richardson CL, Tate WP, Mason SE, Lawlor PA, Dragunow M, Abraham WC. (1992) Correlation between the induction of an immediate early gene, zif/268 and long-term potentiation in the dentate gyrus. *Brain Res.* **580**: 147–154.

Roberts LA, Higgins MJ, O'Shaughnessy CT, Morris BJ. (1995) Changes in hippocampal gene-expression associated with the induction of long-term potentiation. *Br. J. Pharmacol.* **116**: SS pP348.

Ronn LCB, Bock E, Linnemann D, Jahnsen H. (1995) NCAM-antibodies modulate induction of long-term potentiation in rat hippocampal CA1. *Brain Res.* **677**: 145–151.

Sacktor TC, Osten P, Valsamis H, Jiang X, Naik MU, Sublette E. (1993) Persistent activation of the ζ isoform of protein kinase C in the maintenance of long-term potentiation. *Proc. Natl Acad. Sci. USA* **90**: 8342–8346.

Sakimura K, Kutsawada T, Ito I, *et al.* (1995) Reduced hippocampal LTP and spatial learning in mice lacking NMDA ε1 subunit. *Nature* **373**: 151–155.

Seki T, Arai Y. (1991) Expression of highly polysialylated NCAM in the neocortex and piriform cortex of the developing and the adult rat. *Neurosci. Res.* **12**: 503–513.

Shahi K, Baudry M. (1992) Increased binding affinity of agonists to glutamate receptor increases synaptic responses at glutamatergic synapses. *Proc. Natl Acad. Sci. USA* **89**: 6881–6885.

Shinolikar S, Nairn AC. (1991) Protein phosphatases: recent progress. *Adv. Second. Mess. Phosphoprot. Res.* **23**: 1–121.

Silva AJ, Stevens CF, Tonegawa S, Wang Y. (1992) Deficient hippocampal long-term potentiation in α-calcium-calmodulin kinase II mutant mice. *Science* **257**: 201–206.

Smirnova T, Laroche S, Errington ML, Hicks AA, Bliss TVP, Mallet J. (1993a) Transsynaptic expression of a presynaptic glutamate receptor during hippocampal long-term potentiation. *Science* **262**: 433–436.

Smirnova T, Stinnakre J, Mallet J. (1993b) Characterisation of a presynaptic glutamate receptor. *Science* **262**: 430–433.

Sommer B, Kohler M, Sprengel R, Seeburg PH. (1991) RNA editing in brain controls a determinant of ion flow in glutamate-gated channels. *Cell* **67**: 11–19.

Stanton PK, Sarvey JM. (1984) Blockade of long-term potentiation in rat hippocampal CA1 region by inhibitors of protein synthesis. *J. Neurosci.* **4**: 3080–3088.

Storm SM, Cleveland JL, Rapp UR. (1990) Expresssion of raf family proto-oncogenes in normal mouse tissues. *Oncogene* **5**: 345–351.

Sudhof TC. (1995) The synaptic vesicle cycle: a cascade of protein–protein interactions. *Nature* **375**: 645–653.

Sugihara H, Moriyoshi K, Ishii T, Masu M, Nakanishi S. (1992) Structures and properties of seven isoforms of the NMDA receptor generated by alternative splicing. *Biochem. Biophys. Res. Commun.* **185**: 826–832.

Tan SE, Wenthold RJ, Soderling TR. (1994) Phosphorylation of AMPA-type glutamate receptors by calcium/calmodulin-dependent protein kinase II and protein kinase C in cultured hippocampal neurons. *J. Neurosci.* **14**: 1123–1129.

Taylor SS, Knighton DR, Zheng J, Ten Eyck LF, Sowadski JM. (1992) Structural framework for the protein kinase family. *Annu. Rev. Cell Biol.* **8**: 429–462.

Theodosis DT, Rougon G, Poulain DA. (1991) Retention of embryonic features by an adult neuronal system capable of plasticity – polysialylated neural cell-adhesion molecule in the hypothalamo neurohyphophyseal system. *Proc. Natl Acad. Sci. USA* **88**: 5494–5498.

Thiel G, Schoch S, Petersholm D. (1994) Regulation of synapsin 1 gene expression by the zinc-finger transcription factor zif268/egr1. *J. Biol. Chem.* **269**: 15301.

Thoenen H. (1995) Neurotrophins and neuronal plasticity. *Science* **270**: 593–598.

Thomas KL, Hunt SP. (1993) The regional distribution of extracellularly regulated kinase-1 and -2 messenger RNA in the adult rat central nervous system. *Neuroscience* **56**: 741–757.

Thomas KL, Laroche S, Errington ML, Bliss TVP, Hunt SP. (1994a) Spatial and temporal changes in signal transduction pathways during LTP. *Neuron* **13**: 737–745.

Thomas KL, Davis S, Laroche S, Hunt SP. (1994b) Regulation in the expression of NR1 NMDA glutamate receptor subunits during hippocampal LTP. *NeuroReport* **6**: 119–123.

Tingley WG, Roche KW, Thompson AK, Huganir RL. (1993) Regulation of NMDA receptor phosphorylation by alternative splicing of the C-terminal domain. *Nature* **364**: 70–73.

Tocco G, Maren S, Shors TJ, Baudry M, Thompson RF. (1992) Long-term potentiation is associated with increased [^{3}H]-AMPA binding in the rat hippocampus. *Brain Res.* **573**: 228–234.

Urushihara H, Tohda M, Nomura Y. (1992) Selective potentiation of *N*-methyl-D-aspartate-induced current by protein kinase C in *Xenopus* oocytes injected with rat brain RNA. *J. Biol. Chem.* **267**: 11667–11700.

Wan Y, Kurosaki T, Huang X-Y. (1996) Tyrosine kinases in activation of MAP kinase cascade by G-protein-coupled receptors. *Nature* **380**: 541–544.

Wang J-H, Feng D-P. (1992) Postsynaptic protein kinase C essential to induction and maintenance of long-term potentiation in the hippocampal CA1 region. *Proc. Natl Acad. Sci. USA* **89**: 2576–2580.

Wang L-Y, Dudek EM, Browning MD, MacDonald JF. (1994) Modulation of AMPA/karinate receptors in cultured murine hippocampal neurones by protein kinase C. *J. Physiol. (Lond.)* **475**: 431–437.

Wisden W, Errington ML, Williams S, Dunnett SB, Waters C, Hitchcock D, Evan G, Bliss TVP, Hunt SP. (1990) Differential expression of immediate early gene in the hippocampus and spinal cord. *Neuron* **4**: 603–614.

Wood KW, Sarnecki C, Roberts TM, Blenis J. (1992) *ras* mediates nerve growth factor receptor modulation of three signal transducing protein kinases: MAP kinase, Raf-1 and RSK. *Cell* **68**: 11041–11050.

Worley PF, Bhat RV, Baraban JM, Erickson CA, McNaughton BL., Barnes CA. (1993) Thresholds for synaptic activation of transcription factors in hippocampus: correlation with long-term enhancement. *J. Neurosci.* **13**: 4776–4786.

Yamagata K, Kaufmann WE, Lanahan A, Papapalou M, Barnes CA, Andreasson KI, Worley PF. (1994) Egr3/pilot, a zinc-finger transcription factor, is rapidly regulated by activity in brain neurons and colocalizes with Egr1/zif268. *Learn. Mem.* **1**: 140–152.

Yeckel MF, Berger TW. (1990) Feedforward excitation of the hippocampus by afferents from the entorhinal cortex: redefinition of the role of the tri-synaptic pathway. *Proc. Natl Acad. Sci. USA* **87**: 5832–5836.

Yin JCP, Wallach JS, Del Vecchio M, Wilder EL, Zhou H, Quinn WG, Tully T. (1994) Induction of a dominant negative CREB transgene specifically blocks long-term memory in *Drosophila*. *Cell* **79**: 49–58.

Yin JCP, Wallach JS, Wilder EL, Klingensmith J, Dang D, Perrimon N, Zhou H, Tully T, Quinn WG. (1995) A *Drosophila* CREB/CREM homolog encodes multiple isoforms, including a cyclic AMP-dependent protein kinase responsive transcriptional activator and antagonist. *Mol. Cell. Biol.* **15**: 5123–5130.

Neuromodulators of synaptic strength

M. Segal and J.M. Auerbach

7.1 Introduction

Long-term potentiation (LTP) of the efficacy of synaptic transmission (Bliss and Collingridge, 1993; Bliss and Lomo, 1973) is a widespread property of central synapses and has been proposed as a simple cellular model of learning and memory. LTP is proposed to constitute the process underlying learning in complex systems. It can be evoked by tetanic synaptic stimulation, by association of synaptic stimulation with postsynaptic depolarization or by application of chemicals and neurotransmitter substances which affect ionic conductances or second messenger systems to produce long-lasting changes in synaptic efficacy (Aniksztejn and Ben-Ari, 1991; Bortolotto and Collingridge, 1993).

Neuromodulators are substances released from presynaptic terminals which, by themselves, do not produce fast synaptic actions, but are able to modulate reactivity of the affected neurons to stimulation of other afferents. This can be done either by affecting potassium or calcium channels, or via second messenger interactions with the fast neurotransmitter receptors, as will be illustrated below. The intuitive significance of the neuromodulators becomes evident when their action is interrupted by lesion or drug action, and the subsequent loss of memories. Such is the case with acetylcholine (ACh), which is released from septohippocampal terminals and is assumed to have a major role in cognitive processes of hippocampal learning and memory (Bartus *et al.*, 1982; Buresova *et al.*, 1964; Molchan *et al.*, 1992). The role of ACh at the cellular and molecular levels of these processes is still unclear. Likewise, noradrenaline (NA) and serotonin (5-hydroxytryptamine, 5-HT) are assumed to modulate cellular functions associated with long-term plastic properties of neuronal circuits, but their roles in these processes are even less clear (Foote *et al.*, 1983; Levkovitz *et al.*, 1994; Sara, 1985). In this chapter, we will focus on the known actions of ACh, NA and 5-HT and will examine which, if any, of their multiple actions may be related to their putative role in LTP, learning and memory. Other neuromodulators, such as peptides, growth factors and retrograde messengers, will not be dealt with herein.

Cortical Plasticity, edited by M.S. Fazeli and G.L. Collingridge.
© 1996 BIOS Scientific Publishers Ltd, Oxford.

7.2 Acetylcholine

There are five known subtypes of the muscarinic receptor expressed in mammalian brain, M1–M5 (Hulme *et al.*, 1990; Waelbroeck *et al.*, 1990). ACh, acting on muscarinic receptors in the hippocampus, causes blockade of several potassium currents, including a calcium-gated potassium current which underlies the slow afterhyperpolarization (I_{AHP}), a depolarization-evoked sustained potassium current (I_M) and a resting leak potassium current (Cole and Nicoll, 1983; Dutar and Nicoll, 1988; Madison *et al.*, 1987; Muller and Misgeld, 1989). Blockade of these conductances results in depolarization of the cell, increase in input resistance and enhancement of spontaneous activity and reactivity to afferent stimulation. Other reports suggest that ACh may even facilitate a potassium current (I_K; Zhang *et al.*, 1992). In addition, the muscarinic agonist carbachol (CCh) is reported to modulate a calcium-dependent inward current (Fisher and Johnston, 1990) and also to block calcium currents (Segal, 1989). One effect of particular interest is the ACh-induced suppression of the excitatory postsynaptic potential (EPSP) evoked by stimulation of the Schaffer collateral–commissural fibers (Segal, 1982a, 1989; Sheridan and Sutor, 1990). This effect is assumed to be mediated by presynaptic muscarinic receptors and is seen with increasing concentrations of ACh. All of these effects can be seen in most hippocampal neurons with relatively high concentrations of CCh, and are blocked by atropine. The receptor types associated with the various muscarinic effects of CCh are not entirely clear; while most effects involving blockade of potassium currents are likely to be mediated by a muscarinic M1 receptor, the lack of highly specific ligands for most of the M1–M5 receptors hampers progress in this field.

The second messenger systems activated by ACh to produce the effects on the various currents are not clear either. ACh activates second messenger systems involving phospholipases, protein kinase C, inositol trisphosphate (IP_3), and cyclic AMP (McKinney *et al.*, 1991). These, in turn, release calcium from internal stores, which can act on a number of tertiary cellular processes. ACh can also cause a rise in intracellular calcium ($[Ca]_i$) simply by closing potassium channels, thus allowing the membrane to depolarize and activate voltage-gated calcium currents (Muller and Connor, 1991). These actions will trigger several postsynaptic calcium-related processes. One of these processes involves a fast onset potentiation of reactivity of the neurons to application of the glutamate agonist *N*-methyl-D-aspartate (NMDA, Figure 7.1). The cholinergic potentiation of NMDA responses is assumed to involve activation of the IP_3 cascade (Markram and Segal, 1990, 1992), and is mediated by a high affinity, M2 receptor. It is proposed to be one of the main links between ACh, NMDA receptors and plasticity (Markram and Segal, 1992).

The fact that ACh can have different actions on different cell types in the hippocampus complicates the already complex issue of which action of acetylcholine is most relevant to neuronal plasticity. Thus, M2 receptors are found on pyramidal neurons, but also in high concentrations on interneurons located in stratum oriens (Levey *et al.*, 1994). Likewise, all other muscarinic receptors have a selective preferential localization or action on different cell types, and their final effects on hippo-

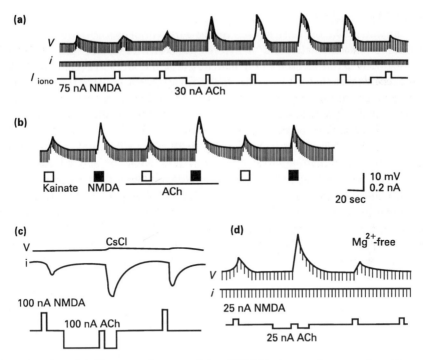

Figure 7.1. Acetylcholine potentiates reactivity of CA1 hippocampal neurons in a slice preparation to NMDA. (a) Iontophoretic application of NMDA near the recorded cell produces a depolarizing response. During the iontophoretic administration of ACh, the depolarizing response to NMDA is enhanced in a transient manner. (b) The response to NMDA but not to iontophoretic application of kainate is enhanced by ACh. (c) The potentiating action of ACh is maintained when the cell is loaded with cesium, and is also seen when [Mg] is reduced in the slice, removing the magnesium block of the NMDA receptor. In the cesium-loaded cell, the current responses to NMDA in a voltage-clamped cell are recorded. In the voltage recordings (a, b, c and d), a hyperpolarizing current pulse is applied every 2 sec, to assess input resistance of the cell. Adapted from Markram and Segal (1990) with permission from The Physiological Society.

campal circuits will be a complex function of their activity and strategic location of the affected cells in the circuit.

Insight into the role of ACh in the cellular and molecular aspects of plasticity has been gained by examining the effects of ACh on LTP. While most cholinergic effects are muscarinic, some nicotinic interactions with LTP have also been reported (Hunter *et al.*, 1994). In the dentate gyrus of rat hippocampus, muscarinic activation facilitates LTP induction (Burgard and Sarvey, 1990) and physostygmine, an inhibitor of acetylcholinesterase, causes potentiation of population spikes resembling LTP (Ito *et al.*, 1988; Levkovitz and Segal, 1994). On the other hand, activation of the medial septum attenuates LTP in the perforant path (Pang *et al.*, 1993). The situation becomes more complex, as in the area CA3–mossy fiber system, low concentrations of CCh suppress tetanus-induced LTP (Maeda *et al.*, 1993,

1994a) via activation of a high affinity M2 receptor [note, however, that Williams and Johnston (1988) obtained similar results with a high concentration of CCh]. In these same neurons, activation of M1 receptors facilitates LTP induction. This effect is opposite to that seen in the dentate gyrus and area CA1 (below), probably related to the fact that LTP in the mossy fiber system is mediated by different mechanisms from those in the other areas, most likely via presynaptic modulation of transmitter release. Another pathway in CA3 where muscarinic actions interact with LTP induction is the fimbria–CA3 pathway (Katsuki et al., 1992).

In area CA1, CCh can enhance LTP (Blitzer et al., 1990) and the muscarinic antagonist atropine can suppress associative LTP (Sokolov and Kleschevnikov, 1995). Cholinergically induced rhythmic activity, obtained with higher concentrations of CCh, can enhance plasticity of neurons in response to afferent stimulation (Huerta and Lisman, 1993), It has also been suggested that anticholinergic drugs suppress the ability of area CA1 to express LTP (Hirotsu et al., 1989). We have found recently that bath application of low concentrations (0.2–0.5 μM) of CCh induces LTP (Auerbach and Segal, 1994), and application of an even lower concentration, while having no observable effect of its own, reduces the threshold for tetanic LTP induction (*Figure 7.2*) This muscarinic LTP (LTP_m) shares similarities with tetanic LTP which lie downstream of the involvement of the NMDA receptor in LTP induction, that is it is dependent on a rise in intracellular calcium, but is independent of activation of the NMDA receptor, or of influx of calcium into the cell during exposure to the drug. This effect is mediated by activation of an M2 muscarinic receptor and is likely to involve protein phosphorylation. This is, by and large, the closest association of ACh to neuronal plasticity in the brain, in that LTP_m is evoked by a very low concentration of the agonist, and is long lasting, as is seen with tetanic LTP. The mechanism underlying LTP_m is still being investigated, but is likely to involve protein phosphorylation, a rise in $[Ca]_i$ and a postsynaptic, long-term change in reactivity of hippocampal neurons to glutamate.

7.3 Noradrenaline

Noradrenaline (NA) is released from terminals of fibers originating in the nucleus locus coeruleus of the brainstem. These fibers innervate the entire forebrain and are assumed to be involved in an array of functions, peripherally related to memory formation and storage (Foote et al., 1983; Sara, 1985). NA activates two classes of receptors in the brain, the α and β receptors, each subdivided in to α-1 and 2, and β-1 and 2. Activation of the β-1 receptor causes a long-lasting blockade of I_{AHP} as seen for the cholinergic muscarinic receptor (Madison and Nicoll, 1982). This response is mediated by a β-1-linked cAMP-generating system. β-2 receptors, found primarily on glial cells, also involve activation of the cAMP second messenger system.

The involvement of NA in plastic processes in the hippocampus has been studied primarily within the dentate gyrus. Early studies suggested that stimulation of the locus coeruleus could augment reactivity of hippocampal neurons to afferent stimulation (Segal and Bloom, 1976). In both intact, awake rats (Harley, 1991) and in brain

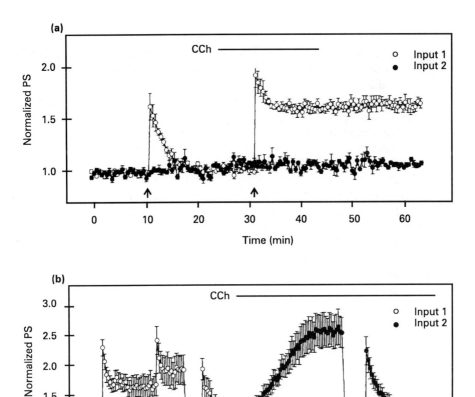

Figure 7.2: LTP$_m$ and tetanus-induced LTP share a common mechanism. (a) Population spike (PS) data recorded from a single site in response to alternating stimuli applied at 0.033 Hz to two separate inputs in stratum radiatum of CA1 area of a rat hippocampal slice. The arrows denote the points at which short tetanic stimulation (35–40 pulses, 100 Hz) was delivered to input 1 only. CCh (0.1 μM) added 10 min before the second tetanus leads to induction of LTP following the second tetanic stimulation. No change in response to the other stimulation was noted throughout the experiment. (b) Population spike data recorded from one site in response to stimulation of two inputs, as above. The mechanism of tetanic LTP was saturated in input 1, the magnitude of stimulation was reduced to verify that the response is not at the saturation level, and a third tetanic stimulation already produced only short-term potentiation. Addition of CCh (0.75 μM) had no effect on the response to stimulation of input 1, but produced the typical potentiation of the response to input 2. Once LTP$_m$ was saturated, suprathreshold tetanus to input 2 failed to further enhance the already potentiated response. Arrows indicate tetanic stimulations (100 pulses, 100 Hz). Reproduced from Auerbach and Segal (1994) with permission from the American Physiological Society.

slices (Stanton and Sarvey, 1985, 1987), stimulation of the locus coeruleus or topical application of NA or the β-agonist isoprotarenol caused a long lasting augmentation of responses of dentate granular cells to stimulation of the perforant path. This augmentation appears to be mediated by a β-receptor, and involves activation of the NMDA receptor, as it is blocked by NMDA antagonists (Burgard et al., 1989; Stanton et. al., 1989; but see Izumi et al., 1992, for another possible interaction between NA and NMDA). This effect is mediated by a change in postsynaptic reactivity to glutamate neurotransmission (Segal, 1982b), although an increase in transmitter release and phosphorylation of presynaptic synapsin I and II proteins has been reported as well (Parfitt et al., 1991), indicating a presynaptic locus of β-adrenergic action. The responses to perforant path stimulation are not enhanced indiscriminately, that is a marked difference between stimulation of the proximal and distal molecular layer (corresponding to the medial and lateral perforant path) has been reported (Pelletier et al., 1994). The NA-enhanced response in the dentate gyrus is genuine LTP in that it lasts for several hours after washout of NA from the tissue. This NA action is not restricted to the dentate gyrus; a β-adrenergic potentiation of some features of LTP by stimulating cAMP production has been reported to take place in region CA3 of the hippocampus (Hopkins and Johnston, 1988).

While the NA innervation of the dentate gyrus appears to be neccesary for the induction of LTP – that is, a markedly reduced tetanus-induced LTP is seen in slices taken from NA-depleted rats, this is not the case in CA1 region of the hippocampus (Stanton and Sarvey, 1985).

The mechanisms underlying the NA-mediated short- and long-term enhancement of reactivity of the dentate to afferent stimulation are not entirely clear. It is not likely that they result from blockade of I_{AHP} since the time course of the AHP is much slower than that of the EPSP or population spike, and the AHP is normally triggered by both depolarization and entry of calcium into the cell. One possible mechanism involves enhancement of calcium currents in hippocampal neurons (Gray and Johnston, 1987). An increase in neurotransmitter release due to phosphorylation of synaptic proteins has been alluded to as another possibility (Parfitt et al., 1991).

The studies mentioned above seem to agree that NA potentiation is mediated by a β-adrenergic receptor. Most of these studies did not consider the possibility that an α-adrenergic receptor may also be involved in LTP. This is a viable possibility since an α-adrenergic receptor is also linked to the phosphatidylinositol (PI) cascade, which causes release of calcium from intracellular stores through activation of the IP_3 receptor (see Segal et al., 1991; *Figure 7.3*). In this respect, the α-adrenergic receptor shares similarities with the metabotropic glutamate receptor, also known to activate the IP_3 receptor and be involved in LTP via release of calcium from intracellular stores (Bortolotto and Collingridge, 1993). This possibility was alluded to in a study which demonstrated that an α-adrenergic receptor potentiated reactivity to NMDA, much like the effect of ACh (Segal et al., 1991). Such potentiation of reactivity to NMDA may interact with synaptic stimulation and facilitate the induction of LTP. It is interesting to note that the α-adrenergic effect on reactivity to glutamate or NMDA is qualitatively different from that of the β-adrenergic effect on the same response (e.g. *Figure 7.3*). Finally, the possibility that some of the effects of NA on

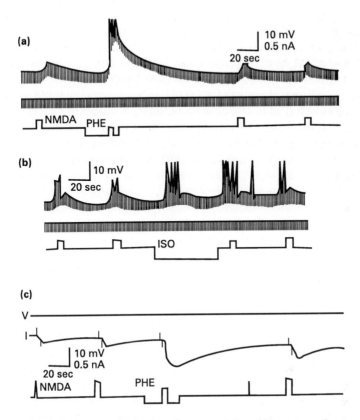

Figure 7.3. Effects of noradrenergic ligands on reactivity of hippocampal neurons of CA1 region in the slice to iontophoretic application of NMDA. Hippocampal neurons were recorded intracellularly with potassium acetate-containing micropipettes, and either current clamped (a and b) or single electrode voltage clamped (c). (a) Voltage (top record) responses to 20 nA current pulses of NMDA (bottom trace) are markedly potentiated during concurrent application of phenylephrine (PHE), a noradrenergic, α-1 agonist. This effect is transient, and the following response to NMDA is actually smaller than the control, pre-PHE response. Input resistance in this cell is estimated by passage of constant current pulses through the recording electrode (middle trace). (b) In another cell, using the same paradigm, isoproterenol (ISO), a β-adrenergic agonist, produces a lasting increase in reactivity to pulse application of NMDA. (c) In a voltage-clamped cell, the current responses to NMDA, middle trace, are potentiated by the presence of PHE. As seen in (a), this effect is short lasting, and a recovery to control response is evident already at the first post-PHE response. Reproduced from Segal *et al.* (1991) Actions of norepinephrine in the rat hippocampus. *Prog. Brain Res.* **88**: 323–330, with permission from Elsevier Science.

plasticity and LTP are mediated by receptors located on interneurons rather than on pyramidal neurons has to be explored further.

7.4 Serotonin

Serotonin-containing fibers originate in the midbrain raphe nuclei and comprise a massive innervation of the hippocampus (Azmitia and Segal, 1978). These fibers terminate preferentially on γ-aminobutyric acid (GABA) interneurons but, in fact, most pyramidal neurons respond to stimulation of the raphe nuclei or local application of serotonin (Segal, 1980). There are many subtypes of serotonergic receptors, grouped into 5-HT_{1-4} types. Most of these receptor types co-exist in the hippocampus, the major receptor being the 5-HT_{1a} subtype, which activates voltage-independent potassium conductance (Segal, 1980). Other major receptor types in the hippocampus are the 5-HT_{2a} receptor, which causes a decrease in I_{AHP} and in a sustained potassium conductance (Colino and Halliwell, 1987). This receptor species is assumed to activate the PI cascade, and to cause release of calcium from intracellular stores (Manor et al., 1994), similar to the effect of the metabotropic glutamate receptors. The activation of 5-HT_2 receptor in the hippocampus results in depolarization and increased firing of the affected neurons. A third receptor type is 5-HT_3, resident primarily on GABAergic interneurons. Activation of the receptor causes a fast depolarization of interneurons and a subsequent burst of inhibitory postsynaptic potentials (IPSPs) in the pyramidal or granule neurons (Ropert, 1988). On a global level, serotonin of the raphe origin is suggested to be involved in generation of hippocampal theta rhythm (see Richter-Levin and Segal, 1991).

The association of serotonin with the induction and maintenance of LTP is less well established than with the other neuromodulators. Unlike NA and ACh, serotonin application by itself is not reported to produce long-lasting changes in reactivity to afferent stimulation. Serotonin depletion does not affect the ability of the dentate gyrus or CA1 area of the hippocampus to express LTP (Richter-Levin and Segal, 1991). Application of serotonin at low concentration does not affect the reactivity to single pulse EPSPs; however, it appears to reduce the ability of the hippocampus to express LTP (Sakai and Tanaka, 1993; Villani and Johnston, 1993). 5HT_3 receptors appear to suppress the induction of LTP (Maeda et al., 1994b) and a selective blockade of the 5-HT_3 receptor in an intact rat brain is reported to increase a theta-burst potentiation in dentate granule cells (Staubli and Otaky, 1994; Staubli and Xu, 1995), most likely through the blockade of the slow IPSPs produced in granule cells following 5-HT_3 activation of interneurons. In the intact brain, removal of 5-HT by lesion of the serotonergic innervation does not affect the ability to express LTP (Richter-Levin and Segal, 1991) but does affect feed-forward inhibition in granule cells. Activation of serotonergic receptors in the intact brain, using drugs known to release serotonin from terminals (e.g. fenfluramine; Richter-Levin and Segal, 1990, 1991), causes a long-lasting increase in population spike response of the dentate gyrus to perforant path stimulation. This effect is caused by serotonin released at its hippocampal terminals, as it is found in brains depleted of their serotonergic innervation, except for the hippocampus, where it is restored with a serotonin-containing raphe graft (Richter-Levin and Segal, 1991). The long-lasting effect of serotonin releasers does not require the continuous presence of serotonin, and in that sense it can be considered a genuine serotonergic LTP. It is different from the typical LTP, though, in that it affects the population spike with little effect on the slope of the population EPSP. The serotonergic potentiation of dentate reactivity to afferent stimulation appears to be mediated by activation of a muscarinic receptor, as atropine blocks the effects of fenfluramine (Levkovitz and Segal, 1994), indicating that serotonin acts on the dentate gyrus by releasing ACh. These interactions may be unique to the intact brain, and not be seen in brain slices, as they may require an intact neuromodulatory system to be detected.

7.5 Conclusions

The biogenic amines, NA, 5-HT and ACh, long thought to modulate reactivity to fast neurotransmitters, may affect the ability of central neurons to express plastic properties via many different channels; they can either themselves produce LTP of reactivity to afferent stimulation, as seen with NA and low concentrations of ACh, or they can modulate the ability of the neurons to react to tetanic stimulation. This comes about either directly through interaction with the receptor in question, for example via phosphorylation of the receptor by activation of a second messenger system, or indirectly, by affecting an interneuron in the hippocampal circuit. The variety of receptor species, and ionic mechanisms activated by the biogenic amine neurotransmitter, as well as the cell type where high affinity receptors are located, makes the analysis of this action rather complicated. The effect on LTP can result from a direct action on LTP-related mechanisms, as well as an action on several mechanisms which are only remotely related to LTP. Such is the case if the action of the modulator is to reduce transmitter release, or to reduce inhibition by hyperpolarizing the interneurons. When the neuromodulator has several actions which can produce opposite effects, for example serotonin acting to hyperpolarize the pyramidal cells but also to block inhibition, the end product may be a delicate balance between the two actions. The presence in the hippocampus of several types of interneurons, containing different receptor types and having different target cells, makes the analysis of the action of the neuromodulators far more complex.

References

Aniksztejn L, Ben-Ari Y. (1991) Novel form of long-term potentiation produced by a K^+ channel blocker in the hippocampus. *Nature* **349**: 67–69.
Auerbach JM, Segal M. (1994) A novel cholinergic induction of long-term potentiation in rat hippocampus. *J. Neurophysiol.* **72**: 2034–2040.
Azmitia EC, Segal M. (1978) An autoradiographic analysis of the differential ascending projections of the dorsal and median raphe nuclei in the rat. *J. Comp. Neurol.* **179**: 641–668.

Bartus RT, Reginald L, Dean RL, Beer B, Lippa AS. (1982) The cholinergic hypothesis of geriatric memory dysfunction. *Science* **217**: 408–417.

Bliss TVP, Collingridge GL. (1993) A synaptic model of memory: long-term potentiation in the hippocampus. *Nature* **361**: 31–39.

Bliss TVP, Lomo T. (1973) Long-lasting potentiation of synaptic transmission in the dentate area of the anaesthetized rabbit following stimulation of the perforant path. *J. Physiol.* **232**: 331–356.

Blitzer RD, Gil O, Landau EM. (1990) Cholinergic stimulation enhances long-term potentiation in the CA1 region of rat hippocampus. *Neurosci. Lett.* **119**: 207–210.

Bortolotto ZA, Collingridge GL. (1993) Characterisation of LTP induced by the activation of glutamate metabotropic receptors in area CA1 of the hippocampus. *Neuropharmacology* **32**: 1–9.

Buresova O, Bures J, Bohdanecky Z, Weiss T. (1964) The effect of atropine on learning, extinction, retention and retrieval in rats. *Psychopharmacologia* **5**: 255–263.

Burgard EC, Sarvey JM. (1990) Muscarinic receptor activation facilitates the induction of long-term potentiation (LTP) in the rat dentate gyrus. *Neurosci. Lett.* **116**: 34–39.

Burgard EC, Decker G, Sarvey JM. (1989) NMDA receptor antagonists block norepinephrine-induced long lasting potentiation in rat hippocampus. *Brain Res.* **428**: 351–355.

Burgard EC, Cote TE, Sarvey JM. (1993) Muscarinic depression of synaptic transmission and blockade of norepinephrine-induced long lasting potentiation in the dentate gyrus. *Neuroscience* **54**: 377–389.

Cole AE, Nicoll RA. (1983) Acetylcholine mediates a slow synaptic potential in hippocampal pyramidal cells. *Science* **221**: 1299–1301.

Colino A, Halliwell JV. (1987) Differential modulation of three separate K conductances in hippocampal CA1 neurons by serotonin. *Nature* **328**: 73–77.

Dutar P, Nicoll R. (1988) Classification of muscarinic responses in hippocampus in terms of receptor subtypes and second-messenger systems: electrophysiological studies *in vitro*. *J. Neurosci.* **8**: 4214–4224.

Fisher R, Johnston D. (1990) Differential modulation of single voltage-gated calcium channels by cholinergic and adrenergic agonists in adult hippocampal neurons. *J. Neurophysiol.* **64**: 1291–1302.

Foote SL, Bloom FE, Aston Jones G. (1983) Nucleus locus coeruleus: new evidence of anatomical and physiological specificity. *Physiol. Rev.* **63**: 844–914.

Gray R, Johnston D. (1987) Noradrenaline and beta adrenoreceptor agonists increase activity of voltage dependent calcium channels in hippocampal neurons. *Nature* **327**: 620–622.

Harley C. (1991) Noradrenergic and locus coeruleus modulation of the perforant path evoked potential in rat dentate gyrus supports a role for the locus coeruleus in attentional and memorial processes. *Prog. Brain Res.* **88**: 307–321.

Hirotsu I, Hori N, Katsuda N, Ishihara T. (1989) Effects of anticholinergic drug on long term potentiation in rat hippocampal slices. *Brain Res.* **482**: 194–197.

Hopkins WF, Johnston D. (1988) Noradrenergic enhancement of long term potentiation at mossy fiber synapses in the hippocampus. *J. Neurophysiol.* **59**: 667–687.

Huerta PT, Lisman JE. (1993) Heightened synaptic plasticity of hippocampal CA1 neurons during a cholinergically-induced rhythmic state. *Nature* **364**: 723–725.

Hulme EC, Birdsall NJM, Buckley NJ. (1990) Muscarinic receptor sybtypes. *Annu. Rev. Pharmacol. Toxicol.* **30**: 633–673.

Hunter BE, de Fiebre CM, Papke RL, Kem WR, Meyer EM. (1994) A novel nicotinic agonist facilitates induction of long term potentiation in the rat hippocampus. *Neurosci. Lett.* **168**: 130–134.

Ito T, Miura Y, Kadokawa T. (1988) Physostigmine induces in rats a phenomenon resembling long-term potentiation. *Eur. J. Pharmacol.* **156**: 351–359.

Izumi Y, Clifford DB, Zorumski CF. (1992) Norepinephrine reverses N-methyl-D-aspartate-mediated inhibition of long term potentiation in rat hippocampal slices. *Neurosci. Lett.* **142**: 163–166.

Katsuki H, Saito H, Satoh M. (1992) The involvement of muscarinic, beta adrenergic and metabotropic glutamate receptors in long term potentiation in the fimbria–CA3 pathway of the hippocampus. *Neurosci. Lett.* **142**: 249–252.

Levey AI, Edmunds SM, Heilman CJ, Desmond TJ, Frey KA. (1994) Localization of muscarinic M3 receptor protein and M3 receptor binding in rat brain. *Neuroscience* **63**: 207–221.

Levkovitz Y, Segal M. (1994) Acetylcholine mediates the effects of fenfluramine on dentate granule cell excitability in the rat. *Eur. J. Pharmacol.* **264**: 279–284.

Levkovitz Y, Richter-Levin G, Segal M. (1994) Effects of 5-HTP on behavior and hippocampal physiology in young and old rats. *Neurobiol. Aging* **15**: 635–641.

Madison DV, Nicoll RA. (1982) Noradrenaline blocks accommodation of pyramidal cell discharge in the hippocampus. *Nature* **299**: 636–638.

Madison DV, Lancaster B, Nicoll RA. (1987) Voltage clamp analysis of cholinergic action in the hippocampus. *J. Neurosci.* **7**: 733–741.

Maeda T, Kaneko S, Satoh M. (1993) Bidirectional modulation of long term potentiation by carbachol via M1 and M2 muscarinic receptors in guinea pig hippocampal mossy fiber CA3 synapses. *Brain Res.* **619**: 324–330.

Maeda T, Kaneko S, Satoh M. (1994a) Roles of endogenous cholinergic neurons in the induction of long term potentiation at hippocampal mossy fiber synapses. *Neurosci. Res.* **20**: 71–78.

Maeda T, Kaneko S, Satoh M. (1994b) Inhibitory influences via 5-HT$_3$ receptors on the induction of LTP in mossy fiber–CA3 system of guinea pig hippocampal slices. *Neurosci. Res.* **18**: 277–282.

Manor D, Moran N, Segal M. (1994) Interactions among calcium compartments in C6 rat glioma cells: involvement of potassium channels. *J. Physiol.* **478**: 251–263.

Markram H, Segal M. (1990) Long lasting facilitation of excitatory postsynaptic potentials in the rat hippocampus by acetylcholine. *J. Physiol.* **427**: 381–393.

Markram H, Segal M. (1992) The inositol 1,4,5-trisphosphate pathway mediates cholinergic potentiation of rat hippocampal neuronal responses to NMDA. *J. Physiol.* **447**: 513–533.

McKinney M, Miller JH, Gibson VA, Nickelson L, Aksoy S. (1991) Interaction of agonists with M2 and M4 muscarinic receptor subtypes mediating cyclic AMP inhibition. *Mol. Pharmacol.* **40**: 1014–1022.

Molchan SE, Martinez RA, Hill JL, Weingartner HJ, Thompson K, Vitiello B, Sunderland T. (1992) Increased cognitive sensitivity to scopolamine with age and a perspective on the scopolamine model. *Brain Res. Rev.* **17**: 215–226.

Muller W, Connor JA. (1991) Cholinergic input uncouples Ca^{2+} changes from K$^+$ conductance activation and amplifies intradendritic Ca^{2+} changes in hippocampal neurons. *Neuron* **6**: 901–905.

Muller W, Misgeld U. (1989) Carbachol and pirenzepine discriminate effects mediated by two muscarinic receptor subtypes on hippocampal neurons *in vitro*. *Experientia* **57**: 114–122.

Pang K, Williams MJ, Olton DS. (1993) Activation of medial septal area attenuates LTP of the lateral perforant path and enhances heterosynaptic LTD of the medial perforant path in aged rats. *Brain Res.* **632**: 150–160.

Parfitt KD, Hoffer BJ, Browning MD. (1991) Norepinephrine and isoproterenol increase the phosphorylation of synapsin I and synapsin II in dentate slices of young but not aged Fisher 344 rats. *Proc. Natl Acad. Sci. USA* **88**: 2361–2365.

Pelletier MR, Kirkby RD, Jones SJ, Corcoran ME. (1994) Pathway specificity of noradrenergic plasticity in the dentate gyrus. *Hippocampus* **4**: 181–188.

Richter-Levin G, Segal M. (1990) Effects of serotonin releasers on dentate granular cell excitability in the rat. *Exp. Brain Res.* **82**: 199–207.

Richter-Levin G, Segal M. (1991) The effects of serotonin depletion and raphe grafts on hippocampal electrophysiology and behavior. *J. Neurosci.* **11**: 1585–1596.

Ropert N. (1988) Inhibitory action of serotonin in CA1 hippocampal neurons *in vitro*. *Neuroscience* **26**: 69–81.

Sakai N, Tanaka C. (1993) Inhibitory modulation of long term potentiation via the 5-HT1A receptor in slices of the rat hippocampal dentate gyrus. *Brain Res.* **613**: 326–330.

Sara SJ. (1985) Noradrenergic modulation of selective attention: its role in memory retrieval. *Ann. NY Acad. Sci.* **444**: 178–193.

Segal M. (1980) The action of serotonin in the rat hippocampal slice preparation. *J. Physiol.* **303**: 423–439.

Segal M. (1982a) Multiple actions of acetylcholine at a muscarinic receptor studied in the rat hippocampal slice. *Brain Res.* **246**: 77–87.

Segal M. (1982b) Norepinephrine modulates reactivity of hippocampal cells to chemical stimulation in vitro. *Exp. Neurol.* **77**: 86–93.
Segal M. (1989) Presynaptic cholinergic inhibition in hippocampal cultures. *Synapse* **4**: 305–312.
Segal M, Bloom FE. (1976) The action of norepinephrine in the rat hippocampus: effects of locus coeruleus stimulation on evoked hippocampal unit activity. *Brain Res.* **107**: 513–525.
Segal M, Greenberger V, Pearl E. (1989) Septal transplants ameliorate spatial deficits and restore cholinergic functions in rats with a damaged septo-hippocampal connection. *Brain Res.* **500**: 139–148.
Segal M, Markram H, Richter-Levin G. (1991) Actions of norepinephrine in the rat hippocampus. *Prog. Brain Res.* **88**: 323–330.
Sheridan RD, Sutor B. (1990) Presynaptic M1 muscarinic cholinoceptors mediate inhibition of excitatory synaptic transmission in the hippocampus *in vitro. Neurosci. Lett.* **108**: 273–278.
Sokolov MV, Kleschevnikov AM. (1995) Atropine suppresses associative LTP in the CA1 region of rat hippocampal slices. *Brain Res.* **672**: 281–284.
Stanton PK, Sarvey JM. (1985) Depletion of norepinephrine but not serotonin reduces long term potentiation in the dentate gyrus of rat hippocampal slices. *J. Neurosci.* **5**: 2169–2176.
Stanton PK, Sarvey JM. (1987) Norepinephrine regulates long term potentiation of both the population spike and dendritic EPSP in hippocampal dentate gyrus. *Brain Res. Bull.* **18**: 115–119.
Stanton PK, Mody I, Heinemann U. (1989) A role for N-methyl D-aspartate receptors in norepinephrine-induced long lasting potentiation in the dentate gyrus. *Exp. Brain Res.* **77**: 517–530.
Staubli U, Otaky N. (1994) Serotonin controls the magnitude of LTP induced by theta bursts via an action on NMDA receptor-mediated responses. *Brain Res.* **643**: 10–16.
Staubli U, Xu FB. (1995) Effects of 5-HT_3 receptor antagonism on hippocampal theta rhythm, memory and LTP induction in freely moving rat. *J. Neurosci.* **15**: 2445–2452.
Villani F, Johnston D. (1993) Serotonin inhibits induction of long term potentiation at commissural synapses in hippocampus. *Brain Res.* **606**: 304–308.
Waelbroeck M, Tastenoy M, Camus J, Christophe J. (1990) Binding of selective antagonists to four muscarinic receptors (M1 to M4) in rat forebrain. *Mol. Pharmacol.* **38**: 267–273.
Williams S, Johnston D. (1988) Muscarinic depression of long term potentiation in CA3 hippocampal neurons. *Science* **242**: 84–87.
Zhang L, Weiner JL, Carlen PL. (1992) Muscarinic potentiation of Ik in hippocampal neurons: electrophysiological characterization of the signal transduction pathway. *J. Neurosci.* **12**: 4510–4520.

8

Epileptogenesis

William W. Anderson

8.1 Introduction

Epileptogenesis is the genesis or creation of a chronic hyperexcitable epileptic state. Epileptogenesis is one of the most dramatic examples of neuronal plasticity, as can be seen by the development of a normal, nonhyperexcitable nervous system into one capable of producing minute long grand mal seizures. Epileptogenesis also has many mechanistic similarities with long-term potentiation (LTP). With the use of modern neuroscience techniques, we are beginning to understand what causes epileptogenesis and, more importantly, how we can stop epileptogenesis from occurring.

The terms *epileptogenesis*, *epilepsy* and *seizures* are often used in slightly different ways by different researchers, and it is important to define how they will be used in this chapter. Epilepsy is a collection of disorders of the nervous system characterized by the chronic occurrence of spontaneous seizures. A seizure is a period of abnormal neuronal hyperexcitability that is of long duration (often lasting 30–60 sec), and consists of an underlying electrographic seizure and often an overt motor seizure. Seizures can be 'nonepileptic' if they occur transiently in normal brain due to some external factor such as inhalation of some compound with convulsant properties. Alternatively, seizures can be 'epileptic' if they occur spontaneously for no apparent reason. Epileptogenesis is the development of a long-lasting or permanent hyperexcitable state that produces spontaneous epileptic seizures. It is important to note that epileptogenesis has also been used by some researchers to describe the creation of transient, 'nonepileptic' seizures – for example, seizures occurring during, but not after, brief application of a convulsant. This chapter will not discuss this type of 'acute epileptogenesis'.

There are three characteristics of epileptogenesis. First, the initial hyperexcitability is often caused by a brain injury or tumor. Microhemorrhaging at the site of injury or tumor causes the local depositing of iron from red blood cell hemoglobin, and this iron then may induce hyperexcitability through the release of free radicals (Rosen and Frumin, 1979; Willmore, 1992; Willmore *et al.*, 1978). Second, a 'kindling' process then spreads this initial hyperexcitability to nearby local neurons. Experimental evidence suggests that this kindling process may

---

Cortical Plasticity, edited by M.S. Fazeli and G.L. Collingridge.
© 1996 BIOS Scientific Publishers Ltd, Oxford.

depend, in part, on the activation of *N*-methyl-D-aspartate (NMDA) receptors. The site of this still localized hyperexcitability is known as the primary focus. The kindling process then further continues to spread the hyperexcitability to other regions of the brain. This produces either secondary 'mirror' hyperexcitable foci (Morrell, 1985), or secondary generalization of hyperexcitability throughout large regions of the brain. Third, this abnormal kindling-induced hyperexcitability is long-lasting and may be permanent. In rats, kindling appears to be permanent. In humans, epilepsy is long-lasting, but in most cases is not permanent. Epilepsy continues for 10 years in approximately 40% of patients, and for 20 or more years in 25% of the patients (Shafer *et al.*, 1988). If epilepsy continues, the seizures may be stopped by neurosurgically removing the hyperexcitable tissue (if it is localized), or by continually administering anticonvulsants. It is therefore important to stop the development of seizures.

This chapter will discuss the *in vivo* kindling model of epileptogenesis and related *in vitro* kindling-like models of epileptogenesis, in which electrical or chemical stimulation produces long-lasting or permanent increases in excitability that result in a hyperexcitable epileptic state. In particular, we will focus on the finding that NMDA receptor antagonists block, or substantially retard, kindling-like epileptogenesis but, once seizures have developed, the NMDA receptor antagonists have little inhibition of seizure expression. This is analogous to the finding that NMDA receptor antagonists block LTP, but have little effect on potentiated excitatory postsynaptic potentials (EPSPs) once LTP has occurred. We will then discuss what mechanisms are responsible for producing kindling-like epileptogenesis, and how these mechanisms are similar to, and different from, the mechanisms producing LTP.

There are many other models of chronic epileptogenesis which will not be discussed in detail here. These include the kainic acid, alumina hydroxide, cobalt, tetanus toxin and lithium–pilocarpine models. For a review of these models, see Fisher (1989) and Jefferys (1990).

8.2 Blockade of kindling-like epileptogenesis by NMDA antagonists

The mechanisms of the kindling-like epileptogenesis have been studied in a wide variety of models ranging from whole animals, to brain slices, to neuronal cultures. A substantial body of evidence obtained from these models supports the hypothesis that activation of NMDA receptors is one of the primary mechanisms producing kindling-like epileptogenesis.

8.2.1 Description of in vivo electrical kindling

The most widely studied animal model of epileptogenesis is the *in vivo* kindling model, which was first carefully characterized by Graham Goddard (Goddard *et al.*, 1969). Kindling is the closest experimental model to complex-partial epilepsy, a type of epilepsy occurring in about half of all epilepsy patients (Löscher, 1993). (Complex means loss of consciousness, partial means focal; therefore complex-partial seizures means focal seizures causing loss of consciousness.) In standard whole animal kindling in rats, stimulating electrodes are usually implanted in the amygdala, hippocampus or piriform cortex, and trains of electrical stimuli (1–2 sec, 60 Hz) are delivered once or twice per day for about 2 weeks. The first stimulus train delivered is subconvulsant and causes only a brief afterdischarge and no motor seizures. However, after 2 weeks this same stimulus train produces a 'full-blown' electrographic and behavioral seizure. Furthermore, if the animal is not stimulated for 3 months, the first stimulus train again often triggers a full-blown seizure, demonstrating that the hyperexcitability is long-lasting and probably permanent (Goddard *et al.*, 1969; see also Wada and Sato, 1974).

A substantial modification of the standard (slow) *in vivo* kindling paradigm is rapid *in vivo* kindling (Lothman *et al.*, 1985). In rapid kindling, the stimulus trains are delivered once every 5–30 min for several hours. A fundamental difference between standard and rapid kindling is that standard kindling produces *fully* kindled seizures which are probably permanent, whereas rapid kindling produces kindled seizures which remain for less than 1 week (Lothman and Williamson, 1994). The first stimulus train after a 1-week stimulus-free period triggers only a very weak seizure indicating a fully kindled state was not present. However, when a second bout of rapid kindling was started 1 week after the stimulus-free period, animals did re-kindle faster. This indicates that 'savings' has occurred and that the first bout of rapid kindling produces signs of epileptogenesis that are still observable 1 week later.

In the *in vivo* kindling studies, the anti-epileptogenic effect of NMDA antagonists is usually measured by how much the NMDA antagonists retard the development of afterdischarge duration and behavioral seizure stage after 10–15 stimulus trains. Behavioral seizure stages in these studies are usually classified on a scale of 0–5: stage 0 = no clinical signs, stage 1 = chewing, stage 2 = chewing and head bobbing, stage 3 = forelimb clonus, stage 4 = forelimb clonus with rearing, stage 5 = forelimb clonus with rearing and falling (Racine, 1972). The anticonvulsant effect of NMDA antagonists is measured by how much NMDA antagonists decrease the afterdischarge duration and behavioral motor seizure stage in fully kindled animals.

8.2.2 Separating anti-epileptogenic versus anticonvulsant effects on kindling

It is not always easy to separate anti-epileptogenic versus anticonvulsant effects of drugs on kindling. Before discussing the anti-epileptogenic effects of NMDA antagonists on kindling, it is first useful to show in a simpler model of synaptic plasticity (LTP) and in idealized models of epileptogenesis how a pure 'anti-epileptogenic' effect (blocking the induction of long-term increases in excitability) differs from a pure 'anticonvulsant' effect (blocking the expression of that excitability).

One of the clearest examples of an 'anti-epileptogenic' but not an 'anticonvulsant' effect is when the NMDA antagonist 2-amino-5-phosphonopentanoate (AP5) blocks LTP of α-amino-3-hydroxy-5-methyl-4-isoxazole propionic acid (AMPA) receptor-mediated EPSPs in CA1 of the hippocampal slice, but has no inhibitory effect on their expression after potentiation has occurred (Collingridge et al., 1983; Figure 8.1a). An NMDA antagonist can be 'anticonvulsant' as well as 'anti-epileptogenic' if the expression depends on NMDA receptor-mediated potentials. One of the clearest examples of this is an NMDA antagonist blocking the NMDA receptor-dependent induction of NMDA receptor-mediated EPSPs, and also the expression of these potentiated NMDA receptor-mediated EPSPs (Bashir et al., 1991; Figure 8.1b). This suggests that a drug can be both 'anti-epileptogenic' and 'anticonvulsant'.

Figure 8.1c–f shows the effects of anti-epileptogenic agents and anticonvulsants on kindling. In contrast to LTP experiments where the measured response (a single stimulus pulse-evoked EPSP) differs from the induction method (a single train), in kindling experiments the stimulus trains cause both the measured response (a triggered seizure) and the method of induction. Figure 8.1c shows an idealized kindling experiment where 10 stimulus trains are required to reach the fully kindled state.

Figure 8.1d shows the effect of a purely anti-epileptogenic agent on kindling. During delivery of 10 stimulus trains in the presence of the anti-epileptogenic agent, there is no increase in seizure state and therefore no kindling. If the anti-epileptogenic agent were an NMDA antagonist, this would indicate NMDA receptor-dependent epileptogenesis. After the anti-epileptogenic drug has cleared, kindling occurs at the normal rate (10 stimulus trains to the fully kindled state in this example). Finally, when the anti-epileptogenic agent is reapplied to the fully kindled seizures, no suppression of seizures occurs, indicating that there is no anticonvulsant effect.

Figure 8.1. Idealized experiments showing 'anti-epileptogenic' and 'anticonvulsant' effects on (a) and (b) LTP and (b–f) kindling. (a) Pure 'anti-epileptogenic' effect of the NMDA antagonist AP5 on LTP of AMPA receptor-mediated EPSPs. AP5 completely blocks the induction of LTP, but has no effect on the expression of the AMPA receptor-mediated EPSPs. (b) An NMDA antagonist can have both 'anti-epileptogenic' and 'anticonvulsant' effects on LTP of NMDA receptor-mediated EPSPs. AP5 completely blocks both the induction of LTP, and also the expression of NMDA receptor-mediated EPSPs. In (a) and (b), the up arrows mark a stimulus train and the closed circles a measured response (the EPSP). The hatched bars mark application of NMDA antagonist. (c) Kindling occurring at a normal rate (10 trains to fully kindled state). (d) The effect of a pure anti-epileptogenic agent on kindling: it completely blocks the development of seizures, but has no effect on the expression of seizures after they have been induced. (e) The effect of a pure anticonvulsant on kindling: it completely masks but does not block the development of seizures (the first drug-free seizure is 'full-blown'), and it completely inhibits the expression of seizures. (f) More realistic kindling where an NMDA antagonist substantially retards, but does not block, kindling (it has a strong but not complete anti-epileptogenic effect), and also has a weak anticonvulsant effect on the expression of fully kindled seizures. In (b–f), the upward triangles mark the delivery of a stimulus train and a measured response (the triggered seizure).

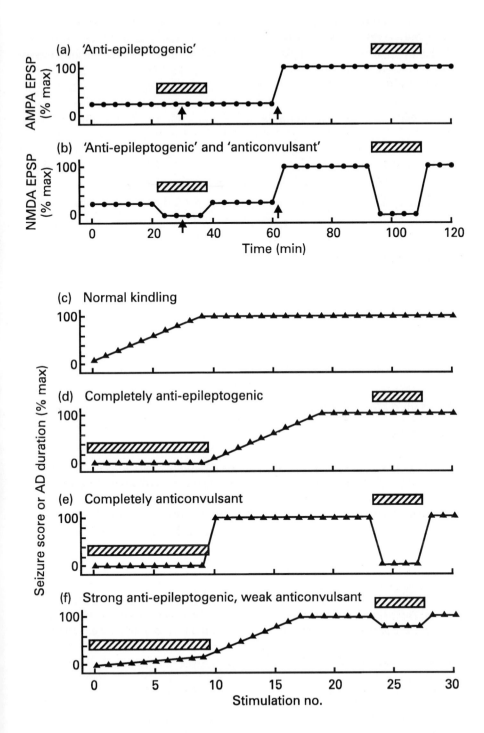

Figure 8.1e shows the effect of a purely anticonvulsant compound on kindling. During delivery of stimulus trains in the presence of the anticonvulsant, there is no increase in seizure stage or afterdischarge duration; however after the anticonvulsant has cleared, the *first* stimulation triggers a fully kindled seizure. This indicates that kindling occurred at the normal kindling rate during the application of the anticonvulsant, but expression of the kindled seizures was masked by the anticonvulsant. When the anticonvulsant agent is reapplied to the fully kindled seizures, complete suppression of seizures occurs, indicating strong anticonvulsant effect.

Sometimes there are 'savings' in the number of drug-free stimulations required to reach the fully kindled state after the normal number of stimulations has been delivered in the presence of the drug. For example, if only five trains were required to reach a fully kindled state in *Figure 8.1d*, then 'savings' has occurred. If savings occurs, this indicates that the drug does not have a complete anti-epileptogenic effect and that some type of induction was occurring in the presence of the drug. A drug having a strong anticonvulsant action and only a moderate anti-epileptogenic action can produce 'savings'.

Figure 8.1f shows a more typical kindling study where application of the drug substantially retards, but does not completely block, the kindling process. This indicates a substantial, but not complete, anti-epileptogenic effect. After the drug has cleared, kindling occurs at the normal rate. When the drug is reapplied to the fully kindled seizures some suppression of seizures occurs, indicating a weak, but clear, anticonvulsant effect.

8.2.3 In vivo *electrical kindling*

The earliest evidence that blockade of the NMDA receptor–channel complex blocks or substantially retards epileptogenesis comes from studies carried out in the early 1980s showing that the dissociative anesthetics ketamine and phencyclidine (PCP, also known as 'Angel Dust') block or substantially retard kindling, but have little inhibitory effect on fully kindled seizures. Callaghan and Schwark (1980) showed that, when ketamine is injected intraperitoneally (i.p.) during amygdala stimulation, the development of afterdischarge duration is retarded by approximately 50%, and more than twice as many stimulations are required to reach stage 5 motor seizures, compared with controls. However, on fully kindled seizures, ketamine causes no change in afterdischarge duration and only a little over one-fifth decrease in motor seizure stage, indicating a slight anticonvulsant effect. Boyer (1982) and Boyer and Winters (1981) also showed that when PCP is injected i.p. during amygdala stimulation, it almost completely blocks the development of motor seizures and substantially retards the increase of afterdischarge duration by about 70%. On fully kindled seizures they found that PCP causes only a one-third decrease in motor seizure stage and an approximately 40% decrease in afterdischarge duration. Although ketamine and PCP have strong anti-epileptogenic effects and weaker anticonvulsant effects, neither completely blocks the development of kindling.

However, at this time, the mechanisms of ketamine and PCP action were not known. Only after Lodge and co-workers (Anis *et al.*, 1983) found that ketamine and PCP block the NMDA channel was the connection made that ketamine and PCP block or retard kindling by blocking the NMDA receptor–channel complex.

Beginning in 1986, a large number of articles began to appear which studied the effects of new noncompetitive NMDA antagonists such as MK-801 and competitive NMDA antagonists such as AP5, CGS19755, 3-[(R)-2-carboxypiperazin-4-yl] propyl-1-phosphonic acid (CPP) and D-CPPene on the development of kindling and on fully kindled seizures (*Table 8.1*). These studies in general show that the anti-epileptogenic effects of various NMDA antagonists on kindling (the ability to retard the development of electrographic afterdischarge duration and motor seizure stage) are generally stronger than the anticonvulsant effects (the ability to decrease fully kindled seizures). Only occasionally is the anti-epileptogenic action of these NMDA antagonists complete so that there is no development of seizures. The rapid kindling study of Trommer and Pasternak (1990) using young animals and 0.1 mg kg^{-1} MK-801 shows the strongest anti-epileptogenic and weakest anticonvulsant effects.

Usually the anti-epileptogenic as well as the anticonvulsant effects of the NMDA antagonists are stronger on motor seizures than on afterdischarge duration. For example, McNamara *et al.* (1988) found that 0.33 mg kg^{-1} MK-801 has a strong anti-epileptogenic effect on motor seizure development (retardation by 85%) (*Figure 8.2a*), but has a weaker anti-epileptogenic effect on the development of afterdischarge duration in the amygdala (retardation by slightly more than half) (*Figure 8.2b*).

There are also some experiments where the anti-epileptogenic effect of the NMDA antagonists on afterdischarge and motor seizure development are not much stronger than the anticonvulsant effect (e.g. Cain *et al.*, 1988; *Table 8.1*). In these cases, it is difficult to separate the anti-epileptogenic and anticonvulsant effects of NMDA antagonists.

In addition to the studies in *Table 8.1*, several other studies show strong anti-epileptogenic effects on kindling, but the anticonvulsant effects were not measured. These include: focal application of AP7 and CPP in the prepiriform cortex (Croucher *et al.*, 1988), intracerebroventricular injection of AP7 during amygdala stimulation (Holmes *et al.*, 1990a) and rapid kindling studies in the hippocampus using i.p. injection of MK-801 (Kapur and Lothman, 1990).

In general, the results show that NMDA antagonists have a stronger anti-epileptogenic and weaker anticonvulsant effect on afterdischarges and motor seizures, but in most studies the NMDA antagonists substantially retard, but do not block, epileptogenesis.

8.2.4 In vivo *chemical kindling*

An alternative method to *in vivo* electrical kindling is *in vivo* chemical kindling – where the repeated administration of a convulsant at initially subconvulsant doses eventually produces full-blown seizures.

Table 8.1. *In vivo* standard and rapid kindling studies comparing various competitive NMDA antagonists (AP5, CPP, D-CPPene, CGS19755) and noncompetitive NMDA antagonists (ketamine, PCP, MK-801) on their anti-epileptogenic effect (blocking the development of kindling) and on their anticonvulsant action (suppressing fully kindled seizures)

	Drug	Anti-epileptogenic on kindling (AD duration)	Anticonvulsant on kindled seizures (AD duration)	Anti-epileptogenic on kindling (motor seizures)	Anticonvulsant on kindled seizures (motor seizures)
Standard kindling					
Callaghan and Schwark (1980)	25 mg kg^{-1} i.p. ketamine	+++	+	ND	++
Boyer (1982); Boyer and Winters (1981)	10 mg kg^{-1} i.p. PCP	++++	+++	+++++	++
McNamara et al. (1988)	0.33 mg kg^{-1} i.p. MK-801	+++	++	+++++	ND
Gilbert (1988)	0.5 mg kg^{-1} i.p. MK-801	++++	++	++	++
Morimoto and Sato (1992)	1 mg kg^{-1} i.p. MK-801	+++++	+	+++++	++++
Holmes and Goddard (1986)	i.c.v. AP5	+	+	++++	++
Cain et al. (1988)	i.c.v. AP5	++++	+++	+++	++
Holmes et al. (1990b)	i.c.v. AP5	+	+	+++	+
Holmes et al. (1990b)	10 mg kg^{-1} i.p. CPP	++	+	+++	+
Morimoto and Sato (1992)	10 mg kg^{-1} i.p. CPP	++	+	+++++	+++
Morimoto and Sato (1992)	10 mg kg^{-1} i.p. CGS19755	++++	+	+++	+++
Dürmüller et al. (1994)	8 mg kg^{-1} i.p. D-CPPene	ND	+	++++	++
Rapid kindling					
Trommer and Pasternak (1990)	10 mg kg^{-1} i.p. ketamine	++++	+++	++	++
Trommer and Pasternak (1990)	0.1 mg kg^{-1} MK-801	+++++	+	+++++	++

Effects were measured on afterdischarge (AD) duration (middle two columns), and behavioral seizures (right two columns). Effects were measured as either percentage retardation of epileptogenesis (anti-epileptogenic effect), or percentage decrease of fully kindled seizures (anticonvulsant effect). '+' = 0–20% retardation or decrease, '++' = 21–40% decrease, '+++' = 41–60% decrease, '++++' = 61–80% decrease, and '+++++' = 81–100% decrease (i.e. complete or almost complete anti-epileptogenic or anticonvulsant effect). Kindling stimulation and recording were done in the amygdala in all these experiments. Drugs were administered by intraperitoneal (i.p.) injection or intracerebroventricular (i.c.v.) injection. ND means the value was not determined. Most studies show that NMDA antagonists had a stronger anti-epileptogenic than anticonvulsant effect on kindling, but the anti-epileptogenic blockade of kindling was usually not complete.

Giorgi et al. (1991) and Corda et al. (1992) used the repeated systemic administration of initially subconvulsant doses of the γ-aminobutyric acid (GABA)$_A$ inhibitory convulsant pentylenetetrazol (PTZ) to chemically kindle rats, and then tested the ability of the NMDA antagonist MK-801 to block this kindling. They administered PTZ at 30 mg kg^{-1} i.p. three times a week for 10 consecutive weeks. During the first week, there was no indication of a seizure. However, after this time, behavioral indications of seizure development began to occur and by the eighth week, 80% of the animals were showing at least stage 4 seizures. However, if MK-801 (0.3 or 1.0 mg kg^{-1} i.p.) was administered just prior to the

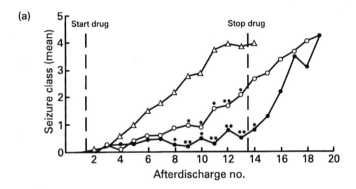

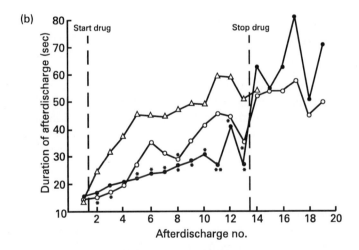

Figure 8.2. The anti-epileptogenic effects of the NMDA antagonist MK-801 on amygdala kindling measured by retardation of the development of motor seizure stage (a) and amygdala afterdischarge duration, in seconds (b). Note that 0.33 mg kg^{-1} MK-801 has a strong anti-epileptogenic effect on the development of motor seizures, but has only a moderate anti-epileptogenic effect on afterdischarge duration. Vehicle = △; 0.1 mg kg^{-1} i.p. MK-801 = O; 0.33 mg kg^{-1} i.p. MK-801 = ●. Modified from McNamara et al. (1988) Anticonvulsant and antiepileptogenic actions of MK-801 in the kindling and electroshock models. *Neuropharmology* **27**: 563–568, with permission from Elsevier Science Ltd.

PTZ, the development of seizures was blocked completely even after 10 weeks. Furthermore, they found that the first PTZ exposure delivered after MK-801 has cleared the system produces no seizures. This indicates that MK-801 produces a true anti-epileptogenic effect, and is not producing an anticonvulsant effect that masquerades as an anti-epileptogenic effect.

Similarly, Stephens and Weidmann (1989) found that repeated systemic administration of the GABA inhibitory convulsant FG 7142 produced chemical kindling, and that administration of MK-801 prior to the FG 7142 also completely blocked the development of kindling. The anticonvulsant effect on FG 7142-kindled seizures was not tested.

8.2.5 In vitro *brain slice electrical kindling*

In 1985, an alternative *in vitro* model of kindling was developed in which kindling-like stimulus trains are delivered to the rat hippocampal slice once every 5–10 min, and this induces triggered and spontaneous epileptiform bursting of about 50 sec duration which continues for many hours with no apparent decrease in activity (Slater *et al.*, 1985; Stasheff *et al.*, 1985). This *in vitro* tetanus delivery rate of once every 5–10 min is similar to the upper rate of tetanus delivery in *in vivo* rapid kindling [once every 5–30 min; Lothman *et al.* (1985)]. Application of the competitive NMDA receptor antagonist AP5 (Anderson *et al.*, 1987; Bawin *et al.*, 1989, 1991; Slater *et al.*, 1985; Stelzer *et al.*, 1987) and the noncompetitive NMDA antagonist MK-801 (Swartzwelder *et al.*, 1989) blocks the stimulus train induction of epileptiform bursting, and reduces, but does not block, the expression of bursting after it has been induced. Stimulus train induction of epileptiform bursting has also been shown recently in the piriform cortex (Pelletier and Carlen, 1995).

It was found subsequently that if the same stimulus trains are delivered to hippocampal slices from young (20- to 35-day-old) rats, electrographic seizures (EGSs) of 20–35 sec duration are induced in physiological medium and continue for several hours with no apparent decrease (Anderson *et al.*, 1988; Stasheff *et al.*, 1989). These EGSs are triggered in an all-or-none manner by stimulus trains (Anderson *et al.*, 1990), and this is also a characteristic of *in vivo* kindled seizures (Goddard *et al.*, 1969; Wada and Sato, 1974). Sometimes, after the initial stimulus train induction, the EGSs in the hippocampal slice will also re-occur spontaneously for several hours (Jones and Hamilton, 1993). Application of the NMDA antagonists AP5 and MK-801 blocks the stimulus train induction of EGSs, and reduces, but does not block, the expression of already induced EGSs (Stasheff *et al.*, 1989; *Figure 8.3*). This indicates that NMDA antagonists are strongly antiepileptogenic and only weakly anticonvulsant in this model.

8.2.6 In vitro *brain slice chemical kindling*

Several chemical manipulations, including bath application of NMDA, high extracellular K^+, kainate, penicillin and low extracellular Mg^{2+}, have been found

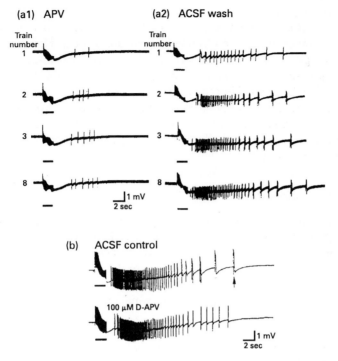

Figure 8.3. The NMDA antagonist AP5 (APV) blocks the stimulus train induction of electrographic seizures (EGSs) in a hippocampal slice bathed in physiological medium, but reduces and does not block the expression of EGSs that have already been induced. (a1) Delivery of eight stimulus trains (2 sec, 60 Hz, once every 10 min) in 50 μM D-AP5 produces little increase in afterdischarge bursting, indicating almost complete block of EGS development. (a2) After AP5 is removed from the physiological medium, the delivery of eight more stimulus trains now results in the gradual development of strong EGSs. (b) In contrast, when 100 μM D-AP5 is applied after EGSs have been induced tetanically, there is almost no decrease in the expression of EGSs (triggered by 2 sec, 60 Hz trains). The EGS duration decreases by less than 10% and the most substantial reduction is in the duration of the individual bursts at the end of the EGS (see arrow). Modified from Stasheff *et al*. (1989) with permission from the American Association for the Advancement of Science.

to induce epileptiform bursting in *in vitro* brain slice preparations which continues long after the chemical manipulation has stopped.

Anderson *et al*. (1987) found that bath application of NMDA to the hippocampal slice induces spontaneous and triggered epileptiform bursting that continues after the NMDA is washed out. If AP5 is applied prior to and during NMDA application, no bursting occurs. However, if AP5 is applied part way through NMDA application after bursting has been induced, bursting is reduced but not blocked. Furthermore, if AP5 is applied in the washout after bursting has been induced, bursting is again reduced but not blocked.

Ben-Ari and Gho (1988) found that exposure of hippocampal slices to high extracellular K^+ or acute kainate application causes spontaneous epileptiform bursting to occur during the high K^+ or kainate exposure, and epileptiform bursts could still be triggered in the physiological medium wash. However, if AP5 or CPP is applied prior to and during application of high K^+ or kainate, spontaneous bursting still continues during high K^+ or kainate exposure, but epileptiform bursts could not be triggered in the wash.

Similarly, Schneidermann et al. (1994) found that when hippocampal slices were bathed in high concentrations of penicillin (2000 IU ml^{-1}), spontaneous bursting occurred in the penicillin, and continued after penicillin was washed out. However, when AP5 was applied along with the penicillin, spontaneous bursting still occurred in the AP5 plus penicillin, but did not continue in the wash.

In each of these chemical manipulations, NMDA antagonists block the induction of long-term epileptiform bursting, and are therefore anti-epileptogenic in these models. Because NMDA antagonists do not block bursting during the application of high K^+, kainate or penicillin, this indicates that the direct blockade of the NMDA receptor, and not the suppression of the bursting *per se*, is the mechanism which blocks the induction of long-term bursting. The results are not consistent with NMDA antagonists having a strong anticonvulsant action on bursting in these models.

Bath application of nominally Mg^{2+}-free (low Mg^{2+}) medium to hippocampal slices produces epileptiform bursting (Coan and Collingridge, 1985; Mody et al., 1987). Low Mg^{2+} is thought to produce this bursting by removing the Mg^{2+} block of the NMDA channel and allowing substantial NMDA-activated synaptic current to occur at depolarized potentials (Nowak et al., 1984). In most low Mg^{2+} studies, exposure to low Mg^{2+} has not been noted to induce long-term increases in excitability (Coan and Collingridge, 1985; Mody et al., 1987). However, there are several studies where exposure to low Mg^{2+} induces long-term, NMDA receptor-dependent increases in several types of excitability, either measured while in low Mg^{2+} medium, or after the return to physiological medium.

Avoli et al. (1988) found that exposure of hippocampal slices to low Mg^{2+} medium causes LTP of CA1 EPSPs when measured after return to physiological medium. Adding AP5 to the low Mg^{2+} medium blocks this induction of the EPSP potentiation.

Lewis et al. (1986) found that exposure of hippocampal slices to low Mg^{2+} medium causes spontaneous epileptiform bursting in CA3 neurons in low Mg^{2+} medium, and this bursting continues after return to physiological medium. Adding AP5 to the low Mg^{2+} medium blocks the appearance of both triggered and spontaneous bursting in low Mg^{2+} medium and no bursting occurs after return to physiological medium. Similarly, Hoffman and Haberly (1989) found that exposure of olfactory cortex slices to low Mg^{2+} medium produces bursting, and induces large, all-or-none EPSPs after return to physiological Mg^{2+} medium. Addition of AP5 or ketamine prior to and during exposure to low Mg^{2+} medium blocks the generation of bursts in low Mg^{2+} medium, and blocks the induction of the all-or-none EPSPs in the washout. However, after induction of the all-or-none

EPSPs has occurred, application of AP5 in the washout has no effect on the generation of these all-or-none EPSPs. Coulter and Lee (1993) also found that low Mg^{2+} induces epileptiform activity in thalamo-cortical slices, and that AP5 application during low Mg^{2+} exposure blocks the induction of this bursting.

Exposure of hippocampal slices to low Mg^{2+} medium also produces spontaneous and triggered EGSs of approximately 50 sec duration (Anderson et al., 1986). *Figure 8.4* shows that the development of these EGSs in low Mg^{2+} medium is due to an NMDA receptor-dependent induction process. If a high concentration

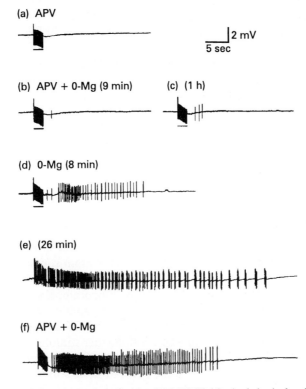

Figure 8.4. The NMDA receptor antagonist AP5 (APV) blocked the induction, but not the expression, of electrographic seizures (EGSs) in the hippocampal slice in low Mg^{2+} medium. (a) In physiological medium containing 500 μM D-AP5, delivery of a stimulus train elicited no after discharge bursts. (b) and (c) When Mg^{2+} was removed from the medium (0-Mg) but the AP5 remained, delivery of six stimulus trains (once every 10 min) still produced only three afterdischarge bursts after 1 h (c). (d) and (e) When AP5 was removed from the 0-Mg medium, strong induction effects were seen after 8 min (d) when a stimulus train triggered an afterdischarge, and spontaneous EGSs occurred by 26 min (e). (f) When 500 μM D-AP5 was added back to the 0-Mg medium, spontaneous EGSs stopped, but stimulus trains could trigger an EGS, indicating that AP5 reduced, but did not block, EGSs. [Methods were similar to those used in Anderson et al. (1990), and similar results were obtained in three slices. Stimulating and recording electrodes were placed in stratum radiatum and stratum pyramidale of CA3, respectively, and stimulus trains were 2 sec, 60 Hz. The rat was 27 days old.]

of AP5 is applied prior to and during low Mg^{2+} exposure, six strong stimulus trains fail to induce EGSs after 1 h in AP5 + low Mg^{2+} medium (*Figure 8.4c*). When AP5 is removed from the low Mg^{2+} medium, EGS induction begins. After a few minutes' stimulus trains can trigger an afterdischarge, and within 30 min the threshold for eliciting EGSs drops so low that spontaneous EGSs occur (*Figure 8.4d* and *e*). If the AP5 is returned to the low Mg^{2+} medium, spontaneous EGSs stop, but stimulus trains can still trigger EGSs (*Figure 8.4f*). However, the EGSs are reduced in duration by one-third, and there is a decrease in the length of individual bursts in the EGS. This shows that a high concentration of AP5 is sufficient to block the stimulus train/low Mg^{2+} medium induction of EGSs in low Mg^{2+} medium, but once induction has occurred AP5 reduces but no longer blocks the EGSs in low Mg^{2+} medium. AP5 is therefore strongly anti-epileptogenic and only moderately anticonvulsant in this low Mg^{2+} EGS model.

Valenzuela and Benardo (1995) recently found that bathing neocortical slices in low Mg^{2+} medium for 2 h induces interictal and ictal activity which continues for 30–240 min after return to physiological medium. Application of the NMDA antagonist CPP during exposure to low Mg^{2+} medium blocks the development of both types of epileptiform activity. However, addition of CPP to physiological medium after the 2 h exposure to low Mg^{2+} medium now fails to block the low Mg^{2+}-induced epileptiform activity. NMDA antagonists are therefore strongly anti-epileptogenic in this low Mg^{2+} model.

8.2.7 Cultured brain slice electrical kindling

In order to have a longer time to study the maintenance of the induced hyperexcitable epileptic state than is possible with acute brain slices, Shin *et al.* (1992) developed a cultured hippocampal slice preparation in which the EGSs could be induced tetanically. After 7–15 days in culture, hippocampal slices were placed in a recording chamber, and a series of (1 sec, 60 Hz) stimulus trains were delivered once every 10 min. After four stimulus trains, the EGS duration approximately doubles, and remains relatively constant with further stimulation.

When 50 μM D-AP5 is applied during the delivery of the stimulus trains, the EGSs in these slices are shorter than the EGSs in slices stimulated in physiological medium. Only a small increase in EGS duration occurs in those slices stimulated in AP5. Significantly, upon washout of the AP5, the first subsequent stimulus train does not produce a full EGS, indicating that AP5 is not simply acting as an anticonvulsant and masking the development of the EGS but is having a true anti-epileptogenic effect (e.g. as in *Figure 8.1d*). Furthermore, when AP5 is applied after the EGSs have been induced, the EGSs are reduced in duration by about half, but are not blocked.

These results indicate that AP5 has a strong anti-epileptogenic effect in cultured hippocampal slices similar to that seen with acute hippocampal slices (Stasheff *et al.*, 1989). However, AP5 has a stronger anticonvulsant action in cultured than in acute hippocampal slices, indicating that NMDA receptor-mediated potentials are probably more involved in generating the EGSs in cultured hippocampal slices.

8.2.8 Cultured neuron chemical kindling

Another *in vitro* method for performing longer duration studies of NMDA receptor-dependent epileptogenesis is to use low Mg^{2+} to induce 'electrographic seizures' in dissociated hippocampal neurons in culture (Sombatti and DeLorenzo, 1995). In this model, the cultured neurons are exposed to low Mg^{2+} for 3 h and then returned to normal medium. After the return to normal medium, spontaneous recurrent EGSs begin almost immediately. The EGSs are approximately 1 min in duration, are similar to some EGSs seen in brain slices and can be blocked by anticonvulsants that block tonic–clonic, but not absence, seizures. The EGSs continue spontaneously for at least 7 days (the longest time measured). When 50 μM AP5 is added during the low Mg^{2+} exposure, no EGSs appear upon return to normal medium. This is consistent with NMDA antagonists blocking low Mg^{2+} induction of EGSs in this model and having an anti-epileptogenic effect.

8.2.9 Dependence of chronic, kainate-induced epileptogenesis on NMDA receptor activation

There are several models of chronic nonkindling epileptogenesis. These include the kainic acid, alumina hydroxide, cobalt, tetanus toxin and lithium–pilocarpine models. One question concerning these models is: to what extent does the activation of NMDA receptor-dependent induction mechanisms contribute to the total induction of hyperexcitability.

Stafstrom *et al.* (1993) have addressed this question recently in the *in vivo* kainate-induced chronic epileptogenesis model by applying NMDA antagonists prior to and during acute kainate exposure. They found that i.p. injection of kainate alone causes an initial period of status epilepticus (continuous seizure activity lasting >30 min), and this is followed by a second period of spontaneous recurrent seizures which continues for at least 4 weeks after kainate exposure. However, if MK-801 (0.2–1 mg kg^{-1} i.p.) is injected 30 min before the kainate, the status epilepticus that is normally present shortly after kainate administration still occurs, but the number of rats showing spontaneous seizures long after kainate exposure is reduced by half, the frequency of the spontaneous seizures is reduced by two-thirds and the severity of the seizures is also reduced. These results suggest that the NMDA receptor-dependent induction of hyperexcitability is *one* of the mechanisms contributing to *in vivo* kainate-induced epileptogenesis, but is not the only one.

8.2.10 Pro-epileptogenic effect of NMDA antagonists during development

Audiogenic seizure susceptibility can be induced permanently in rats by exposure to intense noise during a critical period at age 14 days in postnatal development. Pierson and Swann (1991) hypothesized that application of the NMDA channel blockers MK-801 and PCP during sound exposure on day 14 would suppress the induction of seizures in this model and would, therefore, be anti-epileptogenic.

However, in contrast to expectations, they found that MK-801 actually *increased* the ability of sound to induce seizures. This is in direct contrast to the usual anti-epileptogenic effect of MK-801 and suggests that NMDA antagonists may be pro-epileptogenic in certain developmental conditions. They explain these results with the following hypothesis: if MK-801 and PCP produce a transient 'up-regulation' of NMDA receptors that outlasts the acute NMDA channel blockade by the drug, and if the seizure induction process continues into this later time, then the unblocked, up-regulated NMDA receptors would enhance seizure induction.

8.3 Anticonvulsant action of NMDA antagonists

It sometimes seems contradictory that NMDA antagonists can be weakly anticonvulsant (and strongly anti-epileptogenic) in the kindling model, but strongly anticonvulsant in other seizure models. For example, in clinical trials, the NMDA antagonist D-CPPene had no anticonvulsant effects on complex partial epilepsy (Sveinbjornsdottir *et al.*, 1993). However, whether or not NMDA antagonists are strongly anticonvulsant in a given seizure model can be explained by whether NMDA receptor-mediated potentials are strongly involved in triggering and/or generating the seizure. If this is so, then NMDA antagonists will be strongly anticonvulsant in that seizure model. (For detailed reviews of NMDA antagonists as anticonvulsants, see Clark *et al.*, 1994; Löscher, 1993; Rogawski, 1992; and Rogawski and Porter, 1990.)

This explanation is consistent with our own findings from the *in vitro* hippocampal slice model of epileptogenesis. In this model, NMDA antagonists block the development of seizures (Stasheff *et al.*, 1989; *Figures 8.3* and *8.4*), NMDA antagonists can be anticonvulsant by raising the threshold for triggering seizures (Cohen *et al.*, 1993), and NMDA antagonists are anticonvulsant when seizure generation requires NMDA receptor-mediated potentials, but not when AMPA receptor-mediated potentials are sufficient to generate seizures (Anderson and Coan, 1989; *Figure 8.5*).

In physiological (0.9 mM Mg^{2+}) medium, AMPA but not NMDA receptor-mediated potentials are required for seizure generation. This is indicated by the AMPA antagonist 6-cyano-7-nitroquinoxaline-2,3-dione (CNQX) completely blocking EGSs (*Figure 8.5a4*), and the NMDA antagonist AP5 having little inhibitory effect (*Figures 8.3* and *8.5a2*). In low Mg^{2+} medium, either NMDA or AMPA receptor-mediated potentials are sufficient to generate EGSs. This is indicated by AMPA receptor-mediated EGSs occurring in AP5 + low Mg^{2+} (*Figures 8.4f* and *8.5b2*), and by NMDA receptor-mediated EGSs occurring in CNQX + low Mg^{2+} (*Figure 8.5b4*). [Similar EGSs have been found in low Mg^{2+} plus an AMPA antagonist by Bawin *et al.* (1993) and Traub *et al.* (1994).] In contrast, under conditions which promote NMDA receptor-mediated potentials (by bathing the slice in low Mg^{2+} medium) and block AMPA receptor-mediated potentials (by bathing the slice in CNQX), the subsequent addition of AP5

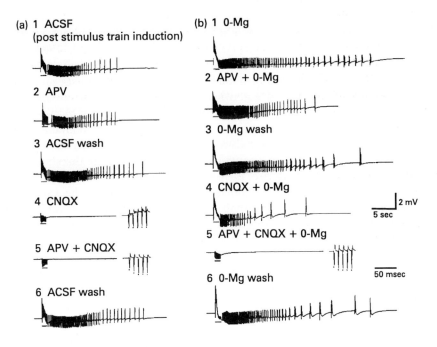

Figure 8.5. Anticonvulsant actions of NMDA and AMPA receptor antagonists in the rat hippocampal slice in (a) physiological (0.9 mM Mg) and (b) nominally Mg^{2+}-free (0-Mg) medium. (a1) Triggering an EGS in physiological medium (or artificial cerebrospinal fluid, ACSF). The EGSs had been induced previously by 10 stimulus trains. (a2) Addition of 100 μM of the NMDA receptor antagonist D-AP5 (APV) to the ACSF had little effect on EGS duration. (a3) Wash in ACSF. (a4) Addition of 10 μM of the AMPA receptor antagonist CNQX completely blocked the EGSs as indicated by the failure of 2 sec, 60 Hz (and 3 sec, 100 Hz, not shown) stimulus trains to produce EGSs. The inset at right is a faster speed recording showing the first five extracellular depolarizing responses in CNQX. (a5) When 100 μM D-AP5 was added to the CNQX solution, EGSs were still blocked. The inset at right shows that the extracellular depolarizing responses have been removed [the remaining deflections are due to antidromically stimulated action potentials which are also seen in 0-Ca^{2+}/high Mg^{2+} medium (Stasheff et al., 1985)]. (a6) Wash in ACSF. (b1) Triggering an EGS in 0-Mg medium. (b2) Addition of 100 μM D-AP5 to 0-Mg medium reduced, but did not block, EGSs. (b3) 0-Mg medium wash. (b4) Addition of 10 μM CNQX to 0-Mg medium also reduced, but did not block, EGSs (compare with a4). (b5) Addition of 100 μM D-AP5 to the CNQX + 0-Mg medium completely blocked EGSs as indicated by the failure of 2 sec, 60 Hz (and 3 sec, 100 Hz, not shown) stimulus trains to produce EGSs. (b6) Recovery in 0-Mg medium wash. [Methods were similar to those used in Anderson et al. (1990) and Figure 8.4. The rat was 27 days old. In (b) 2 μM baclofen was also added to the 0-Mg medium to inhibit interictal burst suppression of the EGSs (Swartzwelder et al., 1987). Each experiment was performed on at least three slices.]

completely blocks the EGSs (*Figure 8.5b5*). In this situation, the NMDA antagonist AP5 is strongly anticonvulsant.

NMDA antagonists can have another more subtle anticonvulsant effect on seizures by raising the threshold for triggering seizures. Cohen *et al.* (1993) found that when EGSs have already been induced in physiological medium and the stimulus merely triggers the all-or-none EGS, addition of AP5 raises the threshold for triggering EGSs twofold (e.g. twice as many stimulus pulses are needed to trigger an EGS in AP5 versus control medium). Similarly, in low Mg^{2+} medium, the EGSs are all-or-none (Anderson *et al.*, 1990) and are either spontaneous (e.g. *Figure 8.4e*) or require just a few pulses to be triggered. However, when AP5 is added to the low Mg^{2+} medium, spontaneous EGSs stop and the threshold number of pulses to trigger an EGS rises to 30–80 (see *Figure 8.4f*). At just suprathreshold stimulations (in medium without an NMDA antagonist), the subsequent addition of an NMDA antagonist will block the triggering of EGSs, and the NMDA antagonist will therefore be anticonvulsant.

Our findings that NMDA antagonists do not block EGSs induced in low Mg^{2+} medium differ from those of Wang and Jensen (1996) (see also Traub *et al.*, 1994). Wang and Jensen found that exposure of hippocampal slices from young (10- to 15-day old) rats to low Mg^{2+} medium produces EGSs of 6–68 sec duration. The addition of 20 μM DL-AP5 to the low Mg^{2+} medium replaces this ictal activity with interictal activity. Wang and Jensen did not use stimulus trains to test if the ictal blockade by AP5 is due to an increase in seizure threshold.

The experiments on the *in vitro* kindling model (*Figures 8.3, 8.4* and *8.5*) are consistent with the following concepts:

(i) If the mechanisms that induce kindling-like epileptogenesis require current flowing through the NMDA channel, then NMDA antagonists will block epileptogenesis and will be strongly anti-epileptogenic.
(ii) If the mechanisms generating seizures depend on NMDA receptor-mediated potentials, NMDA antagonists will block the seizures and will be strongly anticonvulsant.
(iii) Similarly, if mechanisms triggering the seizures depend on NMDA receptor-mediated potentials, NMDA antagonists will tend to prohibit the seizure from occurring and therefore *can* also be anticonvulsant. If neither the seizure-generating mechanism nor the seizure-triggering mechanism depends on NMDA receptor-mediated potentials, NMDA antagonists will not block the seizures and will not be anticonvulsant.

8.4 Mechanisms of epileptogenesis: the initial induction process

8.4.1 Ca^{2+} entry during tetanic stimulation

The concept of how stimulus trains begin the kindling process initially came from LTP studies because of two similarities between LTP and kindling: stimulus trains induce LTP and kindling, and NMDA antagonists block (or substantially retard) the induction of LTP and kindling. The basic concept is that Ca^{2+} enters though NMDA channels during a stimulus train, and this Ca^{2+} then causes subsequent metabolic changes that ultimately produce long-term enhancement of EPSPs for LTP, and additional excitable mechanisms for epileptogenesis. Synaptic transmission in the hippocampus (and presumably in other epileptogenic areas) is comprised of fast excitatory AMPA receptor-mediated potentials, slow excitatory NMDA receptor-mediated potentials and fast and slow inhibitory GABA receptor-mediated potentials. During low frequency synaptic transmission, the GABA inhibitory potentials hyperpolarize the neuron and prohibit the opening of the NMDA channels that would normally occur at depolarized voltages. However, during high frequency synaptic transmission, the GABA inhibition decreases, allowing the neuron to depolarize and the NMDA channels to open (Collingridge et al., 1988). This produces a Ca^{2+} influx through the NMDA channel during the stimulus train (Alford et al., 1993; Lynch et al., 1983). This transient Ca^{2+} signal is detected by intracellular second messengers and ultimately converted into long-term storage. NMDA antagonists block the induction of LTP by blocking the Ca^{2+} entry through NMDA channels during the stimulus train.

Hirayama et al. (1994) have studied the facilitation of tetanic responses during the early stages of amygdala kindling and have found that an NMDA antagonist, CPP, can block the facilitation of EPSPs in the latter part of the stimulus train. This is similar to the effect of AP5 on the tetanic synaptic response during LTP (Collingridge et al., 1988) and consistent with an NMDA receptor-mediated component of synaptic transmission during this time. In an *in vitro* hippocampal slice model of kindling, Bawin et al. (1991) also recorded a similar NMDA receptor-mediated synaptic component in CA2/CA3 neurons during sine wave stimulation that was reduced by AP5.

8.5 Mechanisms of epileptogenesis: changes resulting in seizure expression

Before discussing the intracellular messenger mechanisms responsible for coupling the initial epileptogenic induction process to the mechanisms ultimately responsible for producing seizure expression, we will first discuss what those putative mechanisms of seizure expression are. They include:

(i) increases in excitatory AMPA and NMDA receptor-mediated synaptic transmission.
(ii) decreases in GABAergic inhibitory synaptic transmission; and
(iii) additional physiological changes including increased synaptic connectivity and ectopic potentials.

Other mechanisms such as the 'dormant cell' hypothesis and synaptic rearrangements, which may be very important for establishing the permanence of epileptogenesis (Sloviter, 1994), are beyond the scope of this chapter.

Two basic methods have been used to study the mechanisms causing seizure expression in epileptogenesis: (i) electrophysiological and pharmacological studies *during in vivo* and *in vitro* kindling, and (ii) biochemical and electrophysiological studies on whole tissue, brain slices and dissociated single neurons from *in vivo* kindled animals at various times *after* kindling stimulation has stopped.

8.5.1 LTP and kindling-induced potentiation

There are several reasons why LTP has been proposed as a mechanism of kindling.

(i) Enhanced EPSPs could contribute greatly to the generation of kindled seizures.
(ii) LTP and kindling are both induced by stimulus trains.
(iii) LTP can last for days, weeks or, as early studies have suggested, for 4 months (Bliss and Gardner-Medwin, 1973) (although this was reported in only one animal in their study).
(iv) Kindling-like stimulation causes EPSP potentiation that could last at least 1, 2 or 3 months, respectively (Douglas and Goddard, 1975; Maru and Goddard, 1987; Racine *et al.*, 1991). Also, *in vitro* kindling studies in the hippocampal slice have shown EPSP potentiation occurring in CA1 (Slater *et al.*, 1985) and CA3 synapses (Higashima, 1988).
(v) NMDA antagonists block or substantially retard both LTP and kindling.

However, there are several reasons for suspecting that *conventional* LTP is not a mechanism of kindling (Cain, 1989; Racine *et al.*, 1991). First, some pathways presumably involved in *in vivo* kindling do not show EPSP potentiation during kindling. For example, Giachinno *et al.* (1984) found that when the lateral entorhinal cortex was kindled, seizures developed without the expected potentiation of the entorhinal–granule cell synapse. One answer to this criticism is that EPSP potentiation may occur via other unknown pathways, or may occur undetected in local circuits. Second, conventional LTP usually decays within days or weeks (Racine *et al.*, 1983), even when the LTP-inducing stimulus trains are delivered at the kindling rate of once per day for many consecutive days (de Jonge and Racine, 1985). In contrast, standard *in vivo* kindling lasts for at least 3 months and may be permanent (Goddard *et al.*, 1969; Wada and Sato, 1974). This second criticism is more difficult to refute.

Although it is clear that very long-lasting and perhaps permanent EPSP potentiation can occur during kindling (Douglas and Goddard, 1975; Maru and Goddard, 1987; Racine *et al.*, 1991), the question is whether this EPSP potentiation, called kindling-induced potentiation by Racine, is the same as that occurring during conventional LTP. Conventional LTP can be induced by a stimulus train, and can occur at, and be restricted to, a single synapse. For conventional LTP to occur, the stimulus train does not have to evoke afterdischarge bursting. In contrast, for kindling to occur the stimulus trains must evoke afterdischarge bursting (Racine, 1972). This is partly because, for kindling to occur, the excitability has to spread to many synapses and occur over a comparatively large region of the brain. However, Racine *et al.* (1991) also suggest that stimulus train-evoked afterdischarge bursting is necessary for producing very long-lasting or permanent EPSP potentiation.

It is not at all clear why the afterdischarge bursting following stimulus trains would substantially increase the duration of EPSP potentiation and perhaps make it permanent. The co-activation of many synaptic inputs and substantial cell depolarization that occurs during afterdischarge bursting would not, according to current theory, necessarily make EPSP potentiation permanent. This is an important area of future research. Nevertheless, the data indicate that, in conventional LTP experiments, where stimulus trains are known *not* to evoke afterdischarge bursting, LTP decays within days or weeks (de Jonge and Racine, 1985; Racine *et al.*, 1983). In kindling potentiation experiments, where stimulus trains are known to evoke afterdischarge bursting, EPSP potentiation lasts for at least 1–3 months with little or no decay (Douglas and Goddard, 1975; Maru and Goddard, 1987; Racine *et al.*, 1991). In some early LTP experiments showing very long-lasting potentiation, it is not known whether the stimulus trains evoked afterdischarge bursting (e.g. Bliss and Gardner-Medwin, 1973).

It is clear that kindling stimulations that evoke afterdischarge bursting can produce very long-lasting and perhaps permanent EPSP potentiation. Therefore, it is reasonable to expect that potentiation of excitatory synaptic transmission, and kindling-induced potentiation in particular, could contribute to the expression of permanently kindled seizures. However, it is not clear whether this kindling-induced potentiation produced by afterdischarge bursting is the same as conventional LTP produced without afterdischarge bursting. It is also important to recognize that although NMDA antagonists block or substantially reduce both LTP and kindling, this only indicates that some NMDA receptor-dependent induction process is required for kindling, and it may not be LTP.

8.5.2 Increase of excitatory NMDA receptor-mediated synaptic transmission

There have been several studies showing enhanced NMDA receptor-mediated synaptic transmission during or following *in vivo* or *in vitro* kindling. Mody and Heinemann (1987) found that the NMDA receptor-mediated component of synaptic transmission in the dentate gyrus of hippocampal slices is enhanced in slices obtained from kindled compared with control animals. Similarly, Gean *et al.* (1989) and Rainnie *et al.* (1992) reported an enhancement of NMDA receptor-mediated potentials in amygdala slices from kindled versus control animals. NMDA receptor-dependent induction of NMDA receptor-mediated potentials also appears to occur in the low Mg^{2+} *in vitro* model of kindling. AP5 blocks the induction of EGSs in low Mg^{2+} medium, indicating an NMDA receptor-dependent induction process (*Figure 8.4b* and *c*), but NMDA receptor-mediated synaptic transmission occurs in low Mg^{2+} medium once EGSs have been induced (compare *Figure 8.5b4* with *Figure 8.5b5* which shows AP5 blocking NMDA receptor-mediated synaptic transmission in CNQX + 0-Mg). Therefore, just as NMDA receptor-dependent potentiation of NMDA receptor-mediated synaptic transmission occurs in LTP (Bashir et al., 1991; see *Figure 8.1b*), it also occurs in *in vivo* and *in vitro* kindling.

There have also been several studies showing enhanced NMDA receptor–channel function during or following *in vivo* or *in vitro* kindling. Stelzer et al. (1987) also found that *in vitro* hippocampal slice kindling increased the excitatory effect of iontophoresed NMDA. In grease gap experiments of hippocampal slices obtained from kindled animals, Martin et al. (1992) found that CA3 neurons in slices from kindled animals were five times more sensitive to exogenous NMDA application than those from control animals. Similarly, Wadman et al. (1985) found that exogenous application of NMDA to hippocampal slices produced greater Ca^{2+} entry in slices from kindled versus control animals. Köhr and Mody (1994) used on-cell single channel patch clamp techniques to study NMDA-evoked channel openings in acutely isolated dentate granule neurons from control and kindled animals, 24 h after the last seizure. They found that NMDA channels were opened at lower concentrations of applied NMDA in neurons from kindled animals.

8.5.3 Increase of excitatory AMPA receptor-mediated synaptic transmission

Although many studies have shown that AMPA receptor-mediated potentials are the primary potentials enhanced in LTP, there are few studies that report an enhancement of AMPA receptor-mediated potentials during kindling. However, Gean et al. (1989) and Rainnie et al. (1992) have reported an enhancement of AMPA receptor-mediated potentials in amygdala slices from kindled versus control animals. Valenzuela and Benardo (1995) found that CNQX blocks the epileptiform activity in physiological medium after its induction in low Mg^{2+} medium, suggesting that potentiation of AMPA receptor-mediated potentials occurs in low Mg^{2+} medium and contributes to generation of epileptiform activity in physiological medium. Furthermore, in *in vitro* hippocampal slice kindling, the ability of CNQX to block EGSs in physiological medium after EGSs have been induced is consistent with stimulus train potentiation of AMPA receptor-mediated potentials and their involvement in EGS generation (*Figures 8.5a3* and *8.5a4*).

8.5.4 Decrease of GABAergic inhibition

The fact that many of the common convulsants such as penicillin and PTZ suppress GABA inhibition suggested that a long-term decrease in GABAergic inhibitory synaptic transmission may be a mechanism of kindling. Loss of GABAergic transmission during kindling could theoretically occur in at least two ways: (i) loss of GABAergic neurons occurs during kindling, and (ii) decreasing the strength of the existing GABAergic inhibition. Although loss of GABAergic neurons has been reported to occur during kindling (Kamphuis et al., 1986), we will focus on mechanisms decreasing the strength of existing GABAergic inhibition during kindling.

The earliest data showing that kindling produces a decrease in GABAergic inhibitory synaptic transmission come from the *in vitro* hippocampal slice kindling studies of Stelzer *et al.* (1987). They found that delivery of stimulus trains in region CA1 decreases the frequency of spontaneous inhibitory postsynaptic potentials (IPSPs), decreases the stimulus-evoked $GABA_A$ and $GABA_B$ inhibitory conductances and also reduces the inhibitory response to iontophoresed GABA in CA1 neurons. Addition of AP5 during the delivery of stimulus trains blocks the tetanically induced changes in the GABA receptor-mediated IPSPs and iontophoresed GABA that normally occur during kindling. In a similar *in vitro* hippocampal slice model of kindling, Miles and Wong (1987) detected a decrease in recurrent inhibition after repeated tetanic stimulation when recording from pairs of CA3 neurons, and Merlin and Wong (1993) found that tetanic stimulation produces long-term decreases in inhibitory $GABA_A$ and $GABA_B$ receptor-mediated conductances in CA3 neurons. However, neither Miles and Wong, nor Merlin and Wong tested whether NMDA antagonists could block the long-term reduction in GABA inhibition.

In the *in vivo* hippocampal rapid kindling model, Kapur and Lothman (1990) used measurements of paired-pulse inhibition to measure indirectly decreases in GABA inhibition. They found that a decrease in paired-pulse inhibition occurs during kindling, but systemic administration of MK-801 during delivery of the kindling stimulus trains blocks this decrease in paired-pulse inhibition. In studies testing the properties of slices from kindled versus control animals, Gean *et al.* (1989) and Rainnie *et al.* (1992) reported a decrease in GABA receptor-mediated potentials in amygdala brain slices from kindled animals.

However, some studies showed that GABAergic inhibition does not change much with kindling. In the *in vitro* hippocampal slice kindling model, Higashima (1988) found that repeated tetanic stimulation produced no marked changes in the amplitude and duration of IPSPs in CA3 neurons. Presumably, if there is a sufficient increase in excitatory synaptic transmission during kindling, there need not be a chronic decrease in inhibitory transmission for seizures to occur. This may be particularly true if there is enough excitatory transmission to trigger a seizure, because GABAergic inhibitory transmission transiently 'collapses' during a seizure once a seizure has been triggered (Ben-Ari *et al.*, 1979).

8.5.5 Increased excitatory synaptic connectivity

In addition to enhancing the synaptic strength of excitatory synapses already functioning, another mechanism of producing increased excitability is to make latent excitatory pathways functional. Using the *in vitro* hippocampal slice kindling model, Miles and Wong (1987) used simultaneous intracellular recording from pairs of CA3 neurons and showed that repeated tetanic stimulation causes polysynaptic excitatory pathways between previously unconnected cells to become apparent. This increase in polysynaptic excitatory pathways would be expected to contribute substantially to the generation of epileptiform activity. The tetanically induced increase in polysynaptic excitatory pathways is also associated with a decrease in recurrent inhibition, and a similar increase in polysynaptic excitatory pathways occurs when low concentrations of the GABA$_A$ receptor blocker picrotoxin are applied. Miles and Wong did not test whether NMDA antagonists could block the development of this increased excitatory synaptic connectivity.

8.5.6 Ectopic potentials

One final mechanism found to be involved in epileptogenesis is the NMDA receptor-dependent induction of ectopic action potentials which may contribute to the expression of EGSs in the hippocampal slice. Stasheff *et al.* (1993a,b) found that the stimulus train induction of EGSs in physiological medium is associated with the induction of ectopic ('antidromic') action potentials in CA3 pyramidal axons. These ectopic action potentials go from CA3 axon terminals in region CA1 to the CA3 neuron somata in region CA3. These ectopic action potentials are often correlated with the initiation of individual bursts in the EGS, and could, therefore, contribute to the generation of EGSs.

When the NMDA antagonist AP5 is applied prior to and during the delivery of the stimulus trains, no increase in ectopic action potential frequency occurs. Furthermore, after EGSs and ectopic spikes have been induced tetanically in physiological medium without AP5, subsequent application of AP5 produces little decrease in the EGSs and no decrease in ectopic spike frequency. The ability of AP5 to block the induction, but not the expression, of ectopic spikes is identical to the ability of AP5 to block the induction, but not the expression, of EGSs. These results are consistent with the idea that epileptogenesis may be associated with an NMDA receptor-dependent long-lasting increase in axon terminal excitability which contributes to the generation of seizures.

8.6 Mechanisms of epileptogenesis: intracellular messenger mechanisms

The mechanisms by which the transient Ca^{2+} signal is detected by intracellular messengers and ultimately converted into long-term storage are beginning to be elucidated by LTP studies (see Chapters 3–6), and these findings will be extended to include kindling.

8.6.1 The intracellular second messenger Ca^{2+}

LTP studies indicate that the first step after the Ca^{2+} has entered through the NMDA channel is to amplify this intracellular dendritic Ca^{2+} signal by releasing additional Ca^{2+} from intracellular stores. During a tetanus train, approximately 35% of the intradendritic free Ca^{2+} comes through the NMDA channels, and the rest is released from intracellular stores (Alford *et al.*, 1993). In addition to the release of intracellular Ca^{2+}, glutamate activation of metabotropic glutamate receptors (mGluRs) during a stimulus train may also contribute to the release of intracellular Ca^{2+} (Frenguelli et al., 1993). However, even with intracellular release of Ca^{2+}, the rise in intradendritic Ca^{2+} concentration only outlasts the stimulus train by a few seconds. Malenka et al. (1992) have found that the rise in Ca^{2+} need only last about 2 sec for LTP to occur. Therefore, there must be mechanisms to detect the transient Ca^{2+} signal and extend the storage of this information.

There have been no measurements of intracellular Ca^{2+} during electrical kindling, although similar processes may occur.

8.6.2 Calcium/calmodulin-dependent protein kinase II

The next crucial step is to capture the transient Ca^{2+} signal. One predominant hypothesis is that that this occurs through Ca^{2+} activation of calcium/calmodulin-dependent protein kinase II (CaMKII) (for a detailed review, see Soderling, 1995, and Chapter 3). CaMKII comprises approximately 2% of total hippocampal protein and is concentrated at the postsynaptic site. CaMKII is interesting because a transient rise in Ca^{2+} causes CaMKII to autophosphorylate and turn into a Ca^{2+}-independent, continuously activated kinase which can phosphorylate other substrate proteins continuously (Miller and Kennedy, 1986). In essence, CaMKII is a Ca^{2+}-triggered molecular switch which, at least theoretically, is capable of capturing the transient, tetanically produced Ca^{2+} signal.

In LTP studies, injection of CaMKII peptide inhibitors prior to tetanic stimulation (Malinow et al., 1989), or application of the CaMKII inhibitor KN-62 during the tetanic stimulation (Ito et al., 1991), blocks the induction of LTP. However, if the peptide inhibitor or KN-62 is applied after the tetanic stimulation, the expression of LTP is not blocked. Application of glutamate to cultured hippocampal neurons activates CaMKII and this activation is blocked by the NMDA antagonist AP5 (Fukunaga et al., 1992). Similarly, tetanic stimulation of CA1 in the hippocampal slice produces LTP which is associated with an *increase* in CaMKII activity, and both LTP and CaMKII activation are blocked by application of AP5 (Fukunaga et al., 1993)

Kindling has long been known to produce changes in CaMKII activity lasting several months (Wasterlain and Faber, 1984), but the results have been contradictory as to whether the kindling causes an increase or decrease in CaMKII activity. Studies by Goldenring et al. (1986) indicate that septal kindling produces a long-term decrease in CaMKII activity. However, later studies by Zhou et al. (1994) show that amygdala kindling produces long-term autophosphorylation and an increase of CaMKII activity in the hippocampus, which is similar to Fukunaga's findings on LTP. It is not clear why kindling causes a decrease in CaMKII activity in some studies, and an increase in others.

The effects of the CaMKII inhibitor KN-62 have been studied on one *in vitro* model of epileptogenesis, the low Mg^{2+}, NMDA receptor-dependent induction of epileptiform activity in the piriform cortex (Domroese and Haberly, 1995). They found that KN-62 blocks NMDA receptor-dependent LTP in the piriform cortex, as it does in the hippocampus. However, very interestingly, they found that KN-62 does not block the induction of epileptiform activity. This suggests that there is at least one NMDA receptor-dependent mechanism of epileptogenesis that does not require the activation of CaMKII.

Studies on the role of CaMKII in LTP and kindling were also performed on mutant mice deficient in the CaMKII α-subunit. These α-CaMKII-deficient mice do show impairment of LTP (Silva *et al.*, 1992). However, spontaneous EGSs also occur in the α-CaMKII-deficient mice, and the first tetanic stimulation delivered *in vivo* is capable of triggering strong continuous or repetitive EGSs lasting 0.5–2.5 h, which sometimes result in death (Butler *et al.*, 1995). This status epilepticus-like hyperexcitability indicates an unknown involvement of CaMKII in epilepsy that is far more complicated than previously thought. Whether this abnormal hyperexcitability is due directly to a deficient α-CaMKII in the mature mouse, or is due indirectly to compensatory developmental abnormalities resulting from the deficient α-CaMKII is not known.

8.6.3 Protein kinase C

Protein kinase C (PKC) is also involved in the early stages of LTP, possibly through the activation of mGluRs and the release of intracellular Ca^{2+} (see Chapter 2 and Soderling, 1995). Intracellular injection of peptide inhibitors of PKC block the induction only (Malinow *et al.*, 1989), or the induction and the maintenance (Wang and Feng, 1992), of LTP in the hippocampal slice.

Kindling-induced, long-lasting changes in PKC activity could contribute to the maintenance of seizure expression after epileptogenesis has occurred. Whether long-lasting changes in PKC occur depends on where the animals are kindled. If rats are kindled in the hippocampus, a long-lasting (at least 4 months) increase in membrane-associated PKC activity occurs in the hippocampus, amygdala and piriform cortex (Kohira *et al.*, 1992). However, if rats are kindled in the amygdala, there is only an increase in the ipsilateral hippocampus (Akiyama *et al.*, 1995). Beldhius *et al.* (1993) found that, if rats are kindled in the amygdala, an increase in the γ-isoform of PKC occurs in the contralateral but not ipsilateral piriform cortex. Therefore, although PKC activation can, in theory, contribute to the maintenance of seizure expression in certain situations, it does not appear to be a universal mechanism promoting seizure expression.

The effects of the relatively nonspecific protein kinase C inhibitors W-7 and H-7 were tested on one *in vitro* model of epileptogenesis, the low Mg^{2+} induction of epileptiform activity in the piriform cortex (Domroese and Haberly, 1995). They found that W-7 and H-7 do block the NMDA receptor-dependent LTP in the piriform cortex, but interestingly do not block the low Mg^{2+} induction of epileptiform activity, suggesting that PKC activation is not involved in the induction of epileptiform activity in this model.

8.6.4 Tyrosine kinase

The tyrosine kinase inhibitors lavendustin A and genestein block or reduce the induction of LTP in the hippocampal slice (O'Dell *et al.*, 1991). Cain *et al.* (1995)

recently found that kindling in *fyn* tyrosine kinase-deficient mutant mice is substantially slower than in control wild-type mice. However, kindling is not blocked in the mutant mice and, once the mice are fully kindled, the maintenance of the kindled state is not changed. This suggests the involvement of a *fyn* tyrosine kinase-dependent pathway in the induction, but not the maintenance, of normal kindling.

8.7 Mechanisms of epileptogenesis: gene expression

8.7.1 Immediate-early genes

Protein synthesis inhibitors block the maintenance of LTP (Frey *et al.*, 1988) and *in vitro* kindling in the hippocampal slice (Jones *et al.*, 1992) several hours after stimulus train induction, and also decrease the rate of *in vivo* kindling (Jonec and Wasterlain, 1979). This suggests that changes in gene expression are likely to underlie both LTP and epileptogenesis. The newly synthesized proteins which are involved in epileptogenesis could theoretically include AMPA, NMDA and GABA channels, the protein kinases PKC and CaMKII, and many other structural and regulatory proteins. The transcription of the late response genes which produce these newly synthesized proteins can be altered by immediate-early genes (IEGs), and therefore investigation of alteration of IEGs during LTP and kindling has begun. LTP produces an increase in c-*fos* and NGFI-A mRNA, and this increase is blocked by NMDA receptor antagonists (Cole *et al.*, 1989; Dragunow *et al.*, 1987). However, IEG induction is also produced by ischemia, traumatic brain injury and metrazol convulsant-produced seizures, which require no NMDA receptor-dependent induction process for expression (Morgan *et al.*, 1987). Therefore, it is important to separate NMDA receptor-dependent IEG induction involved in epileptogenesis from IEG induction produced by abnormal, excessive neural activity and that is not NMDA receptor-dependent.

Kindling in the hippocampus, which initially produces hippocampal afterdischarge, causes the induction of c-*fos* and NGFI-A mRNAs, and application of NMDA antagonists reduces the increase of c-*fos* and NGFI-A mRNAs by about half (Dragunow and Robertson, 1987; Labiner *et al.*, 1993; Lerea *et al.*, 1995). This suggests that afterdischarge-mediated induction of IEGs in the hippocampus occurs *in part* through NMDA receptor-dependent pathways.

In contrast, kindling in the amygdala, which initially produces amygdala afterdischarge, causes the induction of c-*fos* and NGFI-A mRNA, but the application of the NMDA antagonist MK-801 does not block the induction of these IEGs (Hughes *et al.*, 1994). This suggests that the afterdischarge-mediated induction of IEGs in the amygdala occurs through NMDA receptor-independent pathways.

8.7.2 NMDA, AMPA and GABA mRNAs

One site that could be expected to store long-term changes in the expression of epileptogenesis would be in the mRNAs that synthesize the NMDA, AMPA and GABA receptors. Interestingly, most changes in NMDA, AMPA and GABA mRNAs occur during or shortly (1 day) after kindling has stopped. However, 4 weeks after standard, slow kindling has been established and the last evoked kindled seizure has occurred, most NMDA, AMPA and GABA mRNAs have returned to control levels.

No long-term changes after 4 weeks were observed in NMDAR1 mRNA in the hippocampus, amygdala and piriform cortex (Hikiji *et al.*, 1993), or in NMDAR1, NR2A, NR2B, NR2C and NR2D mRNAs in the hippocampus (Kamphuis *et al.*, 1995; Kraus *et al.*, 1994). Kindling produces transient but no long-term changes in the AMPA receptor subunit GluR2 mRNA in the hippocampus, amygdala and piriform cortex (Hikiji *et al.*, 1993; Prince *et al.*, 1995), and no changes were observed in GluR1 and GluR4 mRNAs 1 day after the last kindled seizure (Prince *et al.*, 1995). Kindling also produces some transient, but no long-term changes in $GABA_A$ receptor subunit α_1, α_2, α_4 and β_{1-3} mRNAs in the hippocampus (Kamphuis *et al.*, 1994a; Lee *et al.*, 1994). Finally, kindling also produces no long-term change in mGluR1 mRNA in the hippocampus, amygdala or piriform cortex 4 weeks after the last tetanus (Hikiji *et al.*, 1993).

There are, however, exceptions to the finding that NMDA, AMPA and GABA mRNAs return to control levels 4 weeks after kindling has ceased. Kindling has been reported to produce a decrease in the flip variant of GluR1 mRNA in the dentate after 4 weeks (Kamphuis *et al.*, 1994b). Hikiji *et al.* (1993) also detected a small increase in kainate KA-1 mRNA in the hippocampal CA3 region 4 weeks after kindling ceased.

Although Kraus *et al.* (1994) found no changes in known NMDA mRNAs, they did discover a novel NMDA receptor that was induced by kindling and which binds to the NMDA antagonist CPP but, suprisingly, not the similar NMDA antagonist CGS-19755. This may indicate an unknown NMDA mRNA, but may also be due to other reasons including persistent post-translational phosphorylation.

In general, these results suggest that although the induction of the kindled state may be dependent on altered NMDA, AMPA and GABA mRNAs, the long-term maintenance of the kindled state cannot be explained by a long-lasting change in the gene expression of known NMDA, AMPA or GABA receptors.

8.8 Mechanisms of epileptogenesis: post-translational modifications of channels

Earlier, evidence was presented that the changes induced by epileptogenesis that resulted in the expression or generation of seizures were due predominantly to increases in excitatory AMPA and NMDA receptor-mediated synaptic transmission, and decreases in GABAergic inhibitory synaptic transmission (see Section 8.5). One mechanism for this alteration in channel activity could be through post-translational phosphorylation of channel proteins. The phosphorylation and activation of AMPA and NMDA receptor–channels by activated CaMKII and PKC has been reviewed by Soderling (1995) (see also Chapter 5).

There is evidence that phosphorylation and deactivation of $GABA_A$ receptor–channels also occurs. Activation of PKC inhibits $GABA_A$ receptor–channel activity (Leidenheimer *et al.*, 1992), and this could therefore be another mechanism whereby mGluR activation of PKC increases neuronal excitability. Using whole-cell patch-clamped, acutely dissociated CA1 pyramidal cells, Stelzer and Shi (1994) found that application of NMDA causes an approximate 85% decrease in $GABA_A$ currents, and that this decrease is blocked by the addition of a calcineurin inhibitor. This is consistent with the Ca^{2+}/calmodulin-activated protein phosphatase calcineurin dephosphorylating the $GABA_A$ receptor and thereby decreasing $GABA_A$ receptor–channel activity (for a review, see Stelzer, 1992). This, therefore, could be another mechanism whereby NMDA receptor activation leads to calcineurin activation, then to a decrease in $GABA_A$ receptor activity and, finally, to an increase in neuronal excitability. Interestingly, calcineurin inhibitors substantially retard kindling (Moia *et al.*, 1994), and this is consistent with the above mechanism.

8.9 Blockade of kindling by anti-epileptics that are not NMDA antagonists

In addition to NMDA antagonists, several clinically used anti-epileptic drugs have been found to have anti-epileptogenic as well as anticonvulsant effects (Wada, 1977). In the *in vivo* electrical kindling model, Shin *et al.* (1986) found that γ-vinyl-GABA, a GABA transaminase inhibitor which probably enhances GABAergic inhibition, strongly suppresses the rate of kindling. Furthermore, no seizures initially occurred when stimulation resumed after the γ-vinyl-GABA had cleared, indicating that the suppression of kindling was a true anti-epileptogenic effect and not just masked by an anticonvulsant effect. Similarly, Silver *et al.* (1991) and Schmutz *et al.* (1988) found that valproate and pentobarbital are strongly anti-epileptogenic, but only moderately anticonvulsant. In contrast, they found that carbamazepine and phenytoin are strongly anticonvulsant, but not anti-epileptogenic. In *in vivo* chemical kindling by FG 7142, Stephens and Weidmann (1989) found that valproate was strongly anti-epileptogenic, whereas carbamazepine and phenytoin were not.

The mechanism of action of these drugs is not directly on the NMDA receptor, and many of these drugs probably enhance GABAergic inhibition. However, enhancement of GABAergic inhibition may reduce activation of the NMDA system indirectly by helping to maintain the membrane potential more negative than the potential at which Mg^{2+} blocks the NMDA channel. Studies of LTP suggest that enhancement of $GABA_A$ inhibition during the tetanus (by applying a $GABA_B$ antagonist, which blocks presynaptic $GABA_B$ 'autoreceptor'-mediated depression of presynaptic GABA release) can block the induction of LTP (Davies *et al.*, 1991)

Regardless of the mechanisms by which these drugs work, the use of anti-epileptic drugs which have anti-epileptogenic as well as anticonvulsant properties is preferable to the use of those with only anticonvulsant properties, because reducing epileptogenesis may decrease the need for chronic anticonvulsant therapy. The adverse psychotomimetic effects of MK-801 and other NMDA antagonists is another reason why the use of these anti-epileptic drugs to block epileptogenesis may be preferable.

8.10 Are there non-NMDA receptor mechanisms that can induce kindling epileptogenesis?

There is no *a priori* reason to assume that activation of NMDA receptors is the only neurotransmitter receptor mechanism inducing kindling epileptogenesis. The fact that NMDA receptor-dependent epileptogenesis has been demonstrated in the amygdala, hippocampus and piriform cortex, three of the more important sites of epileptogenesis, suggests that it is an extremely important mechanism.

However, there are many reasons for suspecting that non-NMDA transmitter receptor mechanisms may also be important in inducing kindling epileptogenesis, and may act in combination with the NMDA receptor induction system. Failure of NMDA antagonists to block *in vivo* amygdala kindling completely is one reason for suspecting that other receptors are involved. The fact that most *in vitro* studies on epileptogenesis were done on hippocampal tissue and showed almost complete block, and most *in vivo* kindling studies were done on the amygdala and often showed incomplete block, may indicate differences in mechanisms of induction between the amygdala and hippocampus. Finally, there are many examples of non-NMDA receptor induction of long-term changes in neuronal excitability.

The induction of neuronal plasticity events by other receptors could suggest that other receptors are involved in chronic epileptogenesis. For example, induction of LTP in the dorsal lateral septal nucleus is not blocked by the NMDA antagonist AP5, but is blocked by the putative mGluR antagonist DL-AP3 (Zheng and Gallagher, 1992). Also, induction of LTP in the mossy fiber–CA3 pathway is not blocked by NMDA antagonists but is blocked by the mGluR antagonist α-methyl-4-carboxyphenylglycine (MCPG) (Bashir *et al.*, 1993). It was shown recently that induction of LTP in CA1 can be blocked in nonpotentiated slices by the application of MCPG (Bortolotto *et al.*, 1994). However, if the slices are partially potentiated, application of the mGluR antagonist does not block additional LTP. This suggests that an mGluR 'molecular switch' is involved in the induction of LTP and that it acts by operating in series with the NMDA receptor-dependent induction mechanism. Whether a similar system is operating in kindling-like epileptogenesis is not known.

Liu *et al.* (1993) have reported that activation of mGluRs by bath application of (1S, 3R)-1-amino-cyclopentane-1, 3-dicarboxylate (ACPD) produces a long-term decrease in GABAergic inhibition which would contribute to a long-term increase in neuronal hyperexcitability.

Bawin *et al.* (1992) have provided preliminary evidence that activation of an mGluR contributes to *in vitro* kindling in rat hippocampal slices. They found that bath application of the putative mGluR antagonist DL-AP3 during sine-wave stimulation causes a reduction in the frequency of epileptiform bursting compared with physiological medium controls, and that epileptiform bursting continues for only 60–70 min after stimulation has ceased, compared with physiological medium controls where the bursting continues indefinitely. However, if DL-AP3 is added after epileptiform bursting has been induced, the bursting is not affected.

Arvanov *et al.* (1995) have found recently that the mGluR antagonist MCPG blocks the induction of EGSs in the amygdala slice that normally occurs during exposure to 4-aminopyridine. However, if MCPG is applied after the EGSs have already been induced, they are not blocked. This suggests that MCPG has an anti-epileptogenic, but not an anticonvulsant effect in this model.

Early kindling studies also showed that atropine could substantially retard the development of kindling, but had little effect on kindled seizures (Arnold *et al.*, 1973). This suggests that activation of muscarinic acetylcholine receptors could also contribute to epileptogenesis.

Finally, the blockade of kindling by anti-epileptics that are not NMDA antagonists also raises the question of whether these anti-epileptics are blocking kindling by blocking non-NMDA receptors that are directly involved in the induction process. Alternatively, they may be indirectly inhibiting NMDA receptor activation by directly activating GABAergic inhibition.

8.11 Anticonvulsant and anti-epileptogenic therapy

NMDA antagonists have not yet been used in anti-epileptogenic therapy (to block the development of epilepsy after some trauma), and have only been used to a limited extent in anticonvulsant therapy (to suppress seizures after they have developed). One problem is that many NMDA antagonists cause psychotomimetic effects in humans at concentrations that block seizures in several animal models of seizures, and therefore cannot be used for chronic anticonvulsant control of epilepsy (Löscher, 1993). These include the noncompetitive NMDA channel blockers ketamine, PCP and MK-801, and antagonists to the glycine site on the NMDA receptor such as L-687–414. Also, the new competitive NMDA receptor antagonists that cross the blood–brain barrier such as D-CPPene, CGP-37849 or CGP-39551 do not cause PCP-like effects in normal animals, but do cause them in kindled animals (Löscher, 1993).

A second problem is that just as the NMDA antagonists have only weak anticonvulsant action on many models of epilepsy including fully kindled *in vitro* (Stasheff *et al.*, 1989; *Figure 8.3*) and *in vivo* seizures (*Table 8.1*), clinical trials also showed that the NMDA antagonist D-CPPene has no anticonvulsant properties on complex-partial seizures or secondarily generalized seizures (Sveinbjornsdottir *et al.*, 1993). This is not suprising if the seizures are triggered and generated by predominantly non-NMDA receptor-mediated excitatory potentials.

However, the anticonvulsant felbamate, which has NMDA antagonist properties (De Sarro et al., 1994), recently has been in clinical use as an anticonvulsant, and this may indicate that NMDA antagonists can be tolerated if they have the appropriate pharmacodynamics. Furthermore, there are several drugs known as low affinity NMDA channel blockers which do not seem to cause psychotomimetic effects, including dextromethorphan, remacemide and memantine (Rogawski, 1992), and these may be good candidates for clinical anticonvulsant and anti-epileptogenic trials.

NMDA antagonists as anti-epileptogenic agents could be useful in certain instances of acute therapy where protection by NMDA antagonists need only be for a short period (Clark et al., 1994). One situation could be during and shortly after status epilepticus. Status epilepticus is continuous seizure activity lasting 30 min or more, it can be life threatening and it often causes the subsequent development of seizures (Lothman and Bertram, 1993). Another situation could be in neurosurgery where just penetration of the dura causes about a 1% incidence in the development of epilepsy (Wilson et al., 1989). Severe head injury can also lead to the development of seizures, and NMDA antagonists could be useful in blocking seizure development if applied shortly after the injury occurred (Clark et al., 1994). However, usually seizures have developed substantially by the time they are diagnosed, and anti-epileptogenic agents are not useful at this time.

Other non-NMDA antagonist drugs may also be useful as anti-epileptogenic agents. For instance, several anti-epileptic drugs such as γ-vinyl-GABA, valproate and pentobarbital, which are thought to facilitate GABAergic synaptic transmission, block epileptogenesis (Schmutz et al., 1988; Shin et al., 1986; Silver et al. 1991; Wada, 1977). Furthermore, these drugs can be used chronically. Silver et al. (1991) propose that it may be useful to administer drugs that have an anti-epileptogenic as well as an anticonvulsant action when treating severe head injury, because this may lessen the need for chronic anticonvulsant therapy. The initial hyper-excitability occurring during severe head injury may be due to microhemorrhaging, localized deposits of iron from ruptured red blood cells and, possibly, free radical production (Willmore, 1992). Examining drugs which affect free radical production and related processes could also discover useful anti-epileptogenic agents.

8.12 Reversal of permanent epileptogenesis

One of the problems with anticonvulsant and anti-epileptogenic therapy is that, once seizures have developed in humans, they can remain for years, and even be permanent (Shafer et al., 1988). Presently, most seizures can be controlled by chronic administration of anticonvulsants or, alternatively, 'cured' by neurosurgery, if they are localized.

Many researchers have found recently that LTP, although long-lasting, can be reversed rapidly by a process called depotentiation. Because potentiation of AMPA and NMDA receptor-mediated potentials is likely to be one of the mechanisms contributing to permanent epileptogenesis, this suggests that the brain may have the capability of at least partly reversing epileptogenesis by depotentiating these AMPA and NMDA receptor-mediated potentials. One of the next stages in epilepsy research may be to try and reverse 'permanent' epileptogenesis. However, in addition to depotentiating AMPA and NMDA receptor-mediated potentials, other mechanisms that also contribute to epileptogenesis will have to be reversed, such as the long-term decrease in GABAergic inhibition.

8.13 Conclusions

8.13.1 NMDA receptor-dependent epileptogenesis

The main conclusion of this chapter is that there is general agreement that NMDA antagonists either block, or substantially retard, epileptogenesis, and do so in a wide variety of *in vivo* and *in vitro* kindling-like models. This suggests that activation of NMDA receptors could be an important pathophysiological transmitter/receptor mechanism that creates epilepsy.

One notable exception to the findings that NMDA antagonists block or substantially retard epileptogenesis is the report of Pierson and Swann (1991) that MK-801 and PCP are pro-epileptogenic for inducing sound-induced seizures during their critical period of development. This finding is very important because it raises warnings about using NMDA antagonists in anticonvulsant and antiepileptogenic clinical trials, particularly in young patients.

There are many reasons for suspecting that other non-NMDA transmitter receptor induction systems may be involved in inducing kindling epileptogenesis, including failure of NMDA antagonists to completely block *in vivo* amygdala kindling. For example, in the amygdala, the mGluR antagonist MCPG has been found to block 4-aminopyridine-induced epileptogenesis (Arvanov *et al.*, 1995).

8.13.2 NMDA antagonists as anticonvulsants

A substantial amount of research has gone into developing and testing NMDA antagonists as anticonvulsants. In some models of seizures, NMDA antagonists are strongly anticonvulsant, in other models such as the kindling model they are only weakly anticonvulsant (Clark *et al.*, 1994). What is important to realize is that NMDA receptor antagonists will only be strongly anticonvulsant for seizures that occur in brain regions where NMDA receptor-mediated synaptic potentials are strongly involved in triggering and/or generating seizures.

8.13.3 The permanence of slow vs. rapid kindling

Lothman and Williamson (1994) reported that standard, slow *in vivo* kindling (one stimulation a day) produces fully kindled seizures that are probably permanent, whereas rapid *in vivo* kindling (one stimulation every 30 min) produces kindled seizures that decay within 1 week. This raises the question of whether the mechanisms of epileptogenesis determined from rapid *in vivo* and *in vitro* kindling models will be the same as those for slow, but permanent, epileptogenesis.

The finding that NMDA antagonists block or substantially retard slow *in vivo* kindling, rapid *in vivo* kindling and rapid *in vitro* kindling suggests some common mechanisms of seizure induction – at minimum a dependence on the activation of NMDA receptors. The differences in the permanence of the kindled seizures suggest that there are fundamental differences in some of the mechanisms of kindling maintenance, and there also may be differences in the mechanisms of kindling induction. However, the finding that animals re-kindle faster during a second bout of rapid kindling after a 1-week stimulus-free period indicates that the first bout of rapid kindling does produce signs of epileptogenesis that are still observable after 1 week. Long-lasting but impermanent epileptogenesis may also be relevant because it could be the first step to permanent epileptogenesis.

It is not clear why tetanic stimulation that evokes afterdischarge bursting produces permanent EPSP potentiation (kindling-induced potentiation), whereas rapid kindling, which also evokes afterdischarge bursting, does not produce permanent epileptogenesis.

8.13.4 Mechanisms of epileptogenesis

Clearly, the mechanism of kindling epileptogenesis is the area of research that will receive the most future attention. Concerning the mechanisms of induction, the most certain finding is that antagonists of the NMDA receptor–channel complex block or substantially reduce the induction of kindling. However, even this description of the initial induction process is probably incomplete and is likely to include other transmitter receptor induction systems such as the mGluR system. Ca^{2+} entry through the NMDA channel is likely to be an important mechanism. Activation of IEGs is also likely to be important; some of this activation is through an NMDA receptor-dependent pathway, some is through an NMDA receptor-independent pathway.

Even less is known about the mechanisms responsible for the maintenance of kindling. Where do these permanent changes reside? There will probably be a permanent increase in AMPA and NMDA receptor-mediated EPSPs, and a permanent decrease in GABA receptor-mediated IPSPs. However, in some brain regions, the hyperexcitability could be due predominantly to an enhancement of EPSPs, in other regions it could be due predominantly to a decrease in IPSPs. This could involve permanent post-translational phosphorylation of AMPA and NMDA channels, and permanent dephosphorylation of GABA channels. It appears unlikely that the changes will reside in a permanent change in most NMDA, AMPA and GABA mRNAs, because kindling does not cause their permanent modification. Kindling does cause a permanent change in CaMKII, although it is uncertain whether this is an increase or decrease in activity, but this could, nevertheless, be a candidate for a site of permanent storage. PKC activity has been found to undergo long-term changes with kindling in some, but not other, regions of the brain. Although this could be a candidate for permanent storage, it is unlikely to be universal.

Acknowledgments

I would like to thank Dr Suzanne Clark for her very constructive criticisms of this paper. W.W.A. is supported by the Wellcome Trust.

References

Akiyama K, Ono M, Kohira I, Daigen A, Ishihara T, Kuroda S. (1995) Long-lasting increase in protein kinase C activity in the hippocampus of amygdala-kindled rat. *Brain Res*. **679**: 212–220.

Alford S, Frenguelli BG, Schofield JG, Collingridge GL. (1993) Characterization of Ca^{2+} signals induced in hippocampal CA1 neurones by the synaptic activation of NMDA receptors. *J. Physiol*. **469**: 693–716.

Anderson WW, Coan EJ. (1989) Effect of non-NMDA excitatory amino acid receptor blockade on hippocampal slice ictal events in physiological and low Mg^{2+} medium. *Br. J. Pharmacol*. **97**: 588P.

Anderson WW, Lewis DV, Swartzwelder HS, Wilson WA. (1986) Magnesium-free medium activates seizure-like events in the rat hippocampal slice. *Brain Res*. **398**: 215–219.

Anderson WW, Swartzwelder HS, Wilson WA. (1987) The NMDA receptor antagonist 2-amino-5-phosphonovalerate blocks stimulus train-induced epileptogenesis but not epileptiform bursting in the rat hippocampal slice. *J. Neurophysiol*. **57**: 1–21.

Anderson WW, Swartzwelder HS, Wilson WA. (1988) Regenerative, all-or-none electrographic seizures in the rat hippocampal slice in physiological magnesium medium. In: *Synaptic Plasticity of the Hippocampus* (eds Buzaki G, Haas HL), Springer-Verlag, Berlin, pp. 180–183.

Anderson WW, Stasheff SF, Swartzwelder HS, Wilson WA. (1990) Regenerative all-or-none electrographic seizures in the rat hippocampal slice in Mg-free, physiological medium. *Brain Res*. **532**: 288–298.

Anis NA, Berry SC, Burton NR, Lodge D. (1983) The dissociative anesthetics, ketamine and phencyclidine, selectively reduce excitation of central mammalian neurones by N-methyl-aspartate. *Br. J. Pharmacol*. **79**: 565–575.

Arnold PS, Racine RJ, Wise RA. (1973) Effects of atropine, reserpine, 6-hydroxydopamine and handling on seizure development in the rat. *Exp. Neurol*. **40**: 457–470.

Arvanov VL, Holmes KH, Keele NB, Shinnick-Gallagher P. (1995) The functional role of mGluRs in epileptiform activity induced by 4-aminopyridine in the rat amygdala slice. *Brain Res*. **669**: 140–144.

Avoli M, Drapeau C, Kostopoulos G. (1988) Changes in synaptic transmission evoked in the 'in vitro' hippocampal slice by a brief decrease of $[Mg^{++}]_o$: a correlate for long-term potentiation? In: *Synaptic Plasticity of the Hippocampus* (eds Buzaki G, Haas HL), Springer-Verlag, Berlin, pp. 9–11.

Bashir ZI, Alford S, Davies SN, Randall AD, Collingridge GL. (1991) Long-term potentiation of NMDA receptor-mediated synaptic transmission in the hippocampus. *Nature* **349**: 1156–1158.

Bashir ZI, Bortolotto ZA, Davies CH, Beretta N, Irving AJ, Seal AJ, Henley JM, Jane DE, Watkins JC, Collingridge GL. (1993) Induction of LTP in the hippocampus needs synaptic activation of glutamate metabotropic receptors. *Nature* **363**: 347–350.

Bawin SM, Shahhal I, Mahoney MD, Adey WR. (1989) Induction of delayed synchronized bursts in hippocampal slices by weak sine-wave stimulation: role of the NMDA receptor. *Epilepsy Res*. **3**: 41–48.

Bawin SM, Satmary WM, Mahoney MD, Adey WR. (1991) Transition from normal to epileptiform activity in kindled rat hippocampal slices. *Epilepsy Res*. **8**: 107–116.

Bawin SM, Satmany WM, Sheppard AR, Adey WR. (1992) The metabotropic glutamate receptor contributes to kindling of epileptiform events in rat hippocampal slices. *Soc. Neurosci. Abstr*. **18**: 1243.

Bawin SM, Satmary WM, Adey WR. (1993) Roles of the NMDA and quisqualate/kainate receptors in the induction and expression of kindled bursts in rat hippocampal slices. *Epilepsy Res*. **15**: 7–13.

Beldhuis HJA, De Ruiter AJH, Maes FW, Suzuki T, Bohus B. (1993) Long-term increase in protein kinase C-γ, and muscarinic acetylcholine receptor expression in the cerebral cortex of amygdala-kindled rats – a quantitative immunocytochemical study. *Neuroscience* **55**: 965–973.

Ben-Ari Y, Gho M. (1988) Long-lasting modifications of the synaptic properties of rat CA3 hippocampal neurones induced by kainic acid. *J. Physiol.* **404**: 365–384.

Ben-Ari Y, Krnjevic K, Reinhardt W. (1979) Hippocampal seizures and failure of inhibition. *Can. J. Physiol. Pharmacol.* **57**: 1462–1466.

Bliss TVP, Gardner-Medwin AR. (1973) Long-lasting potentiation of synaptic transmission in the dentate area of the unanaesthetized rabbit following stimulation of the perforant path. *J. Physiol.* **232**: 357–374.

Bortolotto ZA, Bashir ZI, Davies CH, Collingridge GL. (1994) A molecular switch activated by metabotropic glutamate receptors regulates induction of long-term potentiation. *Nature* **368**: 740–743.

Boyer JF. (1982) Phencyclidine inhibition of the rate of development of amygdaloid kindled seizures. *Exp. Neurol.* **75**: 173–183.

Boyer JF, Winters WD. (1981) The effects of various anesthetics on amygdaloid kindled seizures. *Neuropharmacol.* **20**: 199–209.

Butler LS, Silva AJ, Abeliovich A, Watanabe Y, Tonegawa S, McNamara JO. (1995) Limbic epilepsy in transgenic mice carrying a Ca^{2+}/calmodulin-dependent kinase II α-subunit mutation. *Proc. Natl Acad. Sci. USA* **92**: 6852–6855.

Cain DP. (1989) Long-term potentiation and kindling: how similar are the mechanisms? *Trends Neurosci.* **12**: 6–10.

Cain DP, Desborough KA, McKitrick DJ. (1988) Retardation of amygdala kindling by antagonism of NMD-aspartate and muscarinic cholinergic receptors: evidence for the summation of excitatory mechanisms. *Exp. Neurol.* **100**: 179–187.

Cain DP, Grant SGN, Saucier D, Hargreaves EL, Kandel ER. (1995) Fyn tyrosine kinase is required for normal amygdala kindling. *Epilepsy Res.* **22**: 107–114.

Callaghan DA, Schwark WS. (1980) Pharmacological modification of amygdaloid-kindled seizures. *Neuropharmacology* **19**: 1131–1136.

Clark S, Stasheff S, Lewis D, Martin D, Wilson WA. (1994) The NMDA receptor in epilepsy. In: *The NMDA Receptor*, 2nd Edn (eds GL Collingridge, JC Wakins). Oxford University Press, Oxford, pp. 395–427.

Coan EJ, Collingridge GL. (1985) Magnesium ions block an *N*-methyl-D-aspartate receptor mediated component of synaptic transmission in rat hippocampus. *Neurosci. Lett.* **53**: 21–26.

Cohen SM, Martin D, Morrisett RA, Wilson WA, Swartzwelder HS. (1993) Proconvulsant and anticonvulsant properties of ethanol: studies of electrographic seizures *in vitro*. *Brain Res.* **601**: 80–87.

Cole AJ, Saffen DW, Baraban JM, Worley PF. (1989) Rapid increase of an immediate-early gene messenger RNA in hippocampal neurons by synaptic NMDA receptor activation. *Nature* **340**: 474–476.

Collingridge GL, Kehl SJ, McLennan H. (1983) Excitatory amino acids in synaptic transmission in the Schaffer collateral–commissural pathway of the rat hippocampus. *J. Physiol.* **334**: 33–46.

Collingridge GL, Herron CE, Lester RAJ. (1988) Frequency-dependent *N*-methyl-D-aspartate receptor-mediated synaptic transmission in rat hippocampus. *J. Physiol.* **399**: 301–312.

Corda MG, Orlandi M, Lecca D, Giorgi O. (1992) Decrease in GABAergic function induced by pentenetetrazol kindling in rats: antagonism by MK-801. *J. Pharmacol. Exp. Ther.* **262**: 792–800.

Coulter DA, Lee CJ. (1993) Thalamocortical rhythm generation *in vitro*: extra- and intracellular recordings in mouse thalamocortical slices perfused with low Mg^{2+} medium. *Brain Res.* **631**: 137–142.

Croucher MJ, Bradford HF, Sunter DC, Watkins JC. (1988) Inhibition of the development of electrical kindling of the pyriform cortex by daily focal injections of excitatory amino acid antagonists. *Eur. J. Pharmacol.* **152**: 29–38.

Davies CH, Starkey SJ, Pozza MF, Collingridge GL. (1991) $GABA_B$ autoreceptors regulate the induction of LTP. *Nature* **349**: 609–611.

de Jonge MC, Racine RJ. (1985) The effects of repeated induction of long-term potentiation in the dentate gyrus. *Brain Res.* **328**: 181–185.

De Sarro G, Ongini E, Bertorelli R, Aguglia U, De Sarro A. (1994) Excitatory amino acid neurotransmission through both NMDA, and non-NMDA receptors is involved in the anticonvulsant activity of felbamate in DBA/2 mice. *Eur. J. Pharmacol.* **262**: 11–19.
Domroese ME, Haberly LB. (1995) NMDA-dependent induction of epileptiform activity in piriform cortex *in vitro* does not involve kinases that are required for LTP. *Soc. Neurosci. Abstr.* **21**: 982.
Douglas RM, Goddard GV. (1975) Long-term potentiation of the perforant path–granule cell synapse in the rat hippocampus. *Brain Res.* **240**: 205–215.
Dragunow M, Robertson HA. (1987) Kindling stimulation induces c-*fos* protein(s) in granule cells of the rat dentate gyrus. *Nature* **329**: 441–442.
Dragunow M, Abraham WC, Goulding M, Mason SE, Robertson HA, Faull RLM. (1987) Long-term potentiation and the induction of c-*fos* mRNA and proteins in the dentate of unanaesthetized rats. *Neurosci. Lett.* **101**: 274–280.
Dürmüller N, Craggs M, Meldrum BS. (1994) The effect of the non-NMDA receptor antagonists GECKO 52466 and NBQX and the competitive NMDA receptor antagonist D-CPPene on the development of amygdala kindling, and on amygdala-kindled seizures. *Epilepsy Res.* **17**: 167–174.
Fisher RS. (1989) Animal models of the epilepsies. *Brain Res. Rev.* **14**: 245–278.
Frenguelli BG, Potier B, Slater NT, Alford S, Collingridge GL. (1993) Metabotopic glutamate receptors and calcium signalling in dendrites of hippocampal CA1 neurones. *Neuropharmacology* **32**: 1229–1237.
Frey U, Krug M, Reymann KG, Matties H. (1988) Anisomycin, an inhibitor of protein synthesis, blocks the late phases of LTP phenomena in the hippocampal CA1 region *in vitro*. *Brain Res.* **452**: 57–65.
Fukunaga K, Soderling T, Miyamoto E. (1992) Activation of Ca^{2+}/calmodulin-dependent protein kinase II and protein kinase C by glutamate in cultured rat hippocampal neurons. *J. Biol. Chem.* **267**: 22527–22533.
Fukunaga K, Stoppini L, Miyamoto E, Muller D. (1993) Long-term potentiation is associated with an increased activity of Ca^{2+}/calmodulin-dependent protein kinase II. *J. Biol. Chem.* **268**: 7863–7867.
Gean PW, Shinnick-Gallagher P, Anderson AC. (1989) Spontaneous epileptiform activity, alteration of GABA- and NMDA-mediated neurotransmission in amygdala neuron kindled *in vivo*. *Brain Res.* **494**: 171–181.
Giacchino JL, Somjen GG, Frush DP, McNamara JO. (1984) Lateral entorhinal cortex kindling can be established without potentiation of the entorhinal–granule cell synapse. *Exp. Neurol.* **86**: 483–492.
Gilbert ME. (1988) The NMDA receptor antagonist, MK-801, suppresses limbic kindling and kindled seizures. *Brain Res.* **463**: 90–99.
Giorgi O, Orlandi M, Lecca D, Corda MG. (1991) MK-801 prevents chemical kindling induced by pentylenetetrazol in rats. *Eur. J. Pharmacol.* **193**: 363–365.
Goddard GV, McIntrye DC, Leech CK. (1969) A permanent change in brain function resulting from daily electrical stimulation. *Exp. Neurol.* **25**: 75–86.
Goldenring JR, Wasterlain CG, Oestreicher AB, de Graan PNE, Farber DB, Glaser G, DeLorenzo RJ. (1986) Kindling induces a long-lasting change in the activity of a hippocampal membrane calmodulin-dependent protein kinase system. *Brain Res.* **377**: 47–53.
Higashima M. (1988) Inhibitory processes in development of seizure activity in hippocampal slices. *Exp. Brain Res.* **72**: 37–44.
Hikiji M, Tomita H, Ono M, Fujiwara Y, Akiyama K. (1993) Increase of kainate receptor mRNA in the hippocampal CA3 of amygdala-kindled rats detected by *in situ* hybridization. *Life Sci.* **53**: 857–864.
Hirayama K, Murata R, Matsuura S. (1994) Effects of an *N*-methyl-D-aspartate antagonist and a GABAergic antagonist on entorhinal tetanic responses during the early stages of amygdala kindling in rats. *Neurosci. Res.* **19**: 397–405.
Hoffman WH, Haberly LB. (1989) Bursting induces persistent all-or-none EPSPs by an NMDA-dependent process in piriform cortex. *J. Neurosci.* **9**: 206–215.
Holmes GL, Thompson JL, Carl GF, Gallagher BS, Hoy J, McLaughlin M. (1990a) Effect of 2-amino-7-phosphonoheptanoic acid (APH) on seizure susceptibility in the prepubescent and mature rat. *Epilepsy Res.* **5**: 125–130.

Holmes KH, Goddard GV. (1986) A role for the *N*-methyl-D-aspartate receptor in kindling. *Proc. Univ. Otago Med. Sch.* **64**: 37–38.

Holmes KH, Bilkey DK, Laverty R, Goddard GV. (1990b) The *N*-methyl-D-aspartate antagonists aminophosphonovalerate and carboxypiperizine phosphonate retard the development and expression of seizures. *Brain Res.* **506**: 227–235.

Hughes P, Singleton K, Dragunow M. (1994) MK-801 does not attenuate immediate-early gene expression following an amygdala afterdischarge. *Exp. Neurol.* **128**: 276–283.

Ito I, Hidaka H, Sugiyama H. (1991) Effects of KN-62, a specific inhibitor of calcium/calmodulin-dependent protein kinase II, on long-term potentiation in the rat hippocampus. *Neurosci. Lett.* **121**: 119–121.

Jefferys JGR. (1990) Basic mechanisms of focal epilepsies. *Exp. Physiol.* **75**: 127–162.

Jonec V, Wasterlain CG. (1979) Effect of inhibitors of protein synthesis on the development of kindled seizures in rats. *Exp. Neurol.* **66**: 524–532.

Jones LS, Hamilton JH. (1993) Electrographic seizures (EGSs) in juvenile rat hippocampal slices in normal ACSF. *Soc. Neurosci. Abstr.* **19**: 1464.

Jones LS, Grooms SY, Lapadula DM, Lewis DV. (1992) Protein synthesis inhibition blocks maintenance but not induction of epileptogenesis in hippocampal slice. *Brain Res.* **599**: 338–344.

Kamphuis W, Wadman WJ, Buijs RM, Lopes da Silva FH. (1986) Decrease in number of hippocampal gamma-aminobutyric acid (GABA) immunoreactive cells in the rat kindling model of epilepsy. *Exp. Brain Res.* **64**: 491–495.

Kamphuis W, De Rijk TC, Lopes da Silva FH. (1994a) $GABA_A$ receptor β_{1-3} subunit gene expression in the hippocampus of kindled rats. *Neurosci. Lett.* **174**: 5–8.

Kamphuis W, De Rijk TC, Talamini LM, Lopes da Silva FH. (1994b) Rat hippocampal kindling induces changes in the glutamate receptor mRNA expression patterns in dentate granule neurons. *Eur. J. Neuroscience.* **6**: 1119–1127.

Kamphuis W, Hendriksen H, Diegenbach PC, Lopes da Silva FH. (1995) *N*-methyl-D-aspartate and kainate receptor gene expression in hippocampal and granular neurons in the kindling model of epileptogenesis. *Neuroscience.* **67**: 551–559.

Kapur J, Lothman EW. (1990) NMDA receptor activation mediates the loss of GABAergic inhibition induced by recurrent seizures. *Epilepsy Res.* **5**: 103–111.

Kohira I, Akiyama K, Daigen A, Otsuki S. (1992) Enduring increase in membrane-associated protein kinase C activity in the hippocampal-kindled rat. *Brain Res.* **593**: 82–88.

Köhr G, Mody I. (1994) Kindling increases *N*-methyl-D-aspartate potency at single *N*-methyl-D-aspartate channels in dentate granule cells. *Neuroscience* **62**: 975–981.

Kraus JE, Yeh GC, Bonhaus DW, Nadler JV, McNamara JO. (1994) Kindling induces the long-lasting expression of a novel population of NMDA receptors in hippocampal region CA3. *J. Neurosci.* **14**: 4196–4205.

Labiner DM, Butler LS, Cao Z, Hosford DA, Shin C, McNamara JO. (1993) Induction of c-*fos* mRNA by kindled seizures: complex relationship with neuronal burst firing. *J. Neurosci.* **13**: 744–751.

Lee S, Miskovsky J, Williamson J, Howells R, Devinsky O, Lothman E, Christakos S. (1994) Changes in glutamate receptor and proenkephalin gene expression after kindled seizures. *Mol. Brain Res.* **24**: 34–42.

Leidenheimer NJ, Mequikin SJ, Hahner LD, Whiting P, Harris RA. (1992) Activation of protein kinase C selectively inhibits the γ-aminobutyric acid a receptor: role of desensitization. *Mol. Pharmacol.* **41**: 1116–1123.

Lerea LS, Carlson NG, McNamara JO. (1995) *N*-methyl-D-aspartate receptors activate transcription of c-*fos* and NGFI-A by distinct phospholipase A_2-requiring intracellular signaling pathways. *Mol. Pharmacol.* **47**: 1119–1125.

Lewis DV, Anderson WW, Swartzwelder HS, Wilson WA. (1986) Zero-magnesium induced epileptiform activity in hippocampal slices: inhibition by an NMDA antagonist. *Soc Neurosci. Abstr.* **12**: 78.

Liu YB, Disterhoft JF, Slater NT. (1993) Activation of metabotropic glutamate induces long-term depression of GABAergic inhibition in hippocampus. *J. Neurophysiol.* **69**: 1000–1004.

Löscher W. (1993) Basic aspects of epilepsy. *Curr. Opin. Neurol. Neurosurg.* **6**: 223–232.

Lothman EW, Bertram EH. (1993) Epileptogenic effects of status epilepticus. *Epilepsia* **34**: 559–570.

Lothman EW, Williamson JM. (1994) Closely spaced recurrent hippocampal seizures elicit two types of heightened epileptogenesis: a rapidly developing, transient kindling and a slowly developing enduring kindling. *Brain Res.* **649**: 71–84.

Lothman EW, Hatelid JM, Zorumski CF, Conry JA, Moon PF, Perlin JB. (1985) Kindling with rapidly recurring hippocampal seizures. *Brain Res.* **360**: 83–91.

Lynch G, Larson J, Kelso S, Barrionuevo G, Schottler F. (1983) Intracellular injections of EGTA block induction of hippocampal long-term potentiation. *Nature* **305**: 719–721.

Malenka RC, Lancaster B, Zucker RS. (1992) Temporal limits on the rise in postsynaptic calcium required for the induction of long-term potentiation. *Neuron* **9**: 121–128.

Malinow R, Schulman H, Tsien RW. (1989) Inhibition of postsynaptic PKC or CaMKII blocks induction but not expression of LTP. *Science* **245**: 862–866.

Martin D, McNamara JO, Nadler JV. (1992) Kindling enhances sensitivity of CA3 hippocampal pyramidal cells to NMDA. *J. Neurosci.* **12**: 1928–1935.

Maru E, Goddard GV. (1987) Alteration in dentate neuronal activities associated with perforant path kindling. *Exp. Neurol.* **96**: 19–32.

McNamara JO, Russell RD, Rigsbee L, Bonhaus DW. (1988) Anticonvulsant and antiepileptogenic actions of MK-801 in the kindling and electroshock models. *Neuropharmacology.* **27**: 563–568.

Merlin LR, Wong RKS. (1993) Synaptic modifications accompanying epileptogenesis *in vitro*: long-term depression of GABA-mediated inhibition. *Brain Res.* **627**: 330–340.

Miles R, Wong RKS. (1987) Latent synaptic pathways revealed after tetanic stimulation in the hippocampus. *Nature* **329**: 724–726.

Miller SG, Kennedy MB. (1986) Regulation of brain type II Ca^{2+}/calmodulin-dependent protein kinase by autophosphorylation: a Ca^{2+}-triggered molecular switch. *Cell* **44**: 861–870.

Mody I, Heinemann U. (1987) NMDA receptors of dentate gyrus granule cells participate in synaptic transmission following kindling. *Nature* **326**: 701–704.

Mody I, Lambert JDC, Heinemann U. (1987) Low extracellular magnesium induces epileptiform activity and spreading depression in rat hippocampal slices. *J. Neurophysiol.* **57**: 869–888.

Moia LJMP, Matsui H, de Barros GAM, Tomizawa K, Miyamoto K, Kuwata Y, Tokuda M, Itano T, Hatase O. (1994) Immunosuppressants and calcineurin inhibitors, cyclosporin A and FK506, reversibly inhibit epileptogenesis in amygdaloid kindled rat. *Brain Res.* **648**: 337–341.

Morgan JI, Cohen DR, Hempstead JL, Curran T. (1987) Mapping patterns of c-*fos* expression in the central nervous system after seizure. *Science* **237**: 192–197.

Morimoto K, Sato M. (1992) NMDA receptor complex and kindling mechanisms. *Epilepsy Res. Suppl.* **9**: 297–305.

Morrell F. (1985) Secondary epileptogenesis in man. *Arch. Neurol.* **42**: 318–335.

Nowak L, Bregestovski P, Ascher P, Herbet A, Prochiantz A. (1984) Magnesium gates glutamate-activated channels in mouse central neurones. *Nature* **214**: 681–683.

O'Dell TJ, Kandel ER, Grant SGN. (1991) Long-term potentiation in the hippocampus is blocked by tyrosine kinase inhibitors. *Nature* **353**: 558–560.

Pelletier MR, Carlen PL. (1995) Persistent spontaneous epileptifrom discharges evoked by kindling-like stimulation in piriform cortex *in vitro*. *Soc. Neurosci. Abstr.* **21**: 2113.

Pierson M, Swann J. (1991) Sensitization to noise-mediated induction of seizure susceptibility by MK-801 and phencyclidine. *Brain Res.* **560**: 229–236.

Prince HK, Conn PJ, Blackstone CD, Huganir RL, Levey AI. (1995) Down-regulation of AMPA receptor subunit GluR2 in amygdaloid kindling. *J. Neurochem.* **64**: 462–465.

Racine RJ. (1972) Modification of seizure activity by electrical stimulation. II. Motor seizure. *Electroencephalogr. Clin. Neurophysiol.* **32**: 281–294.

Racine RJ, Milgram WN, Hafner S. (1983) Long-term potentiation phenomena in the rat limbic forebrain. *Brain Res.* **260**: 217–231.

Racine RJ, Moore KA, Evans C. (1991) Kindling-induced potentiation in the piriform cortex. *Brain Res.* **556**: 218–225.

Rainnie DG, Asprodini EK, Shinnick-Gallagher P. (1992) Kindling-induced long-lasting changes in excitatory and inhibitory transmission in the basolateral amygdala. *J. Neurophysiol.* **67**: 443–454.

Rogawski MA. (1992) The NMDA receptor, NMDA antagonists and epilepsy therapy. *Drugs* **44**: 279–292.

Rogawski MA, Porter RJ. (1990) Antiepileptic drugs: pharmacological mechanisms and clinical efficacy with consideration of promising developmental stage compounds. *Pharmacol. Rev.* **42**: 223–286.

Rosen AD, Frumin NV. (1979) Focal epileptogenesis after intracortical hemoglobin injection. *Exp. Neurol.* **66**: 277–284.

Schmutz M, Klebs K, Baltzer V. (1988) Inhibition or enhancement of kindling evolution by antiepileptics. *J. Neural Transm.* **72**: 245–257.

Schneidermann JH, Sterling CA, Luo R. (1994) Hippocampal plasticity following epileptiform bursting produced by $GABA_A$ antagonists. *Neuroscience.* **59**: 259–273.

Shafer SQ, Hauser WA, Annegers JF, Klass DW. (1988) EEG and other early predictors of epilepsy remission: a community study. *Epilepsia* **29**: 590–600.

Shin C, Rigsbee LC, McNamara JO. (1986) Anti-seizure and anti-epileptogenic effect of γ-vinyl γ-aminobutyric acid in amygdaloid kindling. *Brain Res.* **398**: 370–374.

Shin C, Tamaki Y, Wilson JT, Butler L, Sakaguchi T. (1992) NMDA-receptor mediated electrical epileptogenesis in the organotypic culture of rat hippocampus. *Brain Res.* **589**: 129–134.

Silva AJ, Stevens CF, Tonegawa S, Wang Y. (1992) Deficient hippocampal long-term potentiation in α-calcium-calmodulin kinase II mutant mice. *Science* **257**: 201–206.

Silver JM, Shin C, McNamara JO. (1991) Anti-epileptogenic effects of conventional anticonvulsants in the kindling model of epilepsy. *Ann. Neurol.* **29**: 356–363.

Slater NT, Stelzer A, Galvan M. (1985) Kindling-like stimulus patterns induce epileptiform discharges in the guinea pig *in vitro* hippocampus. *Neurosci. Lett.* **60**: 25–31.

Sloviter RS. (1994) The functional organization of the hippocampal dentate gyrus and its relevance to the pathogenesis of temporal lobe epilepsy. *Ann. Neurol.* **35**: 640–654.

Soderling TR. (1995) Calcium-dependent protein kinases in learning and memory. In: *Advances in Second Messenger and Phosphoprotein Research*, vol. 30 (ed. AR Means). Raven Press, New York, pp. 175–189.

Sombatti S, DeLorenzo RJ. (1995) Recurrent spontaneous seizure activity in hippocampal neuronal networks in culture. *J. Neurophysiol.* **73**: 1706–1711.

Stafstrom CE, Homes GL, Thompson JL. (1993) MK801 pretreatment reduces kainic acid-induced spontaneous seizures in prepubescent rats. *Epilepsy Res.* **14**: 41–48.

Stasheff SF, Bragdon AC, Wilson WA. (1985) Induction of epileptiform activity in hippocampal slices by trains of electrical stimuli. *Brain Res.* **344**: 296–302.

Stasheff S, Anderson WW, Clark S, Wilson WA. (1989) NMDA antagonists differentiate epileptogenesis from seizure expression in an *in vitro* seizure model. *Science* **245**: 648–651.

Stasheff SF, Hines M, Wilson WA. (1993a) Axon terminal hyperexcitability associated with epileptogenesis *in vitro*. I. Origin of ectopic spikes. *J. Neurophysiol.* **70**: 961–975.

Stasheff SF, Mott DD, Wilson WA. (1993b) Axon terminal hyperexcitability associated with epileptogenesis *in vitro*. II. Pharmacological regulation by NMDA and $GABA_A$ receptors. *J. Neurophysiol.* **70**: 976–984.

Stelzer A. (1992) $GABA_A$ receptors control the excitability of neuronal populations. *Int. Rev. Neurobiol.* **33**: 195–287.

Stelzer A, Shi H. (1994) Impairment of $GABA_A$ receptor function by *N*-methyl-D-aspartate-mediated calcium influx in isolated CA1 pyramidal cells. *Neuroscience.* **62**: 813–828.

Stelzer A, Slater NT, ten Bruggencate G. (1987) Activation of NMDA receptors blocks GABAergic inhibition in an *in vitro* model of epilepsy. *Nature* **326**: 698–701.

Stephens DN, Weidmann R. (1989) Blockade of FG 7142 kindling by anticonvulsants acting at sites distant from the benzodiazepine receptor. *Brain Res.* **492**: 89–98.

Sveinbjornsdottir M, Sander JWAS, Upton D, Thompson PJ, Patsalos PN, Hirt D, Emre M, Lowe D, Duncan JS. (1993) The excitatory amino acid antagonist D-CPP-ene (SDZ EAA-494) in patients with epilepsy. *Epilepsy Res.* **16**: 165–174.

Swartzwelder HS, Lewis DV, Anderson WW, Wilson WA. (1987) Seizure-like events in brain slices: suppression by spontaneous interictal spikes. *Brain Res.* **410**: 362–366.

Swartzwelder HS, Ferrari C, Anderson WW, Wilson WA. (1989) The drug MK-801 attenuates the development, but not the expression, of long-term potentiation and stimulus train-induced bursting in hippocampal slices. *Neuropharmacology* **28**: 441–445.
Traub RD, Jefferys JGR, Whittington MA. (1994) Enhanced NMDA conductance can account for epileptiform activity induced by low Mg^{2+} in the rat hippocampal slice. *J. Physiol.* **478**: 379–393.
Trommer BL, Pasternak JF. (1990) NMDA antagonists inhibit kindling epileptogenesis and seizure expression in developing rats. *Devel. Brain Res.* **53**: 248–252.
Valenzuela V, Benardo LS. (1995) An *in vitro* model of persistent epileptiform activity in neocortex. *Epilepsy Res.* **21**: 195–204.
Wada JA. (1977) Pharmacological prophylaxis in the kindling model of epilepsy. *Arch. Neurol.* **34**: 389–395.
Wada J, Sato M. (1974) Generalized convulsive seizures induced by daily electrical stimulation of the amygdala in cats. Correlative electrographic and behavioral features. *Neurology* **24**: 565–574.
Wadman WJ, Heinemann U, Konnerth A, Newhaus S. (1985) Hippocampal slices of kindled rats reveal calcium involvement of epileptogenesis. *Exp. Brain Res.* **57**: 404–407.
Wang JH, Feng DP. (1992) Postsynaptic protein-kinase-C essential to induction and maintenance of long-term potentiation in the hippocampal CA1 region. *Proc. Natl Acad. Sci. USA* **89**: 2576–2580.
Wang C, Jensen FE. (1996) Age dependence of NMDA receptor involvement in epileptiform activity in rat hippocampal slices. *Epilepsy Res.* **23**: 105–113.
Wasterlain CG, Farber DB. (1984) Kindling alters the calcium/calmodulin-dependent phosphorylation of synaptic plasma membrane proteins in rat hippocampus. *Proc. Natl. Acad. Sci. USA* **81**: 1253–1257.
Willmore LF. (1992) Post-traumatic epilepsy – mechanisms, prevention. In: *Recent Advances in Epilepsy* (eds TA Pedley, BS Meldrum). Churchill Livingstone, Edinburgh, pp. 107–117.
Willmore LF, Sypert GW, Munson JB, Hurd RW. (1978) Chronic focal epileptiform discharges induced by injection of iron into rat and cat cortex. *Science* **200**: 1501–1503.
Wilson WA, Stasheff S, Swartzwelder S, Clark S, Anderson WW, Lewis DV. (1989) The NMDA receptor in epilepsy. In: *The NMDA Receptor* (eds JC Wakins, GL Collingridge). Oxford University Press, Oxford, pp. 167–176.
Zheng F, Gallagher JP. (1992) Metabotropic glutamate receptors are required for the induction of long-term potentiation. *Neuron* **9**: 163–172.
Zhou XR, Suzuki T, Shimizu H, Nishino H. (1994) Amygdala kindling activates the phosphorylation of Ca^{2+}/calmodulin-dependent protein kinase II in rat hippocampus. *Neurosci. Lett.* **171**: 45–48.

9
Bidirectional plasticity of cortical synapses

M.F. Bear and A. Kirkwood

9.1 A problem and its theoretical solution

The cerebral cortex is diverse. There are many discrete areas that differ from one another with respect to cytoarchitecture, connectivity and function. Despite this diversity, however, many cortical areas have features in common. For sensory areas of cortex, one common feature is the existence of stimulus-selective neurons. For example, neurons in primary visual cortex show selectivity to particular stimulus attributes, such as which eye is stimulated or the orientation of a contrast border. Neurons in higher order visual areas show selectivity for more complex stimuli, such as faces. Selectivity in many cortical areas can be modified by experience. For example, selectivity of cells in primate inferotemporal cortex to familiar faces shifts as animals learn to recognize new faces. Lasting shifts in selectivity are believed to reflect synaptic changes which, distributed over the cells in the neural network, are the neural basis of memory storage (see Chapter 11). A question of extraordinary interest, therefore, is how synapses adjust their effectiveness to modify neuronal selectivity and store information.

Cortical neurons receive synaptic inputs, both excitatory and inhibitory, from many sources. It is the rich interplay between these inputs that gives rise to the functional properties of cortical neurons, including stimulus selectivity. Nonetheless, it is increasingly clear that the seed for selectivity is the pattern of convergence of excitatory synaptic inputs (Nelson *et al.*, 1994). Thus, it seems reasonable to ask how excitatory synapses might be modified by experience to alter stimulus selectivity.

Serious theoretical analysis of the problem of experience-dependent changes in neuronal selectivity began in the 1970s (Cooper, 1973; Cooper *et al.*, 1979; Grossberg, 1976; Malsburg, 1973; Nass and Cooper, 1975). A number of 'learning rules' were proposed to account for experience-dependent development of stimulus selectivity, each making slightly different assumptions about how synapses

---

Cortical Plasticity, edited by M.S. Fazeli and G.L. Collingridge.
© 1996 BIOS Scientific Publishers Ltd, Oxford.

would modify during various combinations of presynaptic and postsynaptic activity. One such synaptic learning rule, from the work of Cooper and colleagues (Cooper, 1995; Cooper et al., 1979), is illustrated in *Figure 9.1a*. Cooper et al. considered a single neuron receiving an array of excitatory synapses carrying information about the sensory environment. In order to account for the development and plasticity of neuronal stimulus selectivity, they proposed that active synapses are potentiated when the total postsynaptic response exceeds a critical value, called the 'modification threshold' (Θ_m), and that active synapses are depressed when the total postsynaptic response is greater than zero (assumed to be the average 'spontaneous' level), but less than Θ_m. It was later shown by Bienenstock et al. (1982) that if the value of Θ_m was allowed to vary as a nonlinear function of the average integrated postsynaptic activity (*Figure 9.1b*), then the cell would evolve to a stable, selective state in a patterned input environment regardless of the initial condition.

The discovery of *N*-methyl-D-aspartate (NMDA) receptor involvement in hippocampal long-term potentiation (LTP) suggested a physiological basis for this

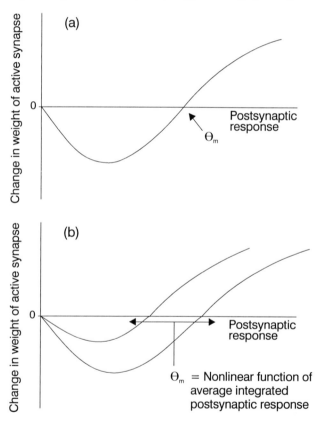

Figure 9.1. Synaptic 'learning rules' devised to account for the acquisition and experience-dependent modification of neuronal selectivity. (a) Function controlling synaptic plasticity at the Cooper synapse. (b) The Cooper synapse learning rule with the variable modification threshold of Bienenstock et al. (1982).

theoretical form of synapse modification. The proposal was made in 1987 that Θ_m corresponds to a critical level of Ca^{2+} entry through the postsynaptic NMDA receptor channel (Bear et al., 1987). Associating Θ_m with a critical level of NMDA receptor activation and postsynaptic Ca^{2+} entry led to two additional hypotheses: *first* that input activity which fails to activate NMDA receptors beyond the level required to trigger LTP should cause long-term depression (LTD) of the active synapses, and *second* that the stimulation threshold for producing synaptic potentiation should depend on the history of postsynaptic activity.

In this chapter, we review experimental studies suggesting that excitatory synapses in the hippocampus and neocortex modify according to principles remarkably consistent with those assumed by Bienenstock et al. (1982). Synapses are bidirectionally modifiable; these modifications can last long enough to contribute to a cortical memory store; and key variables controlling the sign and magnitude of the modifications are the level of NMDA receptor activation and the history of cortical activity.

9.2 Evidence for bidirectional synaptic plasticity in hippocampus

According to Cooper's proposals, synapses become weaker when their activity coincides with postsynaptic activity greater than a lower threshold (operationally defined as zero) and less than a higher threshold (Θ_m); synapses become stronger when their activity coincides with postsynaptic activity greater than Θ_m. The search for 'Cooper synapses' (Bear, 1996) began in the CA1 region of hippocampus, because here it was already well established that strong NMDA receptor activation triggers LTP. Experiments were designed to test the hypothesis that input activity which consistently fails to activate NMDA receptors sufficiently to trigger LTP produces LTD instead (Bear et al., 1987). It was discovered that prolonged low frequency stimulation (1–3 Hz) of the Schaffer collaterals did indeed trigger a long-lasting, homosynaptic LTD in CA1 (Dudek and Bear, 1992). By varying the stimulation frequency, but holding the total number of pulses constant, it was possible to derive a function relating synaptic plasticity to stimulation frequency. Little or no plasticity was observed at frequencies less than 0.1 Hz; robust LTD was observed using 1 Hz stimulation and LTP was observed using stimulation frequencies greater than 10 Hz. Interestingly, 10 Hz stimulation itself produced, in most cases, neither LTD nor LTP. The effects of varying stimulation frequency were interpreted as being due to different values of postsynaptic response [due to temporal summation of excitatory postsynaptic potentials (EPSPs)] during the conditioning stimulation. Indeed, it is now well established that the critical postsynaptic variable is postsynaptic depolarization (reviewed by Bear and Abraham, 1996). For example, while 1 Hz stimulation normally produces LTD, postsynaptic hyperpolarization during 1 Hz conditioning prevents any change, and depolarization leads to induction of LTP (Mulkey and Malenka, 1992). Thus, there exists a postsynaptic 'reversal potential' for synaptic plasticity.

Additional experiments have shown that the same synapses support both LTD and LTP, that is synapses are bidirectionally modifiable (Dudek and Bear, 1993; Heynen *et al.*, 1996; Mulkey *et al.*, 1993). Together, these data show that Cooper synapses exist in hippocampus.

LTD induced by low frequency stimulation in CA1 is blocked by NMDA receptor antagonists (Dudek and Bear, 1992; Heynen *et al.*, 1996; Mulkey and Malenka, 1992; reviewed by Bear and Abraham, 1996). Available data suggest that the information encoded by the pattern or amount of NMDA receptor activation may be sufficient to trigger both forms of synaptic plasticity. The feasibility of this type of regulation was shown in a model in which modest elevations of postsynaptic Ca^{2+} caused LTD by selectively activating protein phosphatases, and large increases in Ca^{2+} caused LTP by activating protein kinases (Artola and Singer, 1993; Lisman, 1989). Mulkey, Malenka and colleagues provided data that support this model in CA1 by showing that LTD is completely blocked by application of several phosphatase inhibitors (Mulkey *et al.*, 1993, 1994).

Forms of LTP and LTD have also been described in hippocampus that are independent of NMDA receptor activation, but they still apparently follow the general rules of the Cooper synapse. For example, in cultured hippocampal neurons, LTD is reduced but not eliminated by application of NMDA receptor antagonists (Goda and Stevens, 1996). Nonetheless, LTD is completely blocked by hyperpolarizing the postsynaptic neuron during conditioning stimulation, showing that the level of postsynaptic response (and presumably Ca^{2+} entry through voltage-gated channels) is a critical variable. Very recently, NMDA receptor-independent LTP and LTD have also been described in the mossy fiber synapses in CA3 (Derrick and Martinez, 1996). Again, however, the modifications appear to be a function of the level of postsynaptic response as assumed in the Cooper synapse learning rule (Reyes *et al.*, 1996).

9.3 Evidence for bidirectional synaptic plasticity in visual cortex

LTP has been relatively well studied in slices of visual cortex. In most cases, the LTP has exhibited properties similar to those found in the CA1 region of hippocampus (Artola and Singer, 1987; Kirkwood and Bear, 1994a; Lee, 1982; Yoshimura and Tsumoto, 1994; but see Komatsu *et al.*, 1991). Evidence for bidirectional plasticity was first provided by Artola *et al.* (1990). They presented evidence suggesting that active synapses are not modified if the level of postsynaptic activation during a high frequency tetanus is low, are depressed if the level of postsynaptic activation is moderate and are potentiated if the level of postsynaptic activation is high. Unlike LTD induced by low frequency stimulation in CA1, however, LTD was not blocked by the NMDA receptor antagonist 2-amino-5-phosphonopentanoate (AP5); in fact, a particularly reliable method for inducing LTD was a high-frequency tetanus in the presence of 25 µM AP5 (and 0.3 µM bicuculline). Very recent work in hippocampus suggests that the AP5 concentration in that study might have been insufficient to block NMDA receptor activation during a strong tetanus (Cummings *et al.*, 1996). In any case, irrespective of the involvement of NMDA receptors in LTD induction, the modifications observed are again consistent with the principles of the Cooper synapse learning rule.

By stimulating layer IV, we have found that synaptic LTP and LTD can be induced in layer III with precisely the same types of stimulation protocols that are effective in CA1 (Kirkwood and Bear, 1994a, b; Kirkwood *et al.*, 1993). Specifically, low frequency stimulation produces LTD and high frequency stimulation produces LTP (*Figure 9.2*). As for hippocampus, neocortical LTP and LTD are specific to the conditioned pathway and are dependent upon activation of NMDA receptors. Furthermore, homosynaptic LTD in visual cortex, like in hippocampus, is blocked by phosphatase inhibitors (Kirkwood and Bear, 1994b; Torii *et al.*, 1995). Using this approach (stimulating layer IV, recording from layer III), we have obtained similar results using slices of mouse, rat and cat visual cortex, prepared from both neonatal and adult animals.

In guinea pig visual cortex, it has been shown that low frequency electrical stimulation of white matter paired with postsynaptic depolarizing or hyperpolarizing current injections can produce transient synaptic enhancements and decrements, respectively (Frégnac *et al.*, 1994). Very recently, by recording from synaptically coupled layer V neurons, Markram and Sakmann beautifully confirmed that neocortical synapses are bidirectionally modifiable, with a critical variable being the pattern or amount of NMDA receptor activation (Markram and Sakmann, 1995). Thus, synapses in superficial and deep layers of visual cortex from many mammalian species modify in a manner consistent with the Cooper synapse learning rule.

9.4 Evidence for bidirectional synaptic plasticity in neocortical areas other than visual cortex

Neurons in the primate inferior temporal cortex show selective responses to faces that are altered as the animal learns to recognize new faces (Rolls *et al.*, 1989). In humans, lesions of this region of cortex lead to an impaired recognition of familiar faces. Are there Cooper synapses in the neocortex of the primate temporal lobe? This question was addressed recently in slices prepared from surgically resected human inferior temporal cortex (Chen *et al.*, 1996). LTP and LTD were evoked in layer III with the same types of layer IV stimulation that are effective in rodent hippocampus and visual cortex (*Figure 9.3*). Specifically, high frequency stimulation produced NMDA receptor-dependent LTP and low frequency stimulation produced LTD that was partially (but not completely) blocked by AP5. The authors suggest that these mechanisms, which follow the Cooper synapse learning rule, are responsible for visual learning and memory in the human temporal cortex.

In the adult somatic sensory cortex, neuronal receptive fields can be altered by a large number of manipulations of the periphery (Donoghue, 1995). For example, simply trimming all but two whiskers of a rat's face will lead to an enhancement

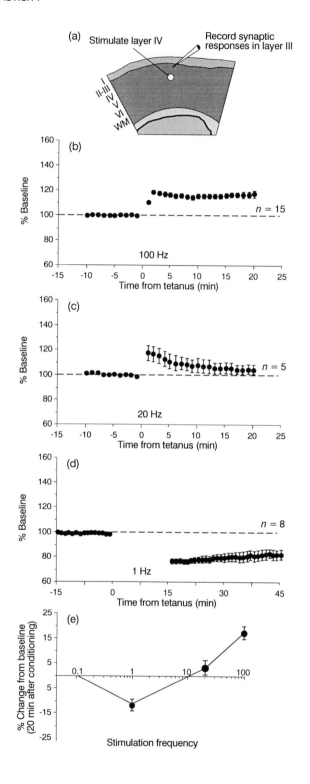

of the cortical responses to the intact whiskers, and a depression of the responses to the trimmed whiskers (Diamond et al., 1993). Computational analysis has shown that these results can be explained by Cooper synapses with a variable modification threshold (Benuskova et al., 1994), and experimental studies confirm that Cooper synapses do exist in somatic sensory cortex. By stimulating layer IV and recording in layer III, the same forms of NMDA receptor-dependent LTP and LTD have been found in adult somatic sensory cortex as have been observed in hippocampus and visual cortex (Castro-Alamancos et al., 1995).

The data from hippocampus, visual cortex, temporal association cortex and somatic sensory cortex suggest the existence of a common set of principles for activity-dependent plasticity of excitatory synapses. These principles appear to apply to motor cortex as well, but with additional constraints imposed by the intracortical circuitry of agranular cortex. While low-frequency stimulation of the middle cortical layers of rat motor cortex does result in NMDA-receptor-dependent LTD in layer III, high frequency stimulation does not reliably elicit LTP (Castro-Alamancos et al., 1995). Similarly, low frequency stimulation reliably produces LTD in the horizontal connections within layer III (Hess and Donoghue, 1996), but little LTP is produced by high-frequency stimulation. However, NMDA receptor-dependent LTP in both vertical and horizontal pathways can be induced by high frequency stimulation if γ-aminobutyric acid $(GABA)_A$ receptors are partially blocked with bicuculline (Hess et al., 1996). A study by Castro-Alamancos and Connors (1996) indicates that facilitation of NMDA receptor-dependent postsynaptic currents normally is required during conditioning to trigger LTP; in motor cortex this usually does not occur with tetanic electrical stimulation because of the strength of inhibitory circuits. Constraints on patterns of activation imposed by intracortical circuits also may contribute to developmental changes in synaptic plasticity in the neocortex, as we discuss below.

9.5 Evidence for a sliding modification threshold

With the addition of a variable modification threshold proposed by Bienenstock et al. (1982), Cooper synapses can account for many aspects of naturally occurring receptive field plasticity. Recently, the issue of a sliding modification threshold was approached by comparing the frequency–response function in visual cortex of animals reared in complete darkness with that of normally reared animals (Kirkwood et al., 1996). In accordance with theoretical predictions, it was found in visual cortex

---

Figure 9.2. Bidirectional plasticity of synaptic responses in adult rat visual cortex. (a) The stimulation–recording arrangement normally used to study LTP and LTD in the superficial layers of visual cortex. (b) After collecting baseline synaptic responses (at 0.07 Hz), a 100-Hz tetanus was delivered (120 pulses total). Resumption of baseline stimulation reveals LTP of the synaptic responses. (c) A 20-Hz tetanus (120 pulses total) yields on average little change in the magnitude of the synaptic responses. (d) 1-Hz stimulation yields stable LTD. (e) The frequency–response function describing the data in b–d.

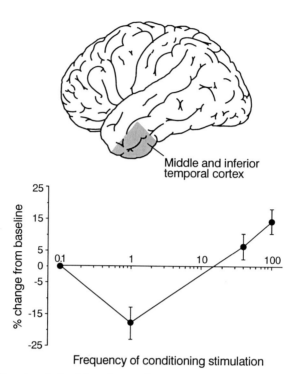

Figure 9.3. Bidirectional plasticity of synaptic responses in adult human inferior temporal cortex. The frequency–response function was derived in a manner similar to that shown in *Figure 9.2*, using slices of human inferior temporal cortex resected during surgery. These data were replotted from Chen *et al.* (1996).

of light-deprived rats that LTP is enhanced, and LTD is diminished, over a range of stimulation frequencies (*Figure 9.4*). Control experiments suggested that the alteration in synaptic plasticity was restricted to visual cortex, as similar changes were not observed in hippocampus. These findings support the concept that Θ_m is set according to the activation history of the cortex.

As a test of the hypothesis that the value of Θ_m actually adjusts to a change in cortical activity, visually deprived rats were exposed to light for various times, and the effects of low frequency stimulation were investigated. It was found that the magnitude of LTD in light-deprived visual cortex returned nearly to control levels after only 2 days of light exposure (Kirkwood *et al.*, 1996). These data are consistent with the hypothesis that Θ_m 'slides' as average cortical activity increases. It is of interest to note that the time course of the observed change in visual cortical LTD closely corresponds to that predicted for Θ_m in modeling studies of visual cortical plasticity (Clothiaux *et al.*, 1991).

The apparently slow rate of change for Θ_m has interesting consequences. First, it suggests that under natural conditions the magnitude of a synaptic change occurring over a period of seconds (the 'step size') is quite small. Large, rapid

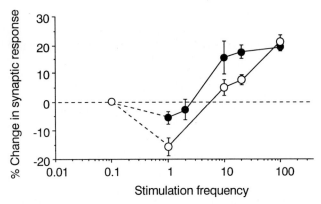

Figure 9.4. Evidence for a sliding modification threshold. Frequency–response functions derived from visual cortex of light-deprived (filled symbols) and normal (open symbols) rats. Figure derived from Kirkwood *et al.* (1996).

synaptic modifications would require a faster moving Θ_m to avoid stability problems, for example to prevent run-away potentiation and saturation of synaptic weights. Thus, the commonly used experimental model of LTP, while very useful for understanding the principles and mechanisms of synaptic plasticity, may be a gross exaggeration of the changes that occur naturally. This may explain why changes equivalent to the magnitude of experimental LTP have not been observed in behaving animals during learning (Moser *et al.*, 1994). The slow time course of Θ_m adjustment, and the fact that it has the same value at all synapses on the same neuron, also places constraints on the possible mechanisms. Activity-dependent regulation of gene expression in the postsynaptic neuron seems ideally suited to account for this form of cellular memory.

9.6 Developmental regulation of LTP and LTD in hippocampus and neocortex

It has been recognized for some time that the mechanisms of NMDA receptor-dependent LTP would be useful for the activity-dependent acquisition of neuronal stimulus selectivity during neonatal and early postnatal life (Artola and Singer, 1987; Bear *et al.*, 1987; Constantine-Paton *et al.*, 1990). Seemingly against this hypothesis, it has been observed repeatedly that little LTP is produced by a high frequency tetanus in the hippocampus from animals less than 2 weeks of age (Dudek and Bear, 1993; Harris and Teyler, 1984; Muller *et al.*, 1989). However, the failure to observe LTP in these studies very likely was due to inadequacies of the stimulation protocol, rather than to the absence of the mechanisms underlying LTP *per se*. Recently, it has been shown that many synapses in neonatal hippocampus have functional NMDA receptors, but no α-amino-3-hydroxy-5-methyl-4-isoxazole proprionic acid (AMPA) receptors (Durand *et al.*, 1996).

Without the contribution of AMPA receptors, tetanic stimulation evidently does not produce sufficient depolarization to cause LTP. However, pairing synaptic activation with intracellular depolarizing current injection does produce LTP in very young hippocampus. Interestingly, the LTP is expressed as the appearance of functional AMPA receptors at the stimulated synapses (Durand *et al.*, 1996; Issac *et al.*, 1995; Liao *et al.*, 1995).

In the neonatal neocortex, LTP not only exists, it apparently is even more robust in some cells than it is in adults. The thalamocortical input to neocortex terminates in layer IV, and it is possible to study synaptic transmission in this projection using a novel slice preparation of rodent somatic sensory thalamus and cortex (Agmon and Connors, 1991). LTP can be elicited in this preparation by pairing stimulation of the thalamus with intracellular depolarizing current injection into layer IV neurons, but only during a critical period of early postnatal development. Layer IV LTP is robust at postnatal day 3 but gradually disappears by about 7 days of age. The decline in LTP correlates with a progressive decrease in the time course of NMDA receptor synaptic currents (Crair and Malenka, 1995).

Critical periods are also characteristic of naturally occurring experience-dependent modifications of cellular response properties in sensory neocortex. The most famous example comes from the work of Hubel and Wiesel (1970) in the visual cortex. They found that visual cortical receptive fields could be modified by manipulations of binocular visual experience during a finite period of postnatal development. It was found later that this critical period could be delayed by rearing animals in complete darkness (Cynader and Mitchel, 1980; Mower, 1991). LTP evoked in layer III by stimulation of the white matter is also much more robust during a critical period (Kato *et al.*, 1991) that can be delayed by dark-rearing (Kirkwood *et al.*, 1995; *Figure 9.5a–c*). Interestingly, LTP evoked in layer III by stimulation of *layer IV* is largely unaffected by age or visual experience (*Figure 9.5d–f*). The different effectiveness of stimulating layer IV and white matter in adults may be explained by the maturation of inhibitory circuits within the cortex, because LTP evoked from white matter can be 'rescued' in adults by bathing slices in low concentrations of $GABA_A$ receptor antagonists (Artola and Singer, 1987; Bear *et al.*, 1992; Kirkwood and Bear, 1994a).

To summarize the data from sensory neocortex, it appears that the mechanism for LTP in layer IV declines during early postnatal life as a result of a change in NMDA receptor properties. In contrast, the mechanism for LTP in the superficial layers is present at all ages, but maturation of inhibition (within or deep to layer IV) during development places increasing constraints on the types of activity that can gain access to it.

Developmental changes in cortical LTD are as dramatic as those in LTP, if not more so. The magnitude of homosynaptic LTD in the CA1 region of hippocampus is much greater in neonates than it is in adults (Dudek and Bear, 1993; Wagner and Alger, 1995). Similarly, LTD of synaptic responses in layer IV can be readily induced with low frequency stimulation of the white matter in visual cortex from young, but not old, animals (Dudek and Friedlander, 1996). Although LTD can be elicited in the superficial layers of visual cortex at all ages (Dudek and

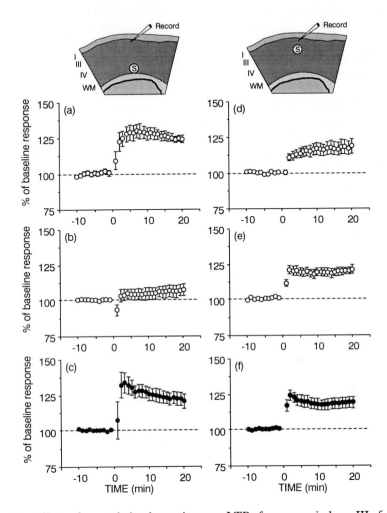

Figure 9.5. Effects of age and visual experience on LTP of responses in layer III of rat visual cortex. (a) Robust LTP may be evoked by high frequency stimulation of the white matter (WM) in visual cortex of rats in the third postnatal week. (b) By the sixth postnatal week, little LTP can be elicited in layer III after stimulation of the WM in normally reared rats. (c) LTP can be evoked using WM stimulation in rats ≥ 5 weeks of age if the animals are reared in complete darkness. (d) LTP evoked from layer IV in visual cortex of rats in the third postnatal week. (e) LTP evoked from layer IV in visual cortex of rats ≥ 5 weeks of age. (f) LTP evoked from layer IV in visual cortex of dark-reared rats ≥ 5 weeks of age.

Friedlander, 1996; Kirkwood and Bear, 1994b), it is also of greater magnitude in neonates (Kirkwood and Bear, unpublished observations).

Together, the data indicate that the mechanisms of bidirectional synaptic plasticity are present in the neonatal cerebral cortex at ages when neurons acquire stimulus selectivity. In most cases, enhanced naturally occurring synaptic plasticity during early postnatal life is accompanied by enhanced LTP and LTD.

9.7 Conclusions

In his lecture at the Clarke University decentennial in 1899, Cajal commented that "while there are very remarkable differences of organization in certain cortical areas, these points of difference do not go so far as to make impossible the reduction of cortical structure to a general plan". The data reviewed in this chapter suggest that, despite the diversity of cortex, there may also be a general plan for cortical synaptic plasticity. Studies have shown in many regions of the cerebral cortex – regions that include sensory, association and motor areas of neocortex as well as the CA1 and CA3 areas of hippocampus – that excitatory synapses are bidirectionally modifiable. In the large majority of cases, the modifications depend upon activation of NMDA receptors but, even when they do not, they conform to the basic principles of the Cooper synapse learning rule.

The Cooper synapse was introduced to account for the experience-dependent modification of neuronal stimulus selectivity. Are the receptive fields of cortical neurons actually modified by this rule and the mechanisms of LTP and LTD? It is possible to manipulate receptive field properties in sensory cortex by pairing sensory stimulation with various imposed patterns of postsynaptic activity (Cruikshank and Weinberger, 1996; Frégnac et al., 1988, 1992). The modifications of receptive fields observed in these studies conform well with theoretical predictions. The challenge now is to see if these changes, and those that actually underlie learning, employ the same mechanisms as LTP and LTD.

Acknowledgments

This research is supported in part by the Howard Hughes Medical Institute, the National Science Foundation, the National Eye Institute and the Charles A. Dana Foundation. We thank Dr Anne Williamson for sharing with us the data on human inferior temporal cortex prior to publication, and Dr Arnold Heynen for helpful comments on the manuscript.

References

Agmon A, Connors BW. (1991) Thalamocortical responses of mouse somatosensory (barrel) cortex *in vitro*. *Neuroscience* **41**: 365–379.

Artola A, Brocher S, Singer W. (1990) Different voltage-dependent thresholds for inducing long-term depression and long-term potentiation in slices of rat visual cortex. *Nature* **347**: 69–72.

Artola A, Singer W. (1987) Long-term potentiation and NMDA receptors in rat visual cortex. *Nature* **330**: 649–652.

Artola A, Singer W. (1993) Long-term depression of excitatory synaptic transmission and its relationship to long-term potentiation. *Trends Neurosci.* **16**: 480–487.

Bear MF. (1996) A synaptic basis for memory storage in the cerebral cortex. *Proc. Natl Acad. Sci. USA* in press.

Bear MF, Abraham WC. (1996) Long-term depression in hippocampus. *Annu. Rev. Neurosci.* **19**: 437–462.

Bear MF, Cooper LN, Ebner FF. (1987) A physiological basis for a theory of synaptic modification. *Science* **237**: 42–48.

Bear MF, Press WA, Connors BW (1992) Long-term potentiation in slices of kitten visual cortex and the effects of NMDA receptor blockade. *J. Neurophysiol.* **67**: 1–11.

Benuskova L, Diamond ME, Ebner FF. (1994) Dynamic synaptic modification threshold: computational model of experience-dependent plasticity in adult rat barrel cortex. *Proc. Natl Acad. Sci. USA* **91**: 4791–4795.

Bienenstock EL, Cooper LN, Munro PW. (1982) Theory for the development of neuron selectivity: orientation specificity and binocular interaction in visual cortex. *J. Neurosci.* **2**: 32–48.

Castro-Alamancos MA, Connors BW. (1996) Short-term synaptic enhancement and long-term potentiation in neocortex. *Proc. Natl Acad. Sci. USA* **93**: 1335–1339.

Castro-Alamancos MA, Donoghue JP, Connors BW. (1995) Different forms of synaptic plasticity in somatosensory and motor areas of the neocortex. *J. Neurosci.* **15**: 5324–5333.

Chen WR, Lee S, Kato K, Spencer DD, Sheperd GM, Williamson A. (1996) Long-term modifications of synaptic efficacy in the human inferior and middle temporal cortex. *Proc. Natl Acad. Sci. USA* **93**: 8011–8015.

Clothiaux EE, Bear MF, Cooper LN. (1991) Synaptic plasticity in visual cortex: comparison of theory with experiment. *J. Neurophysiol.* **66**: 1785–1804.

Constantine-Paton M, Cline HT, Debski E. (1990) Patterned activity, synaptic convergence, and the NMDA receptor in developing visual pathways. *Annu. Rev. Neurosci.* **13**: 129–154.

Cooper LN. (1973) A possible organization for animal learning and memory. In: *Proceedings of the Nobel Symposium on Collective Properties of Physical Systems* (eds B Lindquist, S Lindquist). Academic Press, New York, pp. 252–264.

Cooper LN. (1995) *How We Learn; How We Remember: Toward an Understanding of Brain and Neural Systems. Selected Papers of Leon N Cooper.* World Scientific, London.

Cooper LN, Liberman F, Oja E. (1979) A theory for the acquisition and loss of neuron specificity in visual cortex. *Biol. Cybernet.* **33**: 9–28.

Crair MC, Malenka RC. (1995) A critical period for long-term potentiation at thalamocortical synapses. *Nature* **375**: 325–328.

Cruikshank SJ, Weinberger NM. (1996) Receptive-field plasticity in the adult auditory cortex induced by Hebbian covariance. *J. Neurosci.* **16**: 861–875.

Cummings JA, Mulkey RM, Nicoll RA, Malenka RC. (1996) Ca^{2+} signalling requirements for long-term depression in the hippocampus. *Neuron* **16**: 825–833.

Cynader M, Mitchel DE. (1980) Prolonged sensitivity to monocular deprivation in dark-reared cats. *J. Neurophysiol.* **43**: 1026–1039.

Derrick BE, Martinez JL. (1996) Associative, bidirectional modifications (LTP and LTD) at the hippocampal mossy fiber–CA3 synapse. *Nature* **381**: 429–434.

Diamond ME, Armstrong-James M, Ebner FF. (1993) Experience-dependent plasticity in the adult rat barrel cortex. *Proc. Natl Acad. Sci. USA* **90**: 2082–2086.

Donoghue JP. (1995) Plasticity of adult sensorimotor representations. *Curr. Opin. Neurobiol.* **5**: 749–754.

Dudek SM, Bear MF. (1992) Homosynaptic long-term depression in area CA1 of hippocampus and effects of N-methyl-D-aspartate receptor blockade. *Proc. Natl Acad. Sci. USA* **89**: 4363–4367.

Dudek SM, Bear MF. (1993) Bidirectional long-term modification of synaptic effectiveness in the adult and immature hippocampus. *J. Neurosci.* **13**: 2910–2918.

Dudek SM, Friedlander MJ. (1996) Developmental down-regulation of LTD in cortical layer IV and its independence of modulation by inhibition. *Neuron* **16**: 1–20.

Durand GM, Kovalchuk Y, Konnerth A. (1996) Long-term potentiation and functional synapse induction in developing hippocampus. *Nature* **381**: 71–75.

Frégnac Y, Shultz D, Thorpe S, Bienenstock E. (1988) A cellular analogue of visual cortical plasticity. *Nature* **333**: 367–370.

Frégnac Y, Shulz D, Thorpe S, Bienenstock E. (1992) Cellular analogs of visual cortical epigenesis: I. Plasticity of orientation selectivity. *J. Neurosci.* **12**: 1280–1300.

Frégnac Y, Burke JP, Smith D, Friedlander MJ. (1994) Temporal covariance of pre- and postsynaptic activity regulates functional connectivity in the visual cortex. *J. Neurophysiol.* **71**: 1403–1421.

Goda Y, Stevens CF. (1996) Long-term depression properties in a simple system. *Neuron* **16**: 103–111.

Grossberg S. (1976) Adaptive pattern classification and universal recoding: I. Parallel development and coding of neural feature detectors. *Biol. Cybernetics* **23**: 121–134.

Harris KM, Teyler TJ. (1984) Developmental onset of long-term potentiation in area CA1 of the rat hippocampus. *J. Physiol.* **346**: 27–48.

Hess G, Donoghue JP. (1996) Long-term depression of horizontal connections in rat motor cortex. *Eur. J. Neurosci.* **8**: 658–665.

Hess G, Aizenman CD, Donoghue JP. (1996) Conditions for the induction of long-term potentiation in layer II/III horizontal connections of the rat motor cortex. *J. Neurophysiol.* **75**: 1765–1777.

Heynen AJ, Abraham WC, Bear MF. (1996) Bidirectional modification of CA1 synapses in the adult hippocampus *in vivo*. *Nature* **381**: 163–166.

Hubel DH, Wiesel TN. (1970) The period of susceptibility to the physiological effects of unilateral eye closure in kittens. *J. Physiol.* **206**: 419–436.

Issac JTR, Nicoll RA, Malenka RC. (1995) Evidence for silent synapses: implications for the expression of LTP and LTD. *Neuron* **15**: 427–434.

Kato N, Artola A, Singer W. (1991) Developmental changes in the susceptibility to long-term potentiation of neurones in rat visual cortex slices. *Devel. Brain Res.* **60**: 43–50.

Kirkwood A, Bear MF. (1994a) Hebbian synapses in visual cortex. *J. Neurosci.* **14**: 1634–1645.

Kirkwood A, Bear MF. (1994b) Homosynaptic long-term depression in the visual cortex. *J. Neurosci.* **14**: 3404–3412.

Kirkwood A, Dudek SM, Gold JT, Aizenman CD, Bear MF. (1993) Common forms of synaptic plasticity in the hippocampus and neocortex in vitro. *Science* **260**: 1518–1521.

Kirkwood A, Lee H-K, Bear MF. (1995) Co-regulation of long-term potentiation and experience-dependent plasticity in visual cortex by age and experience. *Nature* **375**: 328–331.

Kirkwood A, Rioult MG, Bear MF. (1996) Experience-dependent modification of synaptic plasticity in visual cortex. *Nature* **381**: 526–528.

Komatsu Y, Nakajima S, Toyama K. (1991) Induction of long-term potentiation without the participation of N-methyl-D-aspartate receptors in kitten visual cortex. *J. Neurophysiol.* **65**: 20–32.

Lee KS. (1982) Sustained enhancement of evoked potentials following brief, high-frequency stimulation of the cerebral cortex *in vitro*. *Brain Res.* **239**: 617–623.

Liao D, Hessler NA, Malinow R. (1995) Activation of postsynaptically silent synapses during pairing-induced LTP in CA1 region of hippocampal slice. *Nature* **375**: 400–404.

Lisman J. (1989) A mechanism for the Hebb and the anti-Hebb processes underlying learning and memory. *Proc. Natl Acad. Sci. USA* **86**: 9574–9578.

Malsburg CVD. (1973) Self-organization of orientation sensitive cells in the striata cortex. *Kybernetik* **14**: 85–100.

Markram H, Sakmann B. (1995) Action potentials propagating back into dendrites trigger changes in efficacy of single-axon synapses between layer V pyramidal cells. *Soc. Neurosci. Abstr.* **21**: 2007.

Moser E, Moser M-B, Andersen P. (1994) Potentiation of dentate synapses initiated by exploratory learning in rats: dissociation from brain temperature, motor activity, and arousal. *Learn. Mem.* **1**: 53–73.

Mower GD. (1991) The effect of dark rearing on the time course of the critical period in cat visual cortex. *Devel. Brain Res.* **58**: 151–158.

Mulkey RM, Malenka RC. (1992) Mechanisms underlying induction of homosynaptic long-term depression in area CA1 of the hippocampus. *Neuron* **9**: 967–975.

Mulkey RM, Herron CE, Malenka RC. (1993) An essential role for protein phosphatases in hippocampal long-term depression. *Science* **261**: 1051–1055.

Mulkey RM, Endo S, Shenolikar S, Malenka RC. (1994) Calcineurin and inhibitor-1 are components of a protein-phosphatase cascade mediating hippocampal LTD. *Nature* **369**: 486–488.
Muller D, Oliver M, Lynch G. (1989) Developmental changes in synaptic properties in hippocampus of neonatal rats. *Devel. Brain Res.* **49**: 105–114.
Nass MM, Cooper LN. (1975) A theory for the development of feature detecting cells in visual cortex. *Biol. Cybernetics* **19**: 1–18.
Nelson S, Toth L, Sheth B, Sur M. (1994) Orientation selectivity of cortical neurons during intracellular blockade of inhibition. *Science* **265**: 774–777.
Reyes JA, Derrick BE, Martinez JL. (1996) Intensity-dependent bidirectional plasticity at the mossy fiber–CA3 synapse. *Soc. Neurosci. Abstr.* in press.
Rolls ET, Baylis GC, Hasselmo ME, Nalwa V. (1989) The effect of learning on the face-selective responses of neurons in the cortex in the superior temporal sulcus of the monkey. *Exp. Brain Res.* **76**: 153–164.
Torii N, Kamishita T, Otsu Y, Tsumoto T. (1995) An inhibitor for calcineurin, FK506, blocks induction of long-term depression in rat visual cortex. *Neurosci. Lett.* **185**: 1–4.
Wagner JJ, Alger BE. (1995) GABAergic and developmental influences on homosynaptic LTD and depotentiation in rat hippocamus. *J. Neurosci.* **15**: 1577–1586.
Yoshimura Y, Tsumoto T. (1994) Dependence of LTP induction on postsynaptic depolarization: a perforated patch–clamp study in visual cortical slices of young rats. *J. Neurophysiol.* **71**: 1638–1645.

10

Synaptic plasticity in the cerebellum

Hervé Daniel, Olivier Blond, Danielle Jaillard and Francis Crepel

10.1 Introduction

The cerebellar cortex has a highly organized architecture. Purkinje cells, which are γ-aminobutyric acid (GABA)ergic inhibitory neurons, represent the single output of the cerebellar cortex. Each of these cells receives two excitatory synaptic inputs displaying distinct characteristics. Single climbing fiber axons, arising from neurons in the inferior olive, establish a very powerful one-to-one synaptic input with Purkinje cells, while about 80 000 parallel fibers, which originate from granule cells, converge on a single Purkinje cell, to form the second input (*Figure 10.1a*) (Ito, 1984).

The participation of the cerebellum in motor learning was postulated by Brindley as early as 1964, and then formalized by Marr (1969) and Albus (1971) in theoretical studies on memory. Taking into account the dual arrangement of the excitatory synaptic inputs, it was proposed that, during motor learning, activity in climbing fibers might serve as an 'external teacher' of synaptic transmission between parallel fibers and Purkinje cells, and thus adjust the cerebellar output to the adapted motor command. In this scheme, repeated coincident activation of climbing fibers and parallel fibers leads to a long-term depression (LTD) of the parallel fiber synaptic inputs.

Direct experimental support for the Marr–Albus theory of motor learning was provided by Ito and co-workers in an initial series of *in vivo* experiments (Ito *et al.*, 1982; see also Ito, 1987, 1989). In rabbit cerebellum, conjunctive stimulation of parallel and climbing fibers leads to LTD of synaptic transmission at parallel fiber–Purkinje cell synapses; whereas activation of climbing fibers alone, or of parallel fibers alone, has no long-term effect. In addition, this change is input specific, since co-activation of climbing fibers and of a group of parallel fibers induces LTD restricted to the activated parallel fiber–Purkinje cell synapses (Ito *et al.*, 1982).

With the experimental advantages of *in vitro* brain slices and culture preparations, cerebellar LTD constitutes a simple and very powerful model of synaptic

---

Cortical Plasticity, edited by M.S. Fazeli and G.L. Collingridge.
© 1996 BIOS Scientific Publishers Ltd, Oxford.

plasticity to further understand biochemical processes involved in long-term changes in synaptic efficacy in the central nervous system. Moreover, parallel fibers and climbing fibers are likely to use an excitatory amino acid 'glutamate', as neurotransmitter (Crepel and Audinat, 1991). Thus, the cerebellum is an interesting region of the brain to study activity-dependent synaptic changes confined to excitatory synapses. Therefore, over the last decade, studies aimed at elucidating the molecular mechanisms that underlie synaptic plasticity in the cerebellum, have outlined a series of events that include rises in intracellular Ca^{2+}, glutamate binding to specific postsynaptic receptors and activation of several second messenger cascades (*Figure 10.1a*).

10.2 Postsynaptic calcium increase is required for LTD induction

It is well known that each climbing fiber possesses multiple synaptic contacts with each Purkinje cell. Climbing fiber activation can potently provide a postsynaptic depolarization of Purkinje cells resulting in the activation of voltage-gated Ca^{2+} channels in the dendrites, and consequently in a rise in postsynaptic calcium level (Konnerth *et al.*, 1992; Miyakawa *et al.*, 1992). Thus, calcium fluorometry studies reveal a large and transient rise in internal Ca^{2+} in Purkinje cell dendrites after climbing fiber activation (Konnerth *et al.*, 1992). In addition, a local rise in postsynaptic Ca^{2+} concentration restricted to the dendritic compartment (spines and adjacent fine dendrites) has been demonstrated recently by calcium imaging studies after activation of at least 20–30 parallel fibers (Ellers *et al.*, 1995).

Initially reported by Crepel and Krupa (1988), the critical role of the postsynaptic Ca^{2+} rise in LTD induction is now largely supported by recent reports. Thus, it has been shown that substitution of climbing fiber stimulation by direct depolarization of Purkinje cells giving rise to calcium spike firing, during a pairing protocol with parallel fiber stimulation (Crepel and Jaillard, 1991; Crepel and Krupa, 1988; Daniel *et al.*, 1992; Glaum *et al.*, 1992; Hirano, 1990; Konnerth *et al.*,

---

Figure 10.1. (a) Simplified neural circuit showing the convergence of the two major inputs, the climbing fiber and the parallel fibers, on to a Purkinje cell. The lower panel shows a more detailed view of the presynaptic parallel fiber and climbing fiber and the postsynaptic voltage-dependent Ca^{2+} channels (VDCC), AMPA receptor (AMPAR) and metabotropic glutamate receptor (mGluR) of the Purkinje cell. (b) Involvement of postsynaptic calcium increase in LTD induction (a diagram of experimental arrangement is shown on the left of the panel). An example of LTD in a cerebellar slice, induced by pairing (P) EPSCs obtained by stimulation of parallel fibers, with depolarization of Purkinje cell giving rise to Ca^{2+} spikes (duration =1 min). Plot of EPSC amplitudes against time before and after the pairing. Inset represents superimposed averaged EPSCs recorded during the pre-pairing period (1) and 15 min after the end of the pairing. Adapted from Hemart *et al.* (1994) Properties of glutamate receptors are modified during long-term depression in cerebellar Purkinje cells. *Neurosci. Res.* **19**: 213–221, with permission from Elsevier Science Ireland Ltd.

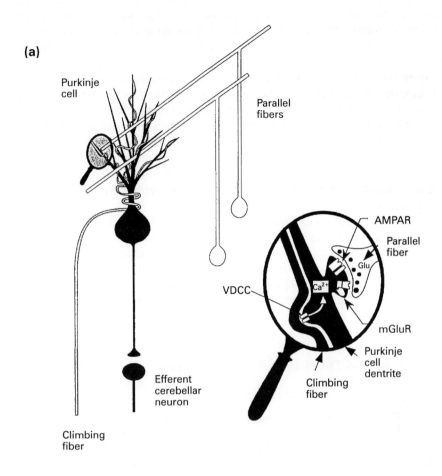

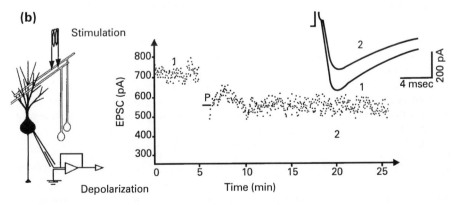

1992; Linden et al., 1991), is sufficient to induce LTD (*Figure 10.1b*). Moreover, similar results were obtained when climbing fiber activation or Purkinje cell depolarization was replaced by a direct rise in internal postsynaptic calcium concentration, following photolysis of calcium-loaded Nitr-5 (Kasono and Hirano, 1994).

The second direct line of evidence supporting a role for postsynaptic Ca^{2+} came from experiments in which a decrease in postsynaptic Ca^{2+} level prevented the generation of LTD. For example, injection into Purkinje cells of Ca^{2+} chelators (EGTA or BAPTA) prevented LTD induction (Kasono and Hirano, 1994; Konnerth et al., 1992; Linden and Connor, 1991; Sakurai, 1990). Finally, the hyperpolarization due to the inhibitory stellate cells also prevents LTD from occurring (Ekerot and Kano, 1985), probably by blocking Ca^{2+}-dependent plateau potentials in Purkinje cell dendrites, after their activation by climbing fibers (Ekerot and Oscarsson, 1981).

The aforementioned results are thus consistent with the fact that a transient rise in postsynaptic internal Ca^{2+} is necessary for LTD induction. However, this rise in cytosolic Ca^{2+} concentration is not sufficient since, in pseudopairing experiments where Purkinje cells are depolarized in the absence of parallel fiber stimulation, no LTD occurs (Daniel et al., 1992). Thus LTD induction also requires the contribution of parallel fiber stimulation and, thus, the activation of postsynaptic glutamate receptors.

10.3 Glutamate receptors are required for LTD induction

10.3.1 Involvement of AMPA receptors

A first target for the glutamate released from parallel fibers is the ionotropic α-amino-3-hydroxy-5-methyl-4-isoxazole proprionic acid (AMPA) receptors of Purkinje cells. These receptors are permeable to monovalent cations and seem to mediate the excitatory action of the parallel fibers, since the responses induced by activation of these fibers are totally blocked by application of 6-cyano-7-nitro-quinoxaline-2,3-dione (CNQX), a selective antagonist of this class of receptors.

Using whole-cell clamped Purkinje cells in acute slices, we have demonstrated that LTD requires activation of AMPA receptors during the pairing protocol (depolarization–parallel fiber stimulation) since application of CNQX prevents induction of LTD. In contrast, the blockade of these receptors after LTD induction does not prevent LTD expression (*Figure 10.2*) (Hemart et al., 1995). This finding is consistent with previous *in vivo* observations in which dendritic application of kynurenate [a glutamate ionotropic receptor antagonist sparing metabotropic glutamate receptors (mGluRs)] blocked LTD produced by parallel fiber–climbing fiber co-activation (Kano and Kato, 1987). Furthermore, it is also in agreement with results obtained in cultured Purkinje neurons, where parallel fiber stimulation was substituted by iontophoretic application of quisqualate (an agonist of both mGluRs and AMPA receptors), and where application of CNQX during depolarization–quisqualate conjunctive stimulation blocked LTD

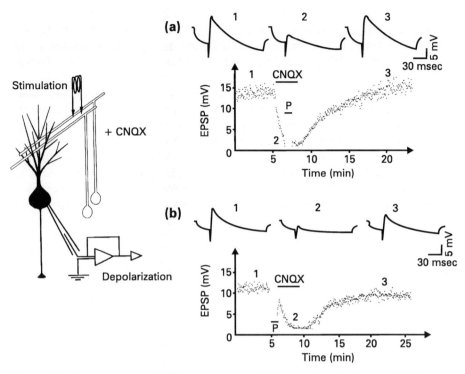

Figure 10.2. Involvement of AMPA receptors in LTD induction. Pairing parallel fiber-mediated EPSPs with Ca^{2+} spikes does not induce LTD in the presence of 4 μM CNQX in the bath (a), whereas the co-activation performed before CNQX application induces a clear LTD, observed after washout of the drug (b). Plot of EPSP amplitudes against time before and after the pairing protocol (P). The bar indicates the duration of CNQX application. Insets: averaged EPSPs at the indicated times. Adapted from Hemart *et al.* (1994) Properties of glutamate receptors are modified during long-term depression in cerebellar Purkinje cells. *Neurosci. Res.* **19**: 213–221, with permission from Elsevier Science Ireland Ltd.

induction (Linden *et al.*, 1991). Thus, these findings also raise the possibility that LTD requires activation of AMPA receptors in addition to a rise in postsynaptic Ca^{2+}. However, although this class of glutamate receptor is necessary, it has also been shown in cultured Purkinje neurons that these two processes are themselves not sufficient to induce LTD (Kasono and Hirano, 1995; Linden *et al.*, 1991).

10.3.2 Involvement of metabotropic receptors

Another target of glutamate seems to be a subtype of glutamate receptors linked to GTP-binding proteins. It is now well established that, besides AMPA receptors mediating fast excitatory synaptic transmission at parallel fiber–Purkinje cell synapses, the dendritic spines of Purkinje cells also bear a subtype of glutamate metabotropic receptors 'mGluR1' (Baude *et al.*, 1993; Masu *et al.*, 1991; Shigemoto *et al.*, 1994). Thus, co-activation of metabotropic and AMPA receptors, using iontophoretic application of glutamate, quisqualate or a mixture of (1*SR*, 3*RS*) 1-amino-cyclopentyl 1,3-dicarboxylate (ACPD) and AMPA, can replace the activation of parallel fibers for inducing LTD, during a pairing protocol with Ca^{2+} spike firing or climbing fiber activation (Crepel and Krupa, 1988; Ito and Karachot, 1990; Kano and Kato, 1987; Kasono and Hirano, 1994; Linden *et al.*, 1991). Moreover, using cultured Purkinje neurons, Linden *et al.* (1991) have demonstrated that G-protein inactivation with pertussis toxin disrupted LTD induced by quisqualate–depolarization conjunction. Furthermore, it has been demonstrated that mGluR1-inactivating antibodies completely prevent the generation of LTD of glutamate-induced currents in cultured Purkinje cells (Shigemoto *et al.*, 1994). However, to establish unambiguously the involvement of these receptors in LTD induction, mGluR1 knockout mice have been generated. It was thus shown that, in acute cerebellar slices from homozygous mutant mice, LTD is significantly impaired (*Figure 10.3*) (Aiba *et al.*, 1994; Conquet *et al.*, 1994), whereas the

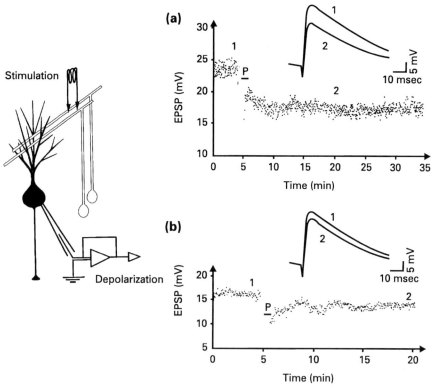

Figure 10.3. Involvement of mGluRs in LTD induction: LTD in a wild-type (a) and in mutant mGluR1$^{-/-}$ mouse Purkinje cell (b). Plots of amplitudes of parallel fiber-mediated EPSPs against time before and after the pairing protocol (P). Insets: averaged EPSPs at the indicated times. Adapted from Conquet *et al.* (1994). Reprinted with permission from *Nature* 372: 237–243. © 1994 Macmillan Magazines Ltd.

intrinsic bioelectrical properties of Purkinje cells, as well as basal synaptic transmission, remain normal.

Taken together, all these findings suggest that three concomitant processes are possibly necessary and sufficient for induction of LTD: activation of voltage-gated Ca^{2+} channels, ionotropic AMPA receptors and mGluR1 of Purkinje cells.

10.4 Second messenger cascades are required for LTD induction

As we have seen before, a rise in postsynaptic calcium seems to be the initial step for generating LTD. Therefore, what are the specific calcium-dependent second messenger cascades triggered by this initial event ? The standard approach to study these biochemical processes has been to apply either extracellularly in the perfusion bath, or intracellularly via the patch recording electrode, compounds that specifically activate or inhibit the activity of enzymes suspected to be involved in these cascades, and to determine whether or not LTD can be generated in such conditions.

10.4.1 Involvement of PKC activation

One isoform of the calcium-dependent protein kinase (PKC I) is known to be very abundant in Purkinje cells (Hidaka *et al.*, 1988; Nishizuka, 1986). It is also established that mGluR1 receptors are coupled to phospholipase C, resulting in production of inositol 1,4,5-trisphosphate (IP_3) and diacylglycerol (DAG) (Aramori and Nakamishi, 1992). It was therefore tempting to postulate that the cascade of events leading to LTD involves a co-activation of PKC by Ca^{2+} entry through voltage-gated ionic channels and by DAG produced by the activation of mGluR1 by glutamate released by parallel fiber activation (Crepel and Krupa, 1988).

Indeed, using bath application of phorbol esters known to directly activate PKC, LTD of Purkinje cell responses to exogenous glutamate or quisqualate was induced, whereas inactive analogs were without any effect (*Figure 10.4a*) (Crepel and Krupa, 1988, 1990; Linden and Connor, 1991). An input-specific LTD might also be induced by local application of 1-oleoyl-2-acetylglycerol, a potent PKC activator, together with direct Purkinje neuron depolarization (Linden, 1994).

The role of PKC in the generation of LTD has been also suggested by the use of PKC inhibitors. Indeed, in acute cerebellar slices, bath application of polymixin B, a potent blocker of PKC and to a lesser extent of calmodulin-dependent kinase, nearly totally prevented LTD induction by a pairing protocol (depolarization–parallel fiber stimulation) (*Figure 10.4b*) (Crepel and Jaillard, 1990). The same results were obtained when the PKC-inhibitory peptide (19–36) was dialyzed directly in the recorded cells (Hemart *et al.*, 1995). These results are in agreement with other studies showing that LTD, induced by iontophoretic glutamate pulses in conjunction with depolarization in cultured Purkinje neurons, can no longer be obtained in the presence of the PKC-inhibitory peptide 19–36 (Linden and Connor, 1991).

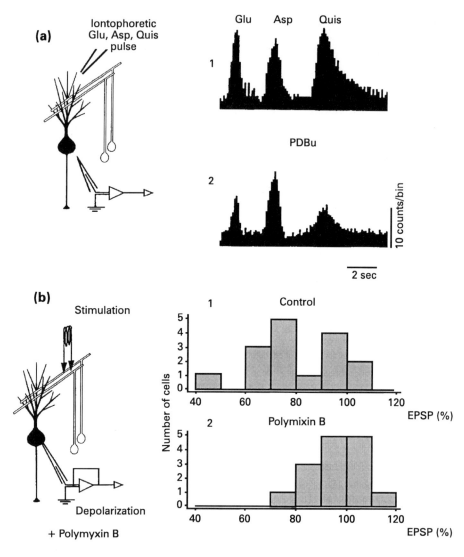

Figure 10.4. Involvement of PKC activation in LTD induction (a) Effects of phorbol esters on excitatory amino acid-induced responses. Poststimulus time histograms of the responses of an extracellularly recorded Purkinje cell to glutamate (Glu), aspartate (Asp) and quisqualate (Quis), before (A1) and after (A2) bath application of 400 μM phorbol 12-13-dibutyrate (PDBu). Adapted from Crepel and Krupa (1990). (b) Distribution of efficacy changes in synaptic transmission (EPSP amplitudes as percentage of control) following a pairing protocol, in standard bathing medium (B1) and in the presence of a potent inhibitor of PKC, polymixin B (B2). Adapted from Crepel and Jaillard (1990) with permission from ITP Journals.

10.4.2 Involvement of nitric oxide formation

In addition to PKC, activation of cGMP-dependent protein kinases via the nitric oxide (NO) pathway could also be implicated in LTD, since it has been demonstrated that Ca^{2+} could induce the formation of NO from arginine by activating a calmodulin-dependent NO synthase (NOS) (Garthwaite et al., 1988, 1989; Ross et al., 1990). NO is highly diffusible and its target is probably the soluble guanylate cyclase (Tremblay et al., 1988) in cells where it is produced, as well as in surrounding cellular elements (Garthwaite et al., 1989; Ross et al., 1990). This transduction cascade could therefore finally activate cGMP-dependent protein kinases.

Intracellular recordings using sharp electrodes or whole cell patch–clamp recordings in cerebellar slices have demonstrated that bath application of N-monomethyl-arginine (L-NMMA), a potent NOS inhibitor (Knowles et al., 1989), almost totally prevented induction of LTD by pairing parallel fiber activation with postsynaptic Purkinje cell Ca^{2+} spikes (*Figure 10.5a*) (Crepel and Jaillard, 1990; Daniel et al., 1993), and this effect was reversed well by co-application of an excess of L-arginine (Daniel et al., 1993). LTD induction can also be prevented by extracellular application of the NO scavenger methylene blue (Crepel and Jaillard, 1990). In contrast, an LTD-like phenomenon, consisting of a long-lasting depression in the amplitude of parallel fiber-mediated responses, was induced by bath application of NO donors such as sodium nitroprusside (SNP; *Figure 10.5b*) (Daniel et al., 1993), or S-nitroso-N-acetyl-D, L-penicillamine (SNAP; unpublished data). In addition, bath application of 8-bromo-cGMP, a membrane-permeant cGMP analog, or intracellular application of cGMP through the patch recording electrode were also able to depress durably the parallel fiber-mediated Purkinje cell responses. This cGMP-mediated decrease in the Purkinje cell response partially occluded the subsequent induction of LTD by a pairing protocol (depolarization–parallel fiber stimulation) (*Figure 10.5c*) (Daniel et al., 1993). Therefore, these results are consistent with a role for NO in LTD induction, and demonstrated that its target is probably the soluble guanylate cyclase of Purkinje cells.

Using the grease-gap recordings from Purkinje cell axons in cerebellar slices, a role for NO has been investigated in another LTD induction paradigm. Indeed, co-application of AMPA and SNP or AMPA and 8-bromo-cGMP in the bath induces an enduring desensitization of AMPA receptors of Purkinje cells, which is likely to require the production of cGMP, via the formation of NO (Ito and Karachot, 1990), since slice incubation with either the NOS inhibitor L-NMMA or the NO scavenger hemoglobin before LTD induction prevents the occurrence of LTD of the responses. In this scheme, Ito and colleagues postulate that PKG activation via cGMP production would allow phosphorylation of its specific endogenous substrate, 'G-substrate', a potent inhibitor of phosphatases (Ito and Karachot, 1992). Accordingly, using an exogenous potent inhibitor of protein phosphatases, it has been demonstrated that LTD involves a cascade including protein phosphatases (Ajima and Ito, 1995), which are also particularly abundant in Purkinje cells (Hashikawa et al., 1995).

Finally, with a NO-sensitive probe inserted in the molecular layer of cerebellar slices, Shibuki and Okada (1991) have demonstrated that protocols used to induce LTD lead to the production of NO.

All these results are in contrast with the findings of Linden and Connor (1992) who have demonstrated, with patch–clamp records in cultured Purkinje neurons, that LTD induced by pairing depolarization of Purkinje cell with iontophoretic glutamate pulses was unaffected by treatment that stimulates or inhibits NO signaling.

Therefore, taken together, and except for the finding of Linden and Connor, these results support the view that NO is an important signal in the cellular events leading to LTD, but it remains to be elucidated how these events are linked to

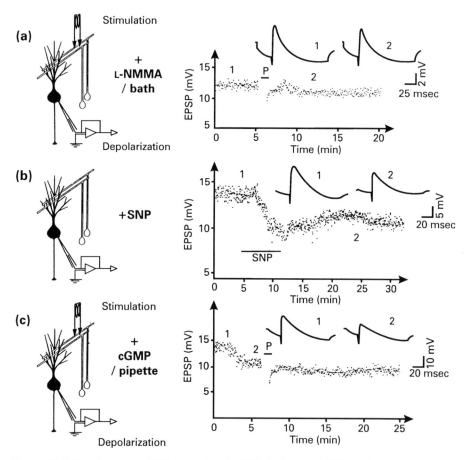

Figure 10.5. Involvement of NO formation in LTD induction (a) Plot of EPSP amplitudes against time in a Purkinje cell before and after a pairing (P) performed with 30 mM NOS inhibitor L-NMMA, in the bath. (b) Plot of EPSP amplitudes against time before, during and after bath application of 8 mM NO donor SNP, for 7 min (heavy bar). (c) Plot of EPSP amplitudes against time in a Purkinje cell where 0.5 μM cGMP was added in the recording pipet. At 7 min after the whole cell recording, a pairing protocol (P) was performed. Insets: averaged EPSPs at the indicated times. Adapted from Daniel *et al.* (1993) Long-term depression requires nitric oxide and guanosine 3' : 5' cyclic monophosphate production in rat cerebellar Purkinje cells. *Eur. J. Neurosci.* **5**: 1079–1083, by permission of Oxford University Press.

each other. Indeed, in the cerebellum, neuronal NOS has not been identified in Purkinje cells, but only in neighboring elements (Bredt et al., 1990; Crepel et al., 1994; Southam et al., 1992). Recent experiments in cerebellar slices (Crepel et al., 1994) support the hypothesis that NO might be produced by the NOS located in parallel fibers and/or in basket cells when these neurons are activated by both parallel fiber stimulation (Eccles et al., 1967) and by the large efflux of potassium which follows the entry of Ca^{2+} in Purkinje cells during pairing experiments (depolarization–parallel fibers stimulation). NO thus produced would then diffuse into Purkinje cells to activate guanylate cyclase and potentially activate cGMP-dependent protein kinases.

10.5 Does LTD require desensitization of AMPA receptors of Purkinje cells?

The fact that co-activation of Purkinje cells by climbing fibers and by iontophoretic application of glutamate in their dendritic fields leads to a long-lasting decrease in their responsiveness to this agonist (Ito et al., 1982) led Ito to propose that induction of LTD might ultimately lead to a long-term desensitization of ionotropic glutamate receptors of Purkinje cells and thus to the observed decrease in synaptic efficacy. Accordingly, it was shown later that pairing iontophoretic application of glutamate and Ca^{2+} spike firing of Purkinje cells induces LTD of their responsiveness to this agonist, both in acute slices (Crepel and Krupa, 1988) and in dissociated cultures (Linden and Connor, 1991). However, the observed decrease in the efficacy of glutamate in activating Purkinje cells might be due to causes other than a true desensitization of glutamate receptors. Recently, the nootropic compound aniracetam has been shown to reduce desensitization of AMPA receptors markedly and/or to decrease the closing rate constant for ion channel gating (Ito et al., 1990; Vyklicky et al., 1991). This compound was therefore used during LTD experiments, to determine whether or not it could interact with induction or expression of LTD. It was found that in whole cell clamped Purkinje cells in acute slices, aniracetam had a larger than normal potentiating effect on parallel fiber-mediated responses during expression of LTD and, moreover, this compound also significantly blocked LTD induction (Hemart et al., 1994). Taken together, these results strongly support the view that this change in synaptic efficacy involves a desensitization of postsynaptic AMPA receptors at parallel fiber–Purkinje cell synapses or, at least, a genuine change in the functional characteristics of these receptors. Futhermore, and in keeping with previous observations of Ito and Karachot (1992), these results also suggest that induction of LTD requires a desensitization of AMPA receptors of Purkinje cells.

Thus, the activation of the abovementioned protein kinases (PKC, PKG) could bring about alterations in the sensitivity of AMPA receptors, by directly phosphorylating or dephosphorylating channel subunits or associated molecules (*Figure 10.6*) (see Chapter 4).

10.6 Conclusions

After one decade of research, we are beginning to unravel the molecular mechanisms underlying LTD induction, but many questions remain unanswered. Although much is known about the role of the different subtypes of the ionotropic and metabotropic glutamate receptor families, the involvement in the LTD induction of the newly discovered δ subunits of this family, which are heavily expressed at parallel fiber–Purkinje cell synapses (Araki *et al.*, 1993; Mayat *et al.*, 1995), remains to be confirmed (Hirano *et al.*, 1994; Kashiwabuchi *et al.*, 1995). On the other hand, recent evidence suggests that insulin-like growth factor (IGF-1) might be released by climbing fibers and contribute to LTD induction, possibly by raising cytosolic Ca^{2+} concentration. Moreover, the fact that deletion of IGF-1 by antisense oligonucleotides injected into the inferior olive effectively impairs eyeblink conditioning in rat suggests a possible role for this factor in motor learning (Castro-Alamancos and Torres-Aleman, 1994).

In spite of remaining uncertainties, and taking into account all experimental evidence mentioned above, we propose the following model, in an attempt to explain the mechanisms of LTD induction (*Figure 10.6*).

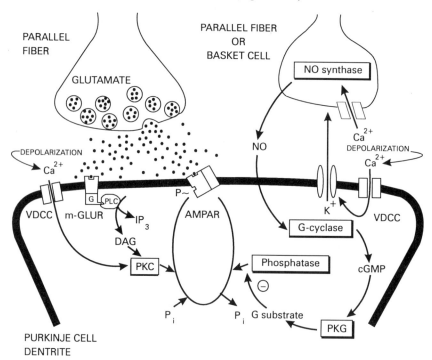

Figure 10.6. Schematic diagram of the signal transduction processes involving the PKC and NO pathway that are presumed to underlie LTD. VDCC, voltage-dependent Ca^{2+} channel; mGluR, metabotropic glutamate receptor; AMPAR, AMPA receptor; G, G-protein; PLC, phospholipase C; IP_3, inositol-1, 4, 5-trisphosphate; DAG, diacylglycerol; PKC, protein kinase C; NO, nitric oxide; cGMP, cyclic GMP; PKG, protein kinase G; P_i, inorganic phosphate; P, phosphorylation.

When the AMPA receptors of Purkinje cells are activated by glutamate released by parallel fibers, their agonist-dependent desensitization could trigger a conformational change in the receptor that makes it susceptible to modulation by cytosolic second messenger pathways. Due to the fact that the phosphorylation site will be exposed after the conformational change, the PKC could directly phosphorylate the AMPA receptor. The signals that activate this kinase could be both Ca^{2+} entry through voltage-gated Ca^{2+} channels and DAG produced by phospholipase C, when mGluRs at parallel fiber–Purkinje cell synapses are also activated by glutamate released by parallel fibers. On the other hand, processes that are reversibly controlled by protein phosphorylation require not only protein kinases, but also protein phosphatases. Thus, during the LTD induction phase, the NO pathway could ultimately inhibit phosphatases. This inhibition might allow the phosphorylated state of AMPA receptors to be maintained at a higher level. As a consequence of this phosphorylation, the kinetics of opening and closing of AMPA receptor-coupled channels could be affected, and/or a larger fraction of glutamate receptors than normally could be stabilized in a desensitized state at rest, thus explaining the maintenance of LTD, at least over a relatively short time.

However, presently there is no evidence concerning possible mechanisms of the maintenance phase of LTD, and it remains now to establish whether LTD is involved in motor learning as suggested by *in vivo* experiments (Nagao and Ito, 1991).

References

Aiba A, Kano M, Chen C, Stanton ME, Fox GD, Herrup K, Zwingman TA, Tonegawa S. (1994) Deficient cerebellar long-term depression and impaired motor learning in mGluR1 mutant mice. *Cell* **79**: 377–388.
Ajima A, Ito M. (1995) A unique role of protein phosphatases in cerebellar long-term depression. *NeuroReport* **6**: 297–300.
Albus JS. (1971) A theory of cerebellar function. *Math. Biosci.* **10**: 25–61.
Araki K, Meguro H, Kushiya E, Takayama C, Inoue Y, Mishina M. (1993) Selective expression of the glutamate receptor δ2 subunit in cerebellar Purkinje cells. *Biochem. Biophys. Res. Commun.* **197**: 1267–1276.
Aramori I, Nakanishi S. (1992) Signal transduction and pharmacological characteristics of a metabotropic glutamate receptor, mGluR1, in transfected CHO cells. *Neuron* **8**: 757–765.
Baude A, Nusser Z, Roberts JDB, Mulvihill E, McIlhinney RAJ, Somogyi P. (1993) The metabotropic glutamate receptor (mGluR1a) is concentrated at perisynaptic membrane of neuronal subpopulations as detected by immunogold reaction. *Neuron* **11**: 771–787.
Bredt S, Hwang PM, Snyder SH. (1990) Localization of nitric oxide synthase indicating a neural role for nitric oxide. *Nature* **347**: 768–770.
Castro-Alamancos MA, Torres-Aleman I. (1994) Learning of the conditioned eye-blink responses is impaired by an antisense insulin-like growth factor I oligonucleotide. *Proc. Natl Acad. Sci. USA* **91**: 10203–10207.
Conquet F, Bashir ZI, Davies CH *et al*. (1994) Motor deficit and impairment of synaptic plasticity in mice lacking mGluR1. *Nature* **372**: 237–243.
Crepel F, Audinat E. (1991) Excitatory amino acid receptors of cerebellar Purkinje cells: development and plasticity. *Prog. Biophys. Mol. Biol.* **55**: 31–46.
Crepel F, Jaillard D. (1990) Protein kinases, nitric oxide and long-term depression of synapses in the cerebellum. *NeuroReport* **1**: 133–136.

Crepel F, Jaillard D. (1991) Pairing of pre- and postsynaptic activities in cerebellar Purkinje cells induces long-term changes in synaptic efficacy. An *in vitro* study. *J. Physiol. (Lond.)* **432**: 123–141.
Crepel F, Krupa M. (1988) Activation of protein kinase C induces a long term depression of glutamate sensitivity of cerebellar Purkinje cells. An *in vitro* study. *Brain Res.* **458**: 397–401.
Crepel F, Krupa M. (1990) Modulation of the responsiveness of cerebellar Purkinje cells to excitatory amino acids. In: *Excitatory Amino Acids and Neuronal Plasticity* (ed. Y Ben-Ari). Plenum Press, New York, pp. 323–329.
Crepel F, Audinat E, Daniel H, Hemart N, Jaillard D, Rossier J, Lambolez B. (1994) Cellular locus of the nitric oxide-synthase involved in cerebellar long-term depression induced by high external potassium concentration. *Neuropharmacology* **33**: 1399–1405.
Daniel H, Hemart N, Jaillard D, Crepel F. (1992) Coactivation of metabotropic glutamate receptors and voltage-gated calcium channels induces long-term depression in cerebellar Purkinje cells *in vitro*. *Exp. Brain Res.* **90**: 327–331.
Daniel H, Hemart N, Jaillard D, Crepel F. (1993) Long-term depression requires nitric oxide and guanosine 3′: 5′ cyclic monophosphate production in rat cerebellar Purkinje cells. *Eur. J. Neurosci.* **5**: 1079–1083.
Eccles JC, Ito M, Szentagothai J. (1967) *The Cerebellum as a Neuronal Machine*. Springer-Verlag, Berlin.
Ekerot CF, Kano M. (1985) Long-term depression of parallel fibre synapses following stimulation of climbing fibres. *Brain Res.* **342**: 357–360.
Ekerot CF, Oscarsson O. (1981) Prolonged depolarization elicited in Purkinje cell dendrites by climbing fibre impulses in the cat. *J. Physiol. (Lond.)* **318**: 207–221.
Ellers J, Augustine GJ, Konnerth A. (1995) Subthreshold synaptic Ca^{2+} signalling in fine dendrites and spines of cerebellar Purkinje neurons. *Nature* **373**: 155–158.
Garthwaite J, Charles SL, Chess-Williams R. (1988) Endothelium-derived relaxing factor release on activation of NMDA receptors suggests a role as intercellular messenger in the brain. *Nature* **336**: 385–388.
Garthwaite J, Southam E, Anderton M. (1989) A kainate receptor linked to nitric oxide synthesis from arginine. *J. Neurochem.* **53**: 1952–1954.
Glaum SR, Slater NT, Rossi DJ, Miller RJ. (1992) Role of metabotropic glutamate (ACPD) receptors at the parallel fiber–Purkinje cell synapse. *J. Neurophysiol.* **68**: 1453–1462.
Hashikawa T, Nakazawa K, Mikawa S, Shima H, Nagao M. (1995) Immunohistochemical localization of protein phosphatase isoforms in the rat cerebellum. *Neurosci. Res.* **22**: 133–136.
Hemart N, Daniel H, Jaillard D, Crepel F. (1994) Properties of glutamate receptors are modified during long-term depression in cerebellar Purkinje cells. *Neurosci. Res.* **19**: 213–221.
Hemart N, Daniel H, Jaillard D, Crepel F. (1995) Receptors and second messengers involved in long-term depression in rat cerebellar slices *in vitro*: a reappraisal. *Eur. J. Neurosci.* **7**: 45–53.
Hidaka H, Tanaka T, Onoda K, Hagiwara M, Watanabe M, Ohta H, Ito Y, Tsurudome M, Yoshida T. (1988) Cell-specific expression of protein kinase C isozymes in the rabbit cerebellum. *J. Biol. Chem.* **263**: 4523–4526.
Hirano T. (1990) Effects of postsynaptic depolarization in the induction of synaptic depression between a granule cell and a Purkinje cell in rat cerebellar culture. *Neurosci. Lett.* **119**: 145–147.
Hirano T, Kasono K, Araki K, Shinozuka K, Mishina M. (1994) Involvement of the glutamate receptor δ2 subunit in the long-term depression of glutamate responsiveness in cultured rat Purkinje cells. *Neurosci. Lett.* **182**: 172–176.
Ito M. (1984) *The Cerebellum and Neural Control*. Raven Press, New York.
Ito M. (1987) Characterization of synaptic plasticity in cerebellar and cerebral neocortex. In: *The Neural and Molecular Bases of Learning* (eds JP Changeux, M Nonishi). John Wiley & Sons, New York, pp. 276–279.
Ito M. (1989) Long term depression. *Annu. Rev. Neurosci.* **12**: 85–102.
Ito M, Karachot L. (1990) Messages mediating long-term desensitization in cerebellar Purkinje cells. *NeuroReport* **1**: 129–132.
Ito M, Karachot L. (1992) Protein kinases and phosphatase inhibitors mediating long-term desensitization of glutamate receptors in cerebellar Purkinje cells. *Neurosci. Res.* **14**: 27–38.

Ito M, Sakurai M, Tongroach P. (1982) Climbing fibre induced depression of both mossy fibre responsiveness and glutamate sensitivity of cerebellar Purkinje cells. *J. Physiol. (Lond.)* 324: 113–134.

Ito I, Tanabe S, Kohda A, Sugiyama H. (1990) Allosteric potentiation of quisqualate receptors by a nootropic drug aniracetam. *J. Physiol. (Lond.)* 424: 533–543.

Kano M, Kato M. (1987) Quisqualate receptors are specifically involved in cerebellar synaptic plasticity. *Nature* 325: 276–279.

Kashiwabuchi K, Ikeda K, Araki K et al. (1995) Impairment of motor coordination, Purkinje cell synapse formation, and cerebellar long-term depression in GluRδ2 mutant mice. *Cell* 81: 245–252.

Kasono K, Hirano T. (1994) Critical role of postsynaptic calcium in cerebellar long-term depression. *NeuroReport* 6: 17–20.

Kasono K, Hirano T. (1995) Involvement of inositol triphosphate in cerebellar long-term depression. *NeuroReport* 6: 569–572.

Knowles RG, Palacios M, Palmer RM, Moncada S. (1989) Formation of nitric oxide from L-arginine in the central nervous system: a transduction mechanism for stimulation of the soluble guanylate cyclase. *Proc. Natl Acad. Sci. USA* 86: 5159–5162.

Konnerth A, Dreesen J, Augustine GJ. (1992) Brief dendritic calcium signals initiate long-lasting synaptic depression in cerebellar Purkinje cells. *Proc. Natl Acad. Sci. USA* 89: 7051–7055.

Linden DJ. (1994) Input-specific induction of cerebellar long-term depression does not require presynaptic alteration. *Learn. Mem.* 1: 121–128.

Linden DJ, Connor JA. (1991) Participation of postsynaptic PKC in cerebellar long-term depression in culture. *Science* 254: 1656–1659.

Linden D, Connor JA. (1992) Long-term depression of glutamate currents in cultured cerebellar Purkinje neurons does not require nitric oxide signalling. *Eur. J. Neurosci.* 4: 10–15.

Linden DJ, Dickinson MH, Smeyne M, Connor JA. (1991) A long-term depression of AMPA currents in cultured cerebellar Purkinje neurons. *Neuron* 7: 81–89.

Marr D. (1969) A theory of cerebellar cortex. *J. Physiol.* 202: 437–470.

Masu M, Tanabe Y, Tsuchida K, Shigemoto R, Nakanishi S. (1991) Sequence and expression of a metabotropic glutamate receptor. *Nature* 349: 760–765.

Mayat E, Petralia RS, Wang YX, Wenthold R. (1995) Immunoprecipitation, immunoblotting, and immunocytochemistry studies suggest that glutamate receptor δ subunits form novel postsynaptic receptor complexes. *J. Neurosci.* 15: 2533–2546.

Miyakawa H, Lev-Ram V, Lasser-Ross N, Ross WN. (1992) Calcium transients evoked by climbing fiber and parallel fiber synaptic inputs in guinea pig cerebellar Purkinje neurons. *J. Neurophysiol.* 4: 1178–1189.

Nagao S, Ito M. (1991) Subdural application of hemoglobin to the cerebellum blocks vestibuloocular reflex adaptation. *NeuroReport* 2: 193–196.

Nishizuka Y. (1986) Studies and perspectives of protein kinase C. *Science* 233: 305–311.

Ross CA, Bredt D, Snyder SH. (1990) Messenger molecules in the cerebellum. *Trends Neurosci.* 13: 216–222.

Sakurai M. (1990) Calcium is an intracellular mediator of the climbing fiber in induction of cerebellar long-term depression. *Proc. Natl Acad. Sci. USA* 87: 8383–8385.

Shibuki K, Okada D. (1991) Endogenous nitric oxide release required for long-term synaptic depression in the cerebellum. *Nature* 349: 326–328.

Shigemoto R, Abe T, Numura S, Nakanishi S, Hirano T. (1994) Antibodies inactivating mGluR1 metabotropic glutamate receptor block long-term depression in cultured Purkinje cells. *Neuron* 12: 1245–1255.

Southam E, Morris R, Garthwaite J. (1992) Sources and targets of nitric oxide in rat cerebellum. *Neurosci. Lett.* 137: 241–244.

Tremblay J, Gerzer R, Hamet P. (1988) Cyclic GMP in cell function. In: *Advances in Second Messengers and Phosphoprotein Research*, Vol. 22 (eds P Greengard, GA Robison). Raven Press, New York, pp. 319–368.

Vyklicky L, Patneau DK, Mayer ML. (1991) Modulation of excitatory synaptic transmission by drugs that reduce desensitization at AMPA/kainate receptors. *Neuron* 7: 971–984.

11

Roles of LTP and LTD in neuronal network operations in the brain

Edmund T. Rolls

11.1 Introduction

Synapse-specific increases and decreases in synaptic strength which depend on the activity of the presynaptic and postsynaptic neuron are central to modern theories of how networks in the brain operate in setting up sensory representations of the world, in memory and in producing appropriate motor responses. The aims of this chapter are to show how different features of these synaptic modifications are crucial to the operation of different types of network, and to the operation of several different brain systems. The types of network considered will be three which are fundamental to brain function, namely pattern associators, autoassociators and competitive networks. Each performs a different type of operation for the brain. Then the ways in which these types of synaptic modification are implicated in the operation of the hippocampus in memory (Section 11.5), and the cerebral neocortex in visual object recognition (Section 11.6), will be described. The points made apply to any synapse-specific modification process in the brain, regardless of whether that process happens to be long-term potentiation (LTP) or long-term depression (LTD). More formal descriptions of the operation of some of the networks introduced here are provided by Rolls and Treves (1997), and by Hertz *et al.* (1991).

11.2 Pattern associators

A fundamental operation of most nervous systems is to learn to associate a first stimulus with a second which occurs at about the same time, and to retrieve the second stimulus when the first is presented. The first stimulus might be the sight

---

Cortical Plasticity, edited by M.S. Fazeli and G.L. Collingridge.
© 1996 BIOS Scientific Publishers Ltd, Oxford.

of food, and the second stimulus the taste of food. After the association has been learned, the sight of food would enable its taste to be retrieved. In classical conditioning, the taste of food might elicit an unconditioned response of salivation and, if the sight of the food is paired with its taste, then the sight of that food would, by learning, come to produce salivation. More abstractly, if one idea is associated by learning with a second, then when the first idea occurs again, the second idea will tend to be associatively retrieved. Areas of the brain in which such pattern associators may be present include the amygdala and orbitofrontal cortex (for learning associations between, e.g. the sight of food and its taste, or more generally in stimulus-reinforcement association learning, see Davis, 1992; Rolls, 1990c; Rolls and Treves, 1997); and in cortico-cortical backprojections (see Section 11.5.3).

11.2.1 Architecture and operation

The prototypical network for pattern association is shown in *Figure 11.1*. What we have called the second or unconditioned stimulus pattern (u) is applied through unmodifiable synapses to produce or force firing of the output neurons (r). (In this notation, u refers to the vector of firing of the unconditioned input stimulus, i.e. the firing rates u_i of the n input neurons i which are indexed by $i = 1, n$. Details of the use of simple linear algebra in defining the operation of such networks are provided by Jordan, 1986; Rolls and Treves, 1996.) The first or conditioned stim-

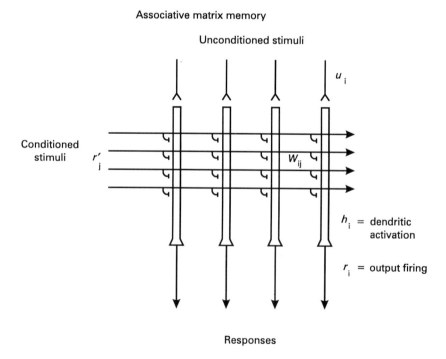

Figure 11.1. The architecture of a network for pattern association (see text).

ulus pattern r' present on the horizontally running axons in *Figure 11.1* is applied through modifiable synapses (w) to the dendrites of the output neurons. The synapses are modifiable in such a way that if there is presynaptic firing on an input axon r'_j paired during learning with postsynaptic activity on neuron r_i, then the strength or weight w_{ij} between that axon and the dendrite increases. This simple learning rule is often called the Hebb rule, after Donald Hebb who in 1949 formulated the hypothesis that if the firing of one neuron was regularly associated with another, then the strength of the synapse or synapses between the neurons should increase. After learning, presenting the pattern r' on the input axons will activate the dendrite through the strengthened synapses. If the cue or conditioned stimulus pattern is similar to that learned, then there will be some activation of the postsynaptic neuron produced by each of the firing axons afferent to a synapse strengthened by the previous learning. The total activation h_i of each postsynaptic neuron i is then the sum of such individual activations. (This is a process intended to model simple summation of inputs by graded depolarization in the postsynaptic neuron.) In this way, just the correct output neurons are strongly activated, and the second or unconditioned stimulus is effectively recalled. The recall is best when only strong activation of the postsynaptic neuron produces firing, that is if there is a threshold for firing, just like real neurons. The reasons for this arise when many associations are stored in the memory, as will soon be shown.

A more precise description of pattern association memory will now be introduced, in order to help us to understand more exactly how pattern associators operate. We have denoted above a conditioned stimulus input pattern as r'. Each of the axons has a firing rate, and if we count or index through the axons using the subscript j, the firing rate of the first axon is r'_1, of the second r'_2, of the jth r'_j, etc. The whole set of axons forms a vector, which is just an ordered (1, 2, 3, etc.) set of elements. The firing rate of each axon r'_j is one element of the firing rate vector r'. Similarly, using i as the index, we can denote the firing rate of any output neuron as r_i, and the firing rate output vector as r. With this terminology, we can then identify any synapse on to neuron i from neuron j as w_{ij} (see *Figure 11.1*). In this chapter, the first index, i, always refers to the receiving neuron (and thus signifies a dendrite), while the second index, j, refers to the sending neuron (and thus signifies an axon in *Figure 11.1*). We can now specify the learning and retrieval operations as follows.

Learning. The firing rate of every output neuron is forced through strong unmodifiable synapses to a value determined by the forcing or unconditioned stimulus input. That is, for any one neuron i, $r_i = u_i$ (see *Figure 11.1*).

The Hebb rule can then be written as follows:

$$\delta w_{ij} = k \cdot r_i \cdot r'_j \qquad (11.1)$$

where dw_{ij} is the change of the synaptic weight w_{ij} which results from the simultaneous (or conjunctive) presence of presynaptic firing r'_j and postsynaptic firing r_i (or strong depolarization), and k is a learning rate constant which specifies how much the synapses alter on any one pairing. The presynaptic and postsynaptic activity must be present approximately simultaneously (to within perhaps 100 msec).

The Hebb rule is expressed in this multiplicative form to reflect the idea that *both* presynaptic and postsynaptic activity must be present for the synapses to increase in strength. The multiplicative form also reflects the idea that strong pre- and postsynaptic firing will produce a larger change of synaptic weight than smaller firing rates. The Hebb rule thus captures what is typically found in studies of associative LTP. It is also assumed for now that, before any learning takes place, the synaptic strengths are small in relation to the changes that can be produced during Hebbian learning. We will see that this assumption can be relaxed later when a modified Hebb rule is introduced that can lead to a reduction in synaptic strength (LTD) under some conditions.

Recall. When the conditioned stimulus is present on the input axons, the total activation h_i of a neuron i is the sum of all the activations produced through each strengthened synapse w_{ij} by each active neuron r'_j. We can express this as:

$$h_i = \Sigma_j r'_j . w_{ij} \qquad (11.2)$$

where Σ_j indicates that the sum is over the n input axons indexed by j. The multiplicative form here indicates that activation should be produced by an axon only if it is firing, and only if it is connected to the dendrite by a strengthened synapse. It also indicates that the strength of the activation reflects how fast the axon r'_j is firing, and how strong the synapse w_{ij} is. The sum of all such activations expresses the idea that summation (of synaptic currents in real neurons) occurs along the length of the dendrite, to produce activation at the cell body, where the activation h_i is converted into firing r_i. This conversion can be expressed as

$$r_i = f(h_i) \qquad (11.3)$$

which indicates that the firing rate is a function of the postsynaptic activation. The function is called the activation function in this case. The function at its simplest could be linear, so that the firing rate would be proportional to the activation (see *Figure 11.2a*). Real neurons have thresholds, with firing occurring only if the activation is above the threshold. A threshold linear activation function is shown in *Figure 11.2b*. This has been useful in formal analysis of the properties of neural networks. Neurons also have firing rates which become saturated at a maximum rate, and we could express this as the sigmoid activation function shown in *Figure 11.2c*. Another simple activation function, used in some models of neural networks, is the binary threshold function (*Figure 11.2d*), which indicates that, if the activation is below threshold, there is no firing, and that if the activation is above threshold, the neuron fires maximally. Some nonlinearity in the activation function is an advantage, for it enables small activations produced by interfering memories to be minimized, and it can enable neurons to perform logical operations, such as fire only if two or more sets of inputs are present simultaneously.

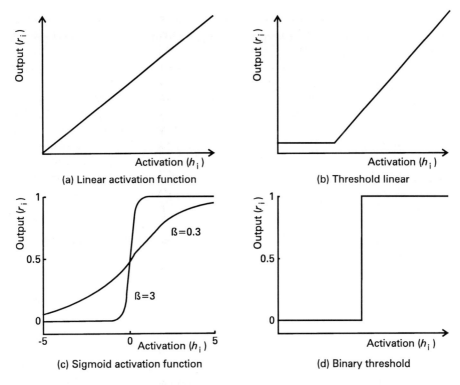

Figure 11.2. A set of different activation functions of neurons (see text). Each activation function shows a relationship between the activation of the neuron by its inputs (abscissa), and the output firing of the neuron (ordinate).

With networks of this type, many associations between different conditioned and unconditioned stimuli can be learned, as will be described shortly. These networks also have many very biologically attractive features, including generalization across similar conditioned input stimuli, and graceful degradation of performance if some of the input axons or the synapses in the network are damaged, or are not available during development. These properties arise only if the input patterns are represented across an ensemble of axons (distributed encoding), because then similarity can be represented by correlations of one pattern of input with another, as described fully by Rolls and Treves, 1997).

11.2.2 Relevance of different aspects of the synaptic modification used

Local learning rule. The learning used in pattern association neural networks (Equation 11.1) is a local learning rule in that the information required to specify the change in synaptic weight is available locally at the synapse, as it is dependent only on the presynaptic firing rate r'_j available at the synaptic terminal, and the postsynaptic activation or firing r_i available on the dendrite of the neuron receiving the synapse. This is a property of LTP and LTD. This makes the learning rule

227

biologically plausible, in that the information about how to change the synaptic weight does not have to be carried from a distant source where it is computed to every synapse. (Such a nonlocal learning rule would not be biologically plausible, in that there are no appropriate connections known in most parts of the brain to bring in the synaptic training signal to every synapse.)

Another useful property of real neurons in relation to Equation 11.1 is that the postsynaptic term, r_i, is available on much of the dendrite of a cell, if (as is the case for many large cells in the brain, e.g. pyramidal cells) the electrotonic length of the dendrite is short. Thus if a neuron is strongly activated with a high value for r_i, then any active synapse on to the cell will be capable of being modified. This enables the cell to learn an association between the pattern of activity on all its axons and its postsynaptic activation. If, in contrast, a group of co-active axons made synapses close together on a small dendrite, then the local depolarization might be intense, and these synapses only would modify on to the dendrite. (A single distant active synapse would not modify in this type of neuron, because of the long electrotonic length of the dendrite.) The computation in this case is described as σ–π, to indicate that there is a local product computed, and then the output of the neuron can reflect the sum of such local multiplications (see Rolls and Treves, 1996).

Capacity (see Rolls and Treves, 1990). The number of different associations that can be stored in a pattern associator is proportional to the number C of independently modifiable synapses on the dendrite of each output neuron. It is for this reason that a synapse-specific modification rule is desirable. If one synapse adjacent to another with the appropriate pre- and postsynaptic activity for synapse modification also showed synapse modification, then the number of different pattern associations that could be stored would be halved. The details of the number of different memories that can be stored in pattern association networks with different types of activation function are described elsewhere (Rolls and Treves, 1996) and summarized next, but the important conclusion is that just stated.

Linear associative neuronal networks have a linear activation function, and are trained by a Hebb (or similar) associative learning rule. The capacity of these networks is C different input vectors (r', where $r'_i = 1,C$) mapped to an output h_i (and thus r_i), or h and thus r. For no interference, the conditioned stimulus input vectors r' must be orthogonal (i.e. they must be uncorrelated). If the conditioned stimulus input pattern vectors are not orthogonal, then the dot product of them is not 0, and an output neuron activated by one of the vectors will be activated by the other. The capacity in this case will be less than C.

If the input firing rates on the axons are in the range $(0,+x)$ (where x is the maximum firing rate in, for example, spikes/sec), as they will be for real neurons, then, the different input vectors will be correlated. Because each neuron effectively calculates the correlation (or more strictly the inner or dot product) of the different input vectors r' with the weight vectors on each dendrite w_i, two different input vectors with positive-only firing rates will always produce correlated outputs (i.e. the correlation between any two such vectors will be positive). This will produce interference between the different memories stored in the network (see Rolls and Treves, 1997). A solution to this issue which arises with input vectors r' which can only assume positive values (e.g. firing rates) is to use a modified learning rule of the following form:

$$\delta w_{ij} = k \cdot r_i \cdot (r'_j - c) \qquad (11.4)$$

where c is a constant. This learning rule includes (in proportion to r_j) increasing the synaptic weight if $(r'_j - c)$ is > 0 (LTP), and decreasing the synaptic weight if $(r'_j - c)$ is <0 (heterosynaptic LTD). If c is the average activity of an input axon r'_j across patterns, then C input vectors can still be learned by the network [provided for linear networks that the resulting vectors $(r'_j - c$, with $j = 1,n)$ are mutually orthogonal]. (In large networks, c can be the average activity of the input vector r', which would be more plausible in the brain, see Rolls and Treves, 1990.)

This modified (Hebbian) learning rule will be described in more detail, in terms of a contingency table showing the synaptic strength modifications produced by different types of learning rule, where LTP indicates an increase in synaptic strength, and LTD indicates a decrease in synaptic strength. In *Table 11.1*, 0 for the post-synaptic activation or the presynaptic firing should be read as low, and 1 as high.

Table 11.1. Changes of synaptic strength produced by different pre- and post- synaptic states

		Postsynaptic activation	
		0	1
Presynaptic firing	0	0	Heterosynaptic LTD
	1	Homosynaptic LTD	LTP

Heterosynaptic LTD is so-called because it is the decrease in synaptic strength that occurs to a synapse which is *other than* that through which the postsynaptic cell is being activated. This heterosynaptic depression is the type of change of synaptic strength that is required (in addition to LTP) for effective subtraction of the average presynaptic firing rate, in order to make the conditioned stimulus vectors appear more orthogonal to the pattern associator. The rule is sometimes called the Singer–Stent rule, after work by Singer (1987) and Stent (1973). Evidence that this type of LTD is found in the hippocampus was described by Levy (Levy, 1985; Levy and Desmond, 1985; see Brown et al., 1990).

Homosynaptic LTD is so called because it is the decrease in synaptic strength that occurs to a synapse which is (the *same as* that which is) active. For it to occur, the postsynaptic neuron must simultaneously have only low activity (which might be below its mean activity, as formalized in Equation 11.8. (This rule is sometimes called the BCM rule after the paper of Bienenstock, Cooper and Munro, 1982.)

The value of heterosynaptic LTD in associative neuronal networks is thus that it enables the correlation between conditioned stimulus input patterns induced by the positive-only firing rates to be removed. This enables interference between different conditioned stimuli induced by this factor to be removed. It enables the theoretical limit for the number of patterns to be associated in such networks to be achieved. This limit is, in the case of linear associative networks, C patterns if the patterns are orthogonal, and somewhat less than C patterns if they are not orthogonal (see Rolls and Treves, 1990, 1996).

With *nonlinear neurons*, for example with a threshold in the output activation function so that the output firing r_i is 0 when the activation h_i is below the threshold, or with a sigmoid activation function, then the capacity can be measured in terms of the number of (conditioned stimulus) input patterns r' that can produce different outputs in the network. As with the linear counterpart, in order to remove the correlation that would otherwise occur between the patterns because the elements can take only positive values, it is useful to use a modified Hebb rule as in Equation 11.4 above.

With fully distributed orthogonal (after subtracting the mean using the term c) input patterns r', it is possible to store, as with linear associative networks, C different patterns where there are C inputs per dendrite.

If sparse (distributed) input patterns r' are used, then many more than C patterns can be stored, even when the patterns are random (e.g. each element is set on or off at random). [A local representation is a representation in which the information that a particular stimulus or event occurred is provided by the activity of one of the elements in the vector, e.g. by one of the nodes or neurons in the network. A (fully) distributed representation is one in which the activity of all the elements in the vector is needed to specify which stimulus or event occurred. A sparse (distributed) representation is one in which only a proportion (a) of the n input axons r'_i are active (>0) at any one time. The sparseness is then a (for binary neurons). The activity of this subset of nodes is sufficient to identify which stimulus or event was present in such a sparse distributed representation.] Indeed, the number of different patterns or prototypes P that can be stored has the following approximate proportionality:

$$P \propto C/a^2 \qquad (11.5)$$

where a is the sparseness, which must hold for both the conditioned stimulus input pattern r' and the unconditioned stimulus pattern r. P can, in this situation, be much larger than C (see Rolls and Treves, 1990, 1997). This is an important result for encoding in pattern associators, for it means that provided that the activation functions are nonlinear (which is the case with real neurons), there is a very great advantage to using sparse encoding, for then many more than C pattern associations can be stored. Sparse representations may well be present in brain regions involved in associative memory for this reason (see Rolls and Treves, 1996).

Overwriting old memories. As additional memories are added to a network, it can be helpful to overwrite old memories. One of the advantages of this is that associative neuronal networks have a limited capacity, noted above, and if this capacity is exceeded, recall of any of the memories from the network can be impaired. Heterosynaptic LTD effectively allows gradual overwriting of old memories, in that synapses strengthened by old memories might be required according to Equation 11.4 to be decreased on to an active dendrite from inactive synapses. If the pattern association must be reversed (e.g. pattern 1 is no longer associated with a taste or reward), then the ability to show heterosynaptic LTD enables the output neurons to stop responding to pattern 1.

Synaptic strength resolution. In the associative networks described, the information that can be utilized per synapse is of the order of 0.2–0.4 bits. (1 bit might correspond to the synapse being either strong or weak/absent; 2 bits might correspond to four levels of synaptic strength.) This implies that the number of different levels of synaptic strength that would need to be implemented by a synaptic modification process such as LTP/LTD when applied to associative networks might be only two to four levels. Much can be achieved if there are only two levels of synaptic strength (absent/weak and strong), as shown in the original analyses of Willshaw and Longuet-Higgins (1969).

Nonlinear operations in the postsynaptic neuron – NMDA receptors. The nonlinear property of the N-methyl-D-aspartate (NMDA) receptors means that synaptic modification would only occur on to the most strongly activated postsynaptic neurons in an associative network. This has led to the notion of cooperativity, in that only postsynaptic neurons with several moderately active presynaptic inputs will have sufficient depolarization for any synaptic modification. This means that effectively the postsynaptic neuron would store the correlation between several conjunctively active inputs (and later respond to any one of the inputs alone). This effect is equivalent to nonlinearity in the output activation function of neurons. It tends (by making the representations stored effectively sparser than the representations received) to increase the number of memories that can be stored in the network (see Equation 11.5 above), at the expense of some (nonlinear) distortion of the inputs stored (see Rolls, 1989c; Rolls and Treves, 1990; Treves and Rolls, 1991).

11.3 Autoassociation memory

Autoassociative memories, or attractor neural networks, store memories, each one of which is represented by a pattern of neural activity. They can then recall the appropriate memory from the network when provided with a fragment of one of

CORTICAL PLASTICITY

the memories. This is called completion. Many different memories can be stored in the network and retrieved correctly. The network can learn each memory in one trial. Because of its 'one-shot' rapid learning, and ability to complete, this type of network is well suited for episodic memory storage, in which each past episode must be stored and recalled later from a fragment, and kept separate from other episodic memories. An autoassociation memory can also be used as a short-term memory, in which iterative processing round the recurrent collateral connection loop keeps a representation active until another input cue is received.

11.3.1 Architecture and operation

The prototypical architecture of an autoassociation memory is shown in *Figure 11.3*. The external input e_i is applied to each neuron i by unmodifiable synapses. This produces firing r_i of each neuron, or a vector of firing on the output neurons r. Each output neuron i is connected by a recurrent collateral connection to the other neurons in the network, via modifiable connection weights w_{ij}. This architecture effectively enables the output firing vector r to be associated during learning with itself. Later on, during recall, presentation of part of the external input will force some of the output neurons to fire but, through the recurrent collateral axons and the modified

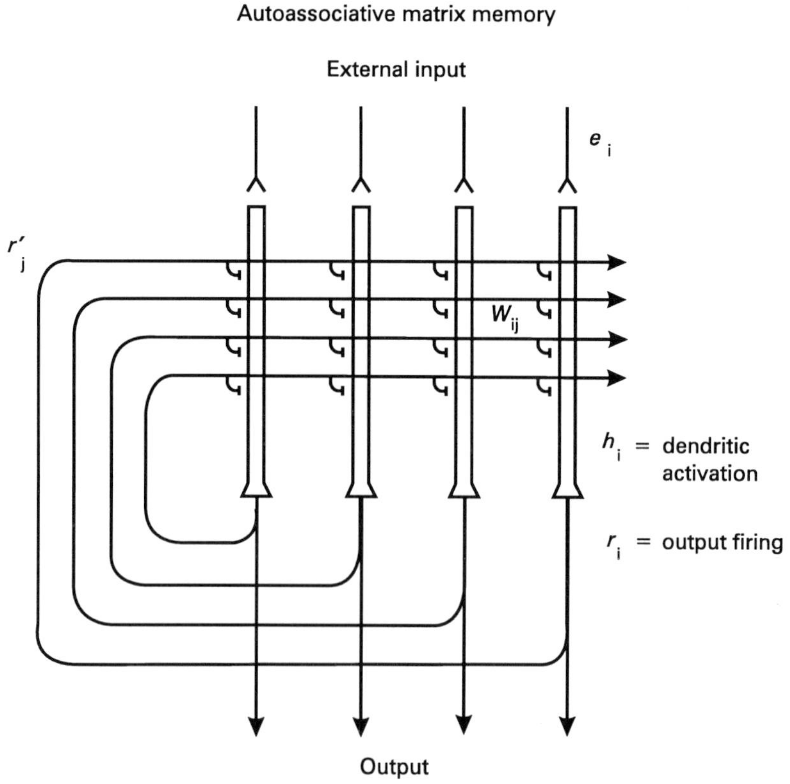

Figure 11.3. The architecture of a network for autoassociation (see text).

synapses, other neurons in *r* can be brought into activity. This process can be repeated a number of times, and recall of a complete pattern may be perfect. Effectively, a pattern can be recalled or recognized because of associations formed between its parts. This of course requires distributed representations.

Next, I introduce a more precise and detailed description of the above, and describe the properties of these networks. A formal description of the operation of these networks is provided in Appendix 3 of Rolls and Treves (1997), and by Hertz *et al.* (1991).

Learning. The firing of every output neuron i is forced to a value r_i determined by the external input e_i. It is sometimes overlooked that there must be a mechanism for ensuring that, during learning, r_i does approximate e_i, and must not be influenced much by the recurrent collateral connections – otherwise the new external pattern *e* will not be stored in the network, but instead something which is influenced by the previously stored memories. It is thought that, in some parts of the brain, such as the hippocampus, there are processes such as the effect of the mossy fiber inputs to the CA3 pyramidal neurons that help the external connections to dominate the firing during learning (see below and Rolls, 1989b,c; Rolls and Treves, 1997; Treves and Rolls, 1992). In addition, synaptic transmission in the recurrent collaterals may be reduced during learning (Hasselmo *et al.*, 1995) (see Equation 11.1)

Recall. During recall, the external input e_i is applied, and produces output firing, operating through the nonlinear activation function described below. The firing is fed back by the recurrent collateral axons, shown in *Figure 11.2*, to produce activation of each output neuron through the modified synapses on each output neuron. The internal activation h_i produced by the recurrent collateral effect on the ith neuron is the sum of the activations produced in proportion to the firing rate of each axon r'_j operating through each modified synapse w_{ij} (see Equation 11.2). The output firing r_i is a function of the activation produced by the recurrent collateral effect (internal recall) and by the external input (e_i):

$$r_i = f(h_i + e_i) \tag{11.6}$$

The output activation function *f* must be nonlinear, and may be, for example, binary threshold, linear threshold, sigmoid, etc. (see *Figure 11.2*). A nonlinear activation function is used to minimize interference between the pattern being recalled and other patterns stored in the network, and to ensure that what is a positive feedback system remains stable. The network can be allowed to repeat this recurrent collateral loop a number of times. Each time the loop operates, the output firing becomes more like the originally stored pattern, and this progressive recall is usually complete within 5–15 iterations.

11.3.2 Properties

The internal recall in autoassociation networks involves multiplication of the firing vector of neuronal activity by the vector of synaptic weights on each neuron.

This inner product vector multiplication allows the similarity of the firing vector to previously stored firing vectors to be provided by the output (as effectively a correlation), if the patterns learned are distributed. As a result of this type of correlation computation performed if the patterns are distributed, many important properties of these networks arise, including pattern completion (because part of a pattern is correlated with the whole pattern), and graceful degradation (because a damaged synaptic weight vector is still correlated with the original synaptic weight vector). These properties are described in more detail by Rolls and Treves (1997).

11.3.3 Relevance of different aspects of the synaptic modification used

Local learning rule. The learning used in autoassociation neural networks, a version of the Hebb rule, is as given in Equation 11.1. The same points apply as described above for pattern associators.

Positive-only firing rates. As with pattern associators, if positive-only firing rates are used, then the correlation this induces between different pattern vectors can be removed by subtracting the mean of the presynaptic activity from each presynaptic term, using a type of long-term depression. This can be specified as in Equation 11.4. This learning rule includes (in proportion to r_i) increasing the synaptic weight if $(r'_j - c)$ is > 0 (LTP), and decreasing the synaptic weight if $(r'_j - c)$ is < 0 (heterosynaptic LTD). This procedure works optimally if c is the average activity of an input axon r'_j across patterns. One implication of this is that, if the postsynaptic activity is high, and the presynaptic activity is below average, then LTD might be predicted if the networks are involved in autoassociation (or pattern association).

Capacity (see Treves and Rolls, 1991). The number of different memories that can be stored in autoassociators is (as with pattern associators) proportional to the number C of independently modifiable synapses on the dendrite of each neuron. It is for this reason that a synapse-specific modification rule is desirable. Some further details are provided next, but what has just been stated is the most important point in relation to synapse specificity.

With the nonlinear neurons used in the network, the capacity can be measured in terms of the number of input patterns r that can be stored in the network and recalled later with a stable basin of attraction. With fully distributed orthogonal (after subtracting the mean) input patterns r (see pattern associators), it is possible to store C different patterns, where there are C inputs to each neuron in the network, each neuron having C synaptic weights. With fully distributed random binary patterns in a fully connected autoassociation network, the number of patterns that can be learned is 0.14C (Hopfield, 1982). Treves and Rolls (1991) have been able to extend this analysis to autoassociation networks which are much more biologically relevant by having diluted connectivity (missing synapses), and neurons with graded (continuously variable) firing rates. The number of different patterns P that can be stored is then:

$$P \approx \frac{C}{a \ln(1/a)} k \qquad (11.7)$$

where C is the number of synapses on each dendrite devoted to the recurrent collaterals from other neurons in the network, and k is a factor that depends weakly on the detailed structure of the rate distribution, on the connectivity pattern, etc., but is roughly of the order of 0.2–0.3. a is the sparseness of the representation, which for binary neurons can be measured by the proportion of neurons that are firing. [For neurons with real firing rates, the corresponding measure is $a = (\Sigma_{i=1,n} r_i/n)^2 / \Sigma_{i=1,n}(r_i^2/n)$ where r_i is the firing rate of the ith neuron in the set of n neurons.]

Overwriting old memories. As additional memories are added to an autoassociative network, it can be helpful to overwrite old memories. One of the advantages of this is that autoassociative neuronal networks have a limited capacity, noted above, and, if this capacity is exceeded, recall of any of the memories from the network is impaired. Heterosynaptic LTD effectively allows gradual overwriting of old memories, in that synapses strengthened by old memories might be required according to Equation 11.4 to be decreased on to an active dendrite from inactive synapses. Homosynaptic LTD may also contribute to this useful forgetting.

Synaptic strength resolution. In the autoassociative networks described, the information that can be utilized per synapse is of the order of 0.2–0.4 bits (Rolls *et al.*, 1996; Treves and Rolls, 1991). This implies that the number of different levels of synaptic strength that would need to be implemented by a synaptic modification process such as LTP/LTD when applied to associative networks might be only two to four (or less) levels.

Nonlinear operations in the postsynaptic neuron – NMDA receptors. The nonlinear property of the NMDA receptors means that synaptic modification would only occur on to the most strongly activated postsynaptic neurons in an autoassociative network. This effect is equivalent to nonlinearity in the output activation function of neurons. It tends (by making the representations stored effectively sparser that the representations received) to increase the number of memories that can be stored in the network (see Equation 11.7), at the expense of some (nonlinear) distortion of the memories stored (see Rolls, 1989c; Treves and Rolls, 1991).

Covariance learning rule vs. LTD. In a covariance learning rule, the means of both the pre- and the postsynaptic factor are subtracted, as indicated in Equation 11.8.

$$\delta w_{ij} = k \cdot (r_i - \mu_i) \cdot (r'_j - \mu_j) \qquad (11.8)$$

where μ_i is the mean of the postsynaptic activity of the output neuron, and μ_j is the mean of the presynaptic terminal j. This rule was effectively used in the original

Hopfield (1982) autoassociation network. An implication of this rule is that, with both the presynaptic and the postsynaptic activity low, the synapse will be strengthened, which seems neurobiologically unnatural. In heterosynaptic LTD, only the mean of the presynaptic activity μ_j is subtracted. Autoassociation networks operate well if only heterosynaptic LTD is present. The term $(r_i - \mu_i)$ in Equation 11.8 would allow homosynaptic LTD if the presynaptic rate is high and the postsynaptic activity r_i is low.

11.4 Competitive networks

11.4.1 Function

Competitive neural nets learn to categorize input pattern vectors. Each category of inputs activates a different output neuron (or set of output neurons – see below). The categories formed are based on similarities between the input vectors. Similar (i.e. correlated) input vectors activate the same output neuron. In that the learning is based on similarities in the input space, and there is no external teacher which forces classification, this is an unsupervised network. The term categorization is used to refer to the process of placing vectors into categories based on their similarity.

The categorization produced by competitive nets is of great potential importance in perceptual systems. Each category formed reflects a set or cluster of active inputs r'_j which occur together. This cluster of co-active inputs can be thought of as a feature, and the competitive network can be described as building feature analyzers, where a feature can now be defined as a correlated set of inputs. During learning, a competitive network gradually discovers these features in the input space, and the process of finding these features without a teacher is referred to as self-organization. Another important use of competitive networks is to remove redundancy from the input space, by allocating output neurons to reflect a set of inputs which co-occur. Another important aspect of competitive networks is that they separate patterns which are somewhat correlated in the input space, to produce outputs for the different patterns which are less correlated with each other, and may indeed easily be made orthogonal to each other. This has been referred to as orthogonalization. Another important function of competitive networks is that partly by removing redundancy from the input information space, they can produce sparse output vectors, without losing information. We may refer to this as sparsification.

These latter operations are useful as preprocessing operations before signals are applied to associative networks, which benefit from sparse, noncorrelated input patterns if they are to store large numbers of memories (see above and Rolls and Treves, 1997).

11.4.2 Architecture and algorithm

Architecture. The basic architecture of a competitive network is shown in *Figure 11.4*. It is a one-layer network with a set of inputs r' which make modifiable excitatory synapses w_{ij} with the output neurons. The output cells compete with each other (e.g. by mutual inhibition) in such a way that the most strongly

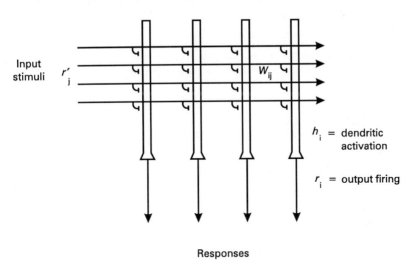

Figure 11.4. The architecture of a competitive network (see text).

activated neuron or neurons win the competition, and are left firing strongly. The synaptic weights, w_{ij}, are initialized to random values. The absence of some of the synapses (e.g. producing randomly diluted connectivity), is not a problem for such networks, and can even help them (see below).

In the brain, the inputs arrive through axons, which make synapses with the dendrites of the output or principal cells of the network. The principal cells are typically pyramidal cells in the cerebral cortex. In the brain, the principal cells are typically excitatory, and mutual inhibition between them is implemented by inhibitory interneurons, which receive excitatory inputs from the principal cells. The inhibitory interneurons then send their axons to make synapses with the pyramidal cells, typically using γ-aminobutyric acid (GABA) as the inhibitory transmitter. An algorithm to represent this follows.

Algorithm.
(i) Apply an input vector r' and calculate the output activation h_i of each neuron as in Equation 11.2. (It is useful to normalize the length of each input vector r'.) The output firing $r1_i$ is a function of the output activation:

$$r1_i = f(h_i) \tag{11.9}$$

This function can be linear, sigmoid, monotonically increasing, etc.

(ii) Allow competitive interaction between the output neurons by a mechanism like lateral or mutual inhibition, to produce a contrast-enhanced version of $r1$:

$$r = f_{comp}(r1) \tag{11.10}$$

This is typically a nonlinear operation and, in its most extreme form, may be a winner-take-all function, in which after the competition one neuron may be 'on', and the others 'off'.

(iii) Apply the Hebb learning rule as in Equation 11.1.

(iv) Normalize the length of the synaptic weight vector on each dendrite to prevent the same few neurons always winning the competition.

$$\sum_j (w_{ij})^2 = 1 \qquad (11.11)$$

(v) Repeat steps (i)–(iv) for each different input stimulus r' in random sequence a number of times.

11.4.3 Properties

Feature discovery by self-organization. Each neuron in a competitive network becomes activated by a set of consistently co-active, that is correlated, input axons, and gradually learns to respond to that cluster of co-active inputs. We can thus think of competitive networks as discovering features in the input space, where features can now be defined by a set of consistently co-active inputs. Competitive networks thus show how feature analyzers can be built, with no external teacher. The feature analyzers respond to correlations in the input space, and the learning occurs by self-organization in the competitive network. Competitive networks are thus well suited to the analysis of sensory inputs (see further Rolls and Treves, 1996).

Capacity. In a competitive net with n output neurons, it is possible to learn up to n output categories, in that the competitive interactions between output neurons are first order (i.e. the competition between the output neurons is uniform in that they are not connected in such a way that pairs, triples, etc. can act together to force themselves apart from other pairs, triples, etc. of output neurons). It is thus the number of output neurons, rather than the number of input connections per neuron, that determines the number of categories that can be formed. The output neurons need not operate in a winner-take-all manner (to produce local or grandmother cell representations), but instead can operate with soft competition to leave a number of neurons still active after the mutual (lateral) inhibition.

Normalization of the length of the synaptic weight vector on each dendrite – a role for heterosynaptic LTD. Normalization is necessary to ensure that one or a few neurons do not always win the competition. (If the weights on one neuron were increased by simple Hebbian learning, and there was no normalization of the weights on the neuron, then it would tend to respond strongly in the future to patterns with some overlap with patterns to which that neuron has learned previously, and gradually that neuron would capture a large number of patterns.) A biologically plausible way to achieve this weight adjustment is to use a modified Hebb rule:

$$\delta w_{ij} = k \cdot r_i \left(r'_j - w_{ij} \right) \qquad (11.12)$$

where k is a constant, and r'_j and w_{ij} are in appropriate units. This implements a Hebb rule which increases synaptic strength according to conjunctive pre- and postsynaptic activity, and also allows the strength of each synapse to decay in proportion to the firing rate of the postsynaptic neuron (as well as in proportion to the existing synaptic strength). This is an important computational use of heterosynaptic LTD. This rule can maintain the sums of the synaptic weights on each neuron to be very similar without any need for explicit normalization of the synaptic strengths, and is useful in competitive nets. (The rule given in Equation 11.12 strictly maintains the sum of the weights on a neuron to be constant if the neurons are linear.) This rule was used by Willshaw and von der Malsburg (1976).

If explicit weight normalization is needed (i.e. keeping the vector length, that is the square root of the sum of the squares of the synaptic strengths, constant), the appropriate form of the modified Hebb rule (again for linear neurons) is:

$$\delta w_{ij} = k \cdot r_i \left(r'_j - r_i w_{ij} \right) \qquad (11.13)$$

This rule, formulated by Oja (1982), makes weight decay proportional to r_i^2, normalizes the synaptic weight vector (see Hertz et al., 1991), is still a local learning rule, and is known as the Oja rule.

In both these rules, the important point is that the degree of LTD and LTP depends on the existing synaptic strength. If the synaptic strength is low, for example before any LTP has been induced, then LTP should be (other things being equal) large. If LTP has already been induced, further LTP might be expected to be smaller. Conversely, LTD should be especially evident when LTP has been induced previously. These effects, which are frequently found in experiments on LTP and LTD, should not therefore necessarily be regarded as spurious effects, but may be related to this fundamental design feature needed in competitive neuronal networks.

Nonlinearity in the learning rule. Nonlinearity in the learning rule can assist competition (Rolls, 1989c). In the brain, LTP typically occurs only when strong activation of a neuron has produced sufficient depolarization for the voltage-dependent NMDA receptors to become unblocked, allowing Ca^{2+} to enter the cell. This means that synaptic modification occurs only on neurons that are strongly activated, effectively assisting competition to select few winners. The learning rule can be written:

$$\delta w_{ij} = k \cdot m_i \left(r'_j - w_{ij} \right) \qquad (11.14)$$

where m_i is a nonlinear function of r_i which mimics the operation of the NMDA receptors in learning.

11.4.4 Brain systems in which competitive networks may be used for orthogonalization and sparsification

One system is the hippocampus, in which the dentate granule cells may operate as a competitive network in order to prepare signals for presentation to the CA3 autoassociative network (see Section 11.5). In this case, the operation is enhanced by expansion recoding, in that (in the rat) there are approximately three times as many dentate granule cells as there are cells in the preceding stage, the entorhinal cortex (see Section 11.4). This expansion recoding will itself tend to reduce correlations between patterns (cf. Marr, 1969, 1970).

Also in the hippocampus, the CA1 neurons are thought to act as a competitive network which recodes the separate representations of each of the parts of an episode which must be separately represented in CA3, into a form more suitable for the recall by pattern association performed by the back-projections from the hippocampus to the cerebral cortex (see Section 11.6).

The granule cells of the cerebellum may perform a similar function, but in this case the principle may be that each of the very large number of granule cells receives a very small random subset of inputs, so that the outputs of the granule cells are decorrelated with respect to the inputs (Marr, 1969; see Rolls and Treves, 1997).

11.4.5 Competitive networks using LTP and homosynaptic LTD

A different learning rule which has been proposed to account for plasticity changes during development of the visual system has, in addition to Hebbian increments (synaptic strength increases when there is high pre- and postsynaptic activity), decrements in synaptic strength if the postsynaptic firing is below a certain threshold T_c and there is presynaptic activity. These decrements would correspond to homosynaptic LTD. In Bienenstock *et al.*'s investigation (1982) of this rule, the threshold (T_c) referred to above was a function of the average activation of the neuron, and varied in such a way that the average activation of different neurons was kept approximately equal. This BCM rule prevents domination by any one dendrite, by adjusting its sensitivity (using T_c) according to its mean response to the input stimulus set. A very active neuron will develop a very high threshold, causing reduction of weights and thus reduced response to future stimuli. It is correspondingly impossible for a neuron not to react to at least one of the stimuli, and so all output neurons become allocated to one category or another. A further analysis of networks taught with this rule is provided by Rolls and Treves (1997).

11.5 The hippocampus

Evidence from the effects of damage to the hippocampus, and from recording the activity of neurons in it, indicates that it is involved in the formation of memories about particular events. These are formed rapidly, and frequently involve associating together a number of spatial inputs (where the event happened) with visual (e.g. the sight of a person) inputs and perhaps auditory, taste and olfactory inputs. It is suggested that the memory for such episodes is formed in the hippocampus, by allowing convergence of all these signals into one network, that of the CA3 pyramidal cells. This network has extensive recurrent connections between the different CA3 pyramidal cells, which are thought to be associatively modifiable. It has, therefore, been suggested that the CA3 cells operate as an autoassociative network, which can store such episodic memories, and then recall each memory from a part of it.

Given these hypotheses, we have developed a computational theory of the operation of the hippocampus (see Rolls, 1987, 1989b,c, 1990a,b,d,e; Treves and Rolls, 1991, 1992, 1994). The ways in which LTP and LTD are thought to contribute to the operation of different parts of the hippocampal circuitry are now described, to show how the three types of network described above may be used in this brain system.

11.5.1 Hippocampal circuitry (see Figure 11.5)

Projections from the entorhinal cortex reach the granule cells (of which there are 10^6 in the rat) in the dentate gyrus via the perforant path. The granule cells project to CA3 cells via the mossy fibers, which provide a *sparse* but possibly powerful connection to the 3×10^5 CA3 pyramidal cells in the rat. Each CA3 cell receives approximately 50 mossy fiber inputs, so that the sparseness of this connectivity is thus 0.005%. By contrast, there are many more – possibly weaker – direct perforant path inputs on to each CA3 cell, in the rat of the order of 4×10^3. The largest number of synapses ($\sim 1.2 \times 10^4$ in the rat) on the dendrites of CA3 pyramidal cells is, however, provided by the (recurrent) axon collaterals of CA3 cells themselves. It is remarkable that the recurrent collaterals are distributed to other CA3 cells throughout the hippocampus (Amaral and Witter, 1989; Amaral *et al.*, 1990; Ishizuka *et al.*, 1990), so that effectively the CA3 system provides a single network, with a connectivity of approximately 2% between the different CA3 neurons given that the connections are bilateral.

11.5.2 CA3 as an autoassociation memory

Many of the synapses in the hippocampus show associative modification as shown by LTP, and this synaptic modification appears to be involved in learning (Morris, 1989). On the basis of the evidence summarized above, Rolls (1987, 1989b,c, 1990a,b, 1991a,b) has suggested that the CA3 stage acts as an autoassociation memory which enables episodic memories to be formed and stored for an intermediate term in the CA3 network, and that subsequently the extensive recurrent collateral connectivity allows for the retrieval of a whole representation to be initiated by the activation of some small part of the same representation (the cue). The hypothesis is that because the CA3 operates effectively as a single network, it can allow arbitrary associations between inputs originating from very different parts of the cerebral cortex to be formed. These might involve associations between information originating in the temporal visual cortex about the presence of an object, and information originating in the parietal cortex about where it is. We have therefore performed quantitative analyses of the storage and retrieval processes in the CA3 network (Treves and Rolls, 1991, 1992). The analysis described in Section 11.3 showed that the number of memories that can be stored in an autoassociation network, such as that believed to be implemented by the

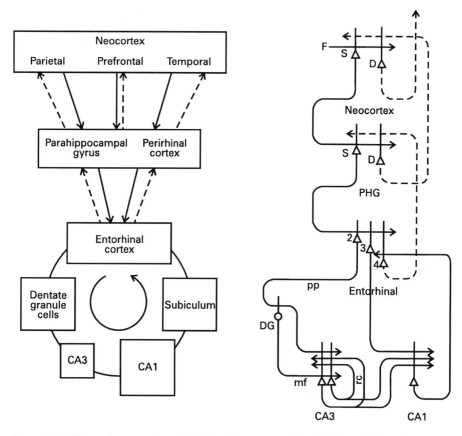

Figure 11.5. Forward connections (solid lines) from areas of cerebral association neocortex via the parahippocampal gyrus and perirhinal cortex and entorhinal cortex, to the hippocampus; and back-projections (dashed lines) via the hippocampal CA1 pyramidal cells, subiculum and parahippocampal gyrus to the neocortex. There is great convergence in the forward connections down to the single network implemented in the CA3 pyramidal cells; and great divergence again in the back-projections. Left: block diagram. Right: more detailed representation of some of the principal excitatory neurons in the pathways. The thick lines above the cell bodies represent the dendrites. D, deep pyramidal cells; DG, dentate granule cells; F, forward inputs to areas of the association cortex from preceding cortical areas in the hierarchy; mf, mossy fibers; PHG, parahippocampal gyrus and perirhinal cortex; pp, perforant path; rc, recurrent collateral of the CA3 hippocampal pyramidal cells; S, superficial pyramidal cells; 2, pyramidal cells in layer 2 of the entorhinal cortex; 3, pyramidal cells in layer 3 of the entorhinal cortex.

CA3 neurons, is proportional to the number of independently modifiable synapses on to each neuron from the other CA3 neurons. The computational requirement is thus for synapse specificity of LTP and LTD if the number of different memories stored in the network is to be maximized. For C = 12 000 modifiable synapses from the recurrent collaterals, and $a = 0.02$ (a realistic estimate of the sparseness of the representation for the rat), P_{max} is calculated to be approximately 36 000 different memories.

For this autoassociation to operate without interference due to the positive-only firing rates of real neurons, it is also predicted that heterosynaptic LTD should be demonstrable in the CA3–CA3 associatively modifiable synapses.

In an argument developed elsewhere, we hypothesize that the mossy fiber inputs force efficient information storage by virtue of their strong and sparse influence on the CA3 cell firing rates (Treves and Rolls, 1992; Rolls, 1989a,b). On the basis of this, we predict that the mossy fibers may be necessary for new learning in the hippocampus, but may not be necessary for recall of existing memories from the hippocampus. The nonassociative LTP of the mossy fiber–CA3 cell synapses is suggested to enhance the signal-to-noise ratio of the mossy fiber input: consistently firing mossy fibers will tend to produce large effects on the postsynaptic neuron (see Treves and Rolls, 1994). Experimental evidence consistent with this prediction about the role of the mossy fibers in learning has been described in mice without mossy fiber LTP associated with a lack of the metabotropic glutamate receptor (Conquet *et al.*, 1994).

We have also presented reasons why the direct perforant path system to CA3 (see *Figure 11.5*), which needs itself to be associatively modifiable, is the one involved in relaying the cues that initiate retrieval (Treves and Rolls, 1992). This set of synapses does show LTP. This perforant path to CA3 projection is hypothesized to operate effectively as a pattern associator, associating whatever firing the mossy fibers produce in the CA3 neurons with the concurrent activity on the direct perforant path inputs. Later, a fragment of the original perforant path input can act as a good retrieval cue to initiate retrieval in the CA3 autoassociation network (Rolls, 1995c; Treves and Rolls, 1992).

The theory is developed elsewhere that the dentate granule cell stage of hippocampal processing acts to produce during learning the sparse yet efficient (i.e. nonredundant) representation in CA3 neurons which is required for the autoassociation to perform well (Rolls, 1989a,c, 1993a; see also Treves and Rolls, 1992). One way in which it may do this is by acting as a competitive network to remove redundancy from the inputs producing a more orthogonal, sparse and categorized set of outputs (Rolls, 1987, 1989b,c, 1990a–e). This is the hypothesized function of LTP on to dentate granule cells from the perforant path input. Heterosynaptic LTD is also predicted to be here, because of its importance in competitive networks (see above).

It is suggested that the CA1 cells, given the separate parts of each episodic memory which must be represented separately in CA3 ensembles, can allocate neurons, by competitive learning, to represent at least larger parts of each episodic memory (Rolls, 1987, 1989b,c, 1990a,b). This implies a more efficient representation, in the sense that when, eventually, after many further stages, neocortical neuronal activity is recalled (as discussed below), each neocortical cell need not be accessed by all the axons carrying each component of the episodic memory as represented in CA3, but instead by fewer axons carrying larger fragments (see Treves and Rolls, 1994).

11.5.3 Back-projections to the neocortex

It is suggested that the hippocampus is able to recall the whole of a previously stored episode for a period of days, weeks or months after the episode, when even a fragment of the episode is available to start the recall. This recall from a fragment of the original episode would take place particularly as a result of completion produced by the autoassociation implemented in the CA3 network. It would then be the role of the hippocampus to reinstate in the cerebral neocortex the whole of the episodic memory. The cerebral cortex would then, with the whole of the information in the episode now producing firing in the correct sets of neocortical neurons, be in a position to use the information to initiate action, or to use the recalled episodic information to contribute to the formation of new structured semantic memories (see Section 11.6).

It is suggested that during recall, the connections from CA3 via CA1 (and the subiculum) would allow activation of at least the pyramidal cells in the deep layers of the entorhinal cortex (see *Figure 11.5*). These neurons would then, by virtue of their back-projections to the parts of cerebral cortex that originally provided the inputs to the hippocampus, terminate in the superficial layers of those neocortical areas, where synapses would be made on to the distal parts of the dendrites of the cortical pyramidal cells (see Rolls, 1989b,c).

Our understanding of the architecture with which this would be achieved is shown in *Figure 11.5*. The feed-forward connections from association areas of the cerebral neocortex (solid lines in *Figure 11.5*), show major convergence as information is passed to CA3, with the CA3 autoassociation network having the smallest number of neurons at any stage of the processing. The back-projections allow for divergence back to neocortical areas. The way in which we suggest that the back-projection synapses are set up to have the appropriate strengths for recall is as follows (see Rolls, 1989b,c; Treves and Rolls, 1994). During the setting up of a new episodic memory, there would be strong feed-forward activity progressing towards the hippocampus. During the episode, the CA3 synapses would be modified and, via the CA1 neurons (and the subiculum), a pattern of activity would be produced on the back-projecting synapses to the entorhinal cortex. Here the back-projecting synapses from active back-projection axons on to pyramidal cells being activated by the forward inputs to entorhinal cortex would be associatively modified. A similar process would be implemented at some at least of the preceding stages of neocortex, that is in the parahippocampal gyrus/perirhinal cortex stage, and in association cortical areas.

How many back-projecting fibers does one need to synapse on any given neocortical pyramidal cell, in order to implement the mechanism outlined above? We have shown that the maximum number of independently generated memory patterns that can be retrieved is given, essentially, by the same formula as Equation 11.7:

$$P \approx \frac{C}{a \ln (1/a)} k' \tag{11.15}$$

where, however, a is now the sparseness of the representation at any given stage, and C is the average number of (back-)projections each cell of that stage receives from cells of the previous one (Treves and Rolls, 1991, 1994). (k' is a similar slowly varying factor to that introduced in Section 11.2.3 above.) This shows that there should be very many back-projection fibers; in fact, there are as many as the forward projecting fibers. It also shows that the back-projection pathway should be multistage, for otherwise each CA3 (or CA1) cell would have to contact an enormous number of neocortical neurons.

This recall process could only operate if there are many back-projection inputs to each neocortical stage. This is the case. There are approximately as many back-projection connections as forward connections between two adjacent cortical areas. The reason for this, it is suggested, is so that as many representations can be recalled by back-projections as can be produced by the forward inputs (Rolls, 1989a–c; Treves and Rolls, 1994). The recall process could also only operate if the back-projections are associatively modifiable, and this is therefore a prediction. It is also predicted that heterosynaptic LTD will be demonstrable in the back-projection pathways, because of its importance in what is a pattern association architecture.

11.6 Synaptic modification rules useful in learning invariant representations in the cerebral cortex

Neurophysiological findings indicate that neurons in the temporal visual cortical areas can respond to objects such as faces relatively independently of where on the retina the face is shown, of its size, and even in some cases relatively independently of the view of the face (see Rolls, 1992, 1994, 1996b). It is suggested that these invariant representations are formed using a multilayer competitive network architecture in which each layer corresponds to one of the visual cortical areas. In order to learn about the invariances which are characteristic of objects in the real world, it is suggested that a modified Hebb rule is used in the competitive networks, which have built in a short-term averaging process, operating over a period of the order of 0.5 sec. This would enable the network to associate together inputs produced by the same object seen in rapid succession in different places on the retina, in different sizes, in different views, etc. Because this trace of previous activation would last for only approximately 0.5 sec, the representations of different objects would not be associated together, due to the statistics of viewing of the world. (One object is typically inspected for a short period, then the eyes move to another, etc.) The prediction is that, in the temporal cortical areas concerned with forming invariant representations, a modified Hebbian learning rule would be present, which would include a trace of previous activation because of its utility in learning invariant representations. An alternative is that ongoing neural activity might implement the short-term memory required.

To test and clarify the hypotheses just described about how the visual system may operate to learn invariant object recognition, we have performed a simulation which implements a specification of how the visual cortex could operate. The network simulated can perform object, including face, recognition in a biologically plausible way and, after training, shows for example translation and view invariance (Wallis and Rolls, 1997; Wallis et al., 1993).

The synaptic learning rule used can be summarized as follows:

$$\delta w_{ij} = k \cdot m_i \cdot r'_j \tag{11.16}$$

and

$$m_i^t = (1-\eta) f_i^{(t)} + \eta m_i^{(t-1)} \tag{11.17}$$

where r'_j is the jth input to the neuron, f_i is the output of the ith neuron, w_{ij} is the jth weight on the ith neuron, η governs the relative influence of the trace and the new input (typically 0.4–0.6), and m_i^t represents the value of the ith cell's memory trace at time t. In our simulations, the neuronal learning was bounded by normalization of each cell's dendritic weight vector, as in standard competitive learning. The alternative, more biologically relevant implementation, using a local weight bounding operation, has in part been explored using a version of the Oja update rule shown in Equation 11.12 above (Kohonen, 1984; Oja, 1982). The results of the simulations show that networks trained with the trace learning rule do have neurons with high values of the discrimination factor which measures how well a neuron discriminates between stimuli independently of location (see *Figure 11.6*). Similar position invariant encoding has been demonstrated for a stimulus set consisting of eight faces. View invariant coding has also been demonstrated for a set

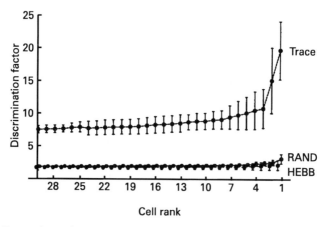

Figure 11.6. Comparison of VisNet network discrimination when trained with the trace learning rule, with a Hebb rule (no trace), and when not trained (random). The discrimination factor, showing how well each neuron discriminated between the stimuli independently of location (see Rolls, 1997b), for the 32 most invariant neurons is shown. The network was trained with three letters (L, T and +) in each of nine different positions on the retina.

of five faces each shown in four views (Rolls, 1997b; Wallis and Rolls, 1996; Wallis et al., 1993).

These results show that the proposed learning mechanism and neural architecture can produce cells with responses selective for stimulus type with considerable position or view invariance. The ability of the network to be trained with natural scenes may also help to advance our understanding of encoding in the visual system.

The hypothesis is that because objects have continuous properties in space and time in the world, an object at one place on the retina might activate feature analyzers at the next stage of cortical processing and, when the object was translated to a nearby position, because this would occur in a short period (e.g. 0.5 sec), the membrane of the postsynaptic neuron would still be in its 'Hebb-modifiable' state (caused, for example, by calcium entry as a result of the voltage-dependent activation of NMDA receptors), and the presynaptic afferents activated with the object in its new position would thus become strengthened on the still-activated postsynaptic neuron. It is suggested that the short temporal window (e.g. 0.5 sec) of Hebb-modifiability helps neurons to learn the statistics of objects moving in the physical world, and at the same time to form different representations of different feature combinations or objects, as these are physically discontinuous and present less regular correlations to the visual system. One idea here is that the temporal properties of the biologically implemented learning mechanism are such that it is well suited to detecting the relevant continuities in the world of real objects. Another suggestion is that a memory trace for what has been seen in the last 300 msec appears to be implemented by a mechanism as simple as continued firing of inferior temporal neurons after the stimulus has disappeared, as was found in the masking experiments described by Rolls and Tovee (1994) and Rolls et al. (1994). I also suggest that other invariances, for example size, spatial frequency and rotation invariance, could be learned by a comparable process. (Early processing in V1 which enables different neurons to represent inputs at different spatial scales would allow combinations of the outputs of such neurons to be formed at later stages. Scale invariance would then result from detecting at a later stage which neurons are almost conjunctively active as the size of an object alters.) It is suggested that this process takes place at each stage of the multiple-layer cortical processing hierarchy, so that invariances are learned first over small regions of space, and then over successively larger regions. This limits the size of the connection space within which correlations must be sought.

The trace process to LTP/LTD described here would be useful in cortical areas involved in learning invariant representations. In the primate brain, this is the ventral visual system projecting via V1, V2 and V4 to the temporal cortical visual areas. It would not be useful in cortical mechanisms involved in motion detection, in which temporal smearing of this type might interfere with the need for precise dynamics in the system. Motion analysis operates especially in the dorsal visual system projecting from V1 via MT to MST, etc. It is therefore predicted that a temporal trace in the synaptic modification rules may be less evident in the dorsal than the ventral visual system. The requirement for a different type of synaptic modification rule might indeed be one of the fundamental computational reasons for separation of visual processing into these two partially separate processing streams.

11.7 Conclusions

The importance of some of the different properties of LTP and LTD for understanding *how* different parts of the brain compute has been described. A number of predictions about the properties of LTP and LTD in biological systems have been made.

Acknowledgments

The author has worked with Dr Alessandro Treves, Dr G Wallis and Dr P Foldiak on some of the investigations referred to here, and their contribution is sincerely acknowledged. The author thanks Dr R Baddeley and S Schultz for reading an earlier version of the chapter. Different parts of the research described were supported by the Medical Research Council, PG8513790; by a Human Frontier Science Program grant; by an EC Human Capital and Mobility grant; by the MRC Oxford Interdisciplinary Research Centre in Brain and Behaviour; and by the Oxford McDonnell–Pew Centre in Cognitive Neuroscience.

References

Amaral DG, Witter MP. (1989) The three-dimensional organization of the hippocampal formation: a review of anatomical data. *Neuroscience* **31**: 571–591.
Amaral DG, Ishizuka N, Claiborne B. (1990) Neurons, numbers and the hippocampal network. *Prog. Brain Res.* **83**: 1–11.
Bienenstock EL, Cooper LN, Munro PW. (1982) Theory for the development of neuron selectivity: orientation specificity and binocular interaction in visual cortex. *J. Neurosci.* **2**: 32–48.
Brown TH, Kairiss EW, Keenan CL. (1990) Hebbian synapses: biophysical mechanisms and algorithms. *Annu. Rev. Neurosci.* **13**: 475–511.
Conquet F, Bashir ZI, Davies CH et al. (1994) Motor deficit and impairment of synaptic plasticity in mice lacking mGluR1. *Nature* **372**: 237–243.
Davis M. (1992) The role of the amygdala in conditioned fear. In: *The Amygdala* (ed. JP Aggleton). Wiley-Liss, New York, pp. 255–305.
Hasselmo ME, Schnell E, Barkai E. (1995) Dynamics of learning and recall at excitatory recurrent synapses and cholinergic modulation in rat hippocampal region CA3. *J. Neurosci.* **15**: 5249–5282.
Hertz J, Krogh A, Palmer RG. (1991) *Introduction to the Theory of Neural Computation*. Addison-Wesley, Wokingham.
Hopfield JJ. (1982) Neural networks and physical systems with emergent collective computational abilities. *Proc. Natl Acad. Sci USA* **79**: 2554–2558.
Ishizuka N, Weber J, Amaral DG. (1990) Organization of intrahippocampal projections originating from CA3 pyramidal cells in the rat. *J. Comp. Neurol.* **295**: 580–623.

Jordan MI. (1986) An introduction to linear algebra in parallel distributed processing. In *Parallel Distributed Processing*, Vol. 1, *Foundations* (eds DE Rumelhart, JL McClelland). MIT Press, Cambridge, MA, pp. 365–442.

Kohonen T. (1984) *Self-Organization and Associative Memory*. Springer-Verlag, Berlin.

Levy WB. (1985) Associative changes in the synapse: LTP in the hippocampus. In: *Synaptic Modification, Neuron Selectivity, and Nervous System Organization* (eds WB Levy, JA Anderson, S Lehmkuhle). Erlbaum, Hillsdale, NJ, pp. 5–33.

Levy WB, Desmond NL. (1985) The rules of elemental synaptic plasticity. In: *Synaptic Modification, Neuron Selectivity, and Nervous System Organization* (eds WB Levy, JA Anderson, S Lehmkuhle). Erlbaum, Hillsdale, NJ, pp. 105–121.

Marr D. (1969) A theory of cerebellar cortex. *J. Physiol.* 202: 437–470.

Marr D. (1970) A theory for cerebellar cortex. *Proc. R. Soc. B* 76: 161–234.

Marr D. (1971) Simple memory: a theory for archicortex. *Phil. Trans. R. Soc. B* 262: 23–81.

Morris RGM. (1989) Does synaptic plasticity play a role in information storage in the vertebrate brain? In: *Parallel Distributed Processing: Implications for Psychology and Neurobiology* (ed RGM Morris). Oxford University Press, Oxford, pp. 248–285.

Oja E. (1982) A simplified neuron model as a principal component analyzer. *J. Math. Biol.* 15: 267–273.

Rolls ET. (1987) Information representation, processing and storage in the brain: analysis at the single neuron level. In: *The Neural and Molecular Bases of Learning* (eds J-P Changeux, M Konishi). John Wiley & Sons, Chichester, pp. 503–540.

Rolls ET. (1989a) Information processing in the taste system of primates. *J. Exp. Biol.* 146: 141–164.

Rolls ET. (1989b) Functions of neuronal networks in the hippocampus and neocortex in memory. In: *Neural Models of Plasticity: Experimental and Theoretical Approaches* (eds JH Byrne, WO Berry). Academic Press, San Diego, pp. 240–265.

Rolls ET. (1989c) Parallel distributed processing in the brain: implications of the functional architecture of neuronal networks in the hippocampus. In: *Parallel Distributed Processing: Implications for Psychology and Neurobiology* (ed RGM Morris). Oxford University Press, Oxford, pp. 286–308.

Rolls ET. (1990a) Principles underlying the representation and storage of information in neuronal networks in the primate hippocampus and cerebral cortex. In: *An Introduction to Neural and Electronic Networks* (eds SF Zornetzer, JL Davis, C Lau). Academic Press, San Diego, pp. 73–90.

Rolls ET. (1990b) Theoretical and neurophysiological analysis of the functions of the primate hippocampus in memory. *Cold Spring Harbor Symp. Quant. Biol.* 55: 995–1006.

Rolls ET. (1990c) A theory of emotion, and its application to understanding the neural basis of emotion. *Cognit. Emot.* 4: 161–190.

Rolls ET. (1990d) Functions of neuronal networks in the hippocampus and of backprojections in the cerebral cortex in memory. In: *Brain Organization and Memory: Cells, Systems and Circuits* (eds JL McGaugh, NM Weinberger, G Lynch). Oxford University Press, New York, pp. 184–210.

Rolls ET. (1990e) Functions of the primate hippocampus in spatial processing and memory. In: *Neurobiology of Comparative Cognition* (eds DS Olton, RP Kesner). Erlbaum, Hillsdale, NJ, pp. 339–362.

Rolls ET. (1990f) Functions of different regions of the basal ganglia. In: *Parkinson's Disease* (ed. GM Stern). Chapman and Hall, London, pp. 151–184.

Rolls ET. (1991a) Functions of the primate hippocampus in spatial and non-spatial memory. *Hippocampus* 1: 258–261.

Rolls ET. (1991b) Functions of the primate hippocampus in spatial processing and memory. In: *Brain and Space* (ed. J Paillard). Oxford University Press, Oxford, pp. 353–376.

Rolls ET. (1992) Neurophysiological mechanisms underlying face processing within and beyond the temporal cortical visual areas. *Phil. Trans. R. Soc.* 335: 11–21.

Rolls ET. (1993a) Neurophysiological and neuronal network analysis of how the hippocampus functions in memory. In: *The Memory System of the Real Brain* (ed. J Delacour). World Scientific Publishing, London, pp. 713–744.

Rolls ET. (1993b) The neural control of feeding in primates. In: *Neurophysiology of Ingestion* (ed. DA Booth). Pergamon Press, Oxford, pp. 137–169.

Rolls ET. (1994) Brain mechanisms for invariant visual recognition and learning. *Behav. Process.* **33**: 113–138.

Rolls ET. (1995a) Central taste anatomy and neurophysiology. In: *Handbook of Olfaction and Gustation* (ed. RL Doty). Dekker, New York, pp. 549–573.

Rolls ET. (1995b) Learning mechanisms in the temporal lobe visual cortex. *Behav. Brain Res.* **66**: 177–185.

Rolls ET. (1995c) A model of the operation of the hippocampus and entorhinal cortex in memory. *Int. J. Neural Systems* **6** (Suppl.): 51–70.

Rolls ET. (1997a) Brain mechanisms involved in perception and memory, and their relation to consciousness. In: *Cognition, Computation, and Consciousness* (eds M Ito, Y Miyashita, ET Rolls). Oxford University Press, Oxford.

Rolls ET. (1997b) A neurophysiological and computational approach to the functions of the temporal lobe cortical visual areas in invariant object recognition. In: *Computational and Biological Mechanisms of Visual Coding* (eds L Harris, M Jenkin). Cambridge University Press, Cambridge.

Rolls ET, Tovee MJ. (1994) Processing speed in the cerebral cortex and the neurophysiology of visual masking. *Proc. R. Soc. Lond. B* **257**: 9–15.

Rolls ET, Tovee MJ, Purcell DG, Stewart AL, Azzopardi P. (1994) The responses of neurons in the temporal cortex of primates, and face identification and detection. *Exp. Brain Res.* **101**: 474–484.

Rolls ET, Treves A. (1990) The relative advantages of sparse versus distributed encoding for associative neuronal networks in the brain. *Network* **1**: 407–421.

Rolls ET, Treves A. (1997) *Neural Networks and the Brain.* Oxford University Press, Oxford.

Rolls ET, Treves A, Foster D, Perez-Vicente C. (1996) Simulation studies of the CA3 hippocampal subfield modelled as an attractor neural network. In: *Neural Networks*, in press.

Singer W. (1987) Activity-dependent self-organization of synaptic connections as a substrate for learning. In: *The Neural and Molecular Bases of Learning* (eds J-P Changeux, M Konishi). John Wiley & Sons, Chichester, pp. 301–335

Stent GS. (1973) A psychological mechanism for Hebb's postulate of learning. *Proc. Natl Acad. Sci. USA* **70**: 997–1001.

Treves A, Rolls ET. (1991) What determines the capacity of autoassociative memories in the brain? *Network* **2**: 371–397.

Treves A, Rolls ET. (1992) Computational constraints suggest the need for two distinct input systems to the hippocampal CA3 network. *Hippocampus* **2**: 189–199.

Treves A, Rolls ET. (1994) A computational analysis of the role of the hippocampus in memory. *Hippocampus* **4**: 374–391.

Wallis G, Rolls ET. (1997) A model of invariant object recognition in the visual system.

Wallis G, Rolls ET, Foldiak P. (1993) Learning invariant responses to the natural transformations of objects. *Int. Joint Conf. Neur. Networks* **2**: 1087–1090.

Willshaw DJ, Longuet-Higgins HC. (1969) The holophone – recent developments. In: *Machine Intelligence 4* (ed. D Michie). Edinburgh University Press, Edinburgh.

Willshaw DJ, von der Malsburg C. (1976) How patterned neural connections can be set up by self-organization. *Proc. R. Soc. Lond. B* **194**: 431–445.

Index

Acetylcholine, 137, 138–140
Arachidonic acid, 48, 50
Autoassociation, 231–236, 241

Basic leucine zipper transcription factors, *see* Immediate-early genes
Brain-derived neurotrophic factor, 126

'Caged' glutamate, 65
Calmodulin, 93
Calmodulin kinase
    activity after LTP, 41
    autophosphorylation, 44
    changes in expression after LTP, 112
    composition of, 37
    mice deficient in α-CaMKII, 40, 111, 174
    phosphorylation of glutamate receptors, 90–92
    role in kindling/seizures, 173–174
    transient activation in E-LTP, 38
Calpain, 48
Carbachol, 138–140
Carbon monoxide, 48, 49–50
Cell adhesion molecules, 127–128
CREB, 54
Cyclic AMP (cAMP), 215
Cyclic AMP-dependent kinase
    changes in expression after LTP, 119–120
    composition of, 44
    in LTP, 45, 51, 138
    mice lacking, 111
    phosphorylation of glutamate receptors, 89–91

E-LTP, 41–45, 48, 103
Ectopic potentials, 172
Epilepsy, 149, 166–179
Extracellulary regulated kinases (ERKs), 111, 114–119

GABA receptors
    changes in expression after kindling/seizures, 176
    $GABA_A$ in LTD, 27
    $GABA_A$ IPSPs, 17–18
    $GABA_B$ IPSPs, 18
    $GABA_B$ LTD of, 28
    $GABA_B$ in LTD, 27

GABAergic transmission
    changes after kindling, 170–171
    effects of 5-HT on, 144
Glutamate receptors, 3, 84–95
Glutamate receptors, AMPA
    cellular localization, 94–95
    changes in gene expression after kindling/seizures, 175
    changes in gene expression after LTP, 121
    functional modification after kindling, 170
    functional modification after LTP, 10, 96–97
    LTD of, 23–27
    modulatory protein, 94
    phosphorylation, 86, 89–91, 176
    role in LTD, 24–25
    subunits, 85
Glutamate receptors, kainate
    subunits, 85
Glutamate receptors, metabotropic receptors
    changes in expression, 124–126
    -deficient mice, 21, 27, 124, 212
    in LTD, 26–27
    palmitoylation, 93
    role in kindling/seizures, 178–179
    role in LTP, 211–213
    role in LTP induction, 13–17, 21
    subunits, 3, 85–86
Glutamate receptors, NMDA
    calcium influx through, 12–13
    cellular localization, 94–95
    changes in expression after kindling/seizures, 176
    changes in expression after LTP, 122–124
    distribution of, 20
    glycine site, 11
    interaction with actin, 93
    interaction with PSD-95, 94
    LTD of, 27–28, 208–211
    LTP of, 9, 21–23, 65–66, 71–72
    modification after kindling, 169–170
    modification after LTP, 97–98, 138
    noncompetitive antagonists, 11
    phosphorylation, 87–88, 91–93, 122, 177
    subunits, 85

Heme oxygenase, 49
5-Hydroxytryptamine, 137, 144–145

I_{AHP}, 138–145
Immediate-early genes (IEGs), 66, 104, 105–110
    CREB, 54, 107–110, 119
    Erg3, 106
    expression after kindling/seizures, 175
    fos, 67, 107, 175
    jun, 107
    zip268/NGFI-A, 67, 105, 175
Isoproterenol, 143

Kindling
    chemical, 155–162
    electrical, 150–151, 158

L-LTP, 45–46, 103–104
Learning and memory, 6, 53, 128, 225–226, 231–233
Learning rules, 191–192, 227–228, 234–235, 239, 245–248
Long-term depression, 3, 193–197
    calcium requirement, 208–210
    heterosynaptic, 26, 229
    in developing CNS, 199–202
    of AMPA currents, 23–27, 210–213, 217
    of $GABA_A$ currents, 28
    of NMDA currents, 27–28
    phosphatases in 51
Long-term potentiation
    calcium-dependence, 12–13, 21
    changes in cytoskeletal gene expression, 128
    changes in growth factor gene expression, 126
    changes in IEG expression, 105–110
    changes in receptor gene expression, 120–126
    glutamate-induced, 12
    in developing CNS, 199–202
    locus of expression, 4, 52, 61, 69, 77
    neocortex, NMDA-dependent, 11
    NMDA-dependent increase in AMPA, 11–13, 18–19, 64–65, 126
    NMDA-dependent increase in NMDA, 21–23, 65–66, 71–72
    NMDA-independent, 20–21
    quantal analysis of, 67–71
    relation to kindling/epilepsy, 166–169
    striatum, NMDA-dependent, 11
    transmitter release, 62–64

MAP kinase, 47, 52
Mossy fiber–CA3 synapse, 20–21
Motor learning, 207–208
Muscarinic receptors, 138–140

Networks
    capacity of, 228, 234, 238
    competitive, 236–240
    linear, 228–230
    nonlinear, 230–231, 235, 239
Nitric oxide
    in LTD, 215–217
    in LTP, 48, 49
Nitric oxide synthase, 49, 215–217
Noradrenaline, 137, 140–144
Neurotrophin 3, 126

PAF, 48, 50
Paired-pulse depression, 72–76
Paired-pulse facilitation, 72–76
Pattern association, 223–231
Pattern completion, 6
Phenylephrine, 143
Phosphatases
    1, 2A and 2B, 51
    calcineurin, 51
    changes in expression after LTP, 120
    dephosphorylation of glutamate receptors, 91–93
    inhibitor protein 1, 51
Phospholipase A_2, 24, 50 138
Phospholipase C, 24, 50, 124–125, 138, 213
    activity after LTD, 213
    activity after LTP, 41, 138
Protein kinase C
    autophosphorylation, 41–44
    isozymes, 37
    mice lacking δ PKC, 39, 111
    phorbol esters, 40–41
    phosphorylation of glutamate receptors, 90, 91–92, 122–123
    role in kindling/seizures, 174–175
    substrates
        changes in gene expression of, 112–114
        GAP43, 43, 52, 66
        MAP2, 111
        neurogranin, 43, 67
    transient activation in E-LTP, 38
Protein tyrosine kinases, 45
    in LTP, 46–47, 65, 67
    phosphorylation of CREB, 107
    phosphorylation of glutamate receptors, 91–92

Raf-1 and Raf-B, *see* Extracellularly regulated kinases
Retrograde messenger, 4, 48, 76–77

Serotonin, *see* 5-Hydroxytryptamine
Short-term potentiation, 3,10
    NMDA-induced, 12
Silent receptors, 69–71
Synaptic vesicle proteins,
    66, 106, 125, 142

Tissue plasminogen activator, 126–127
Transgenic/knockout mice, 6, 53, 111–124
Transmitter release during LTP, 62–64

Zinc finger transcription factors, *see* Immediate-early genes

ALSO AVAILABLE FROM BIOS SCIENTIFIC PUBLISHERS LTD

Glial Cell Development
Basic principles and clinical relevance

K.R. Jessen & W.D. Richardson (Eds)
University College London, UK

This book reviews recent advances in our understanding of glial cell development and how this basic knowledge is being harnessed for the treatment of human disease.

"I recommend throwing away all those other glial textbooks and reading this one. This book presents an excellent collection of essays from leaders in the field, who have made considerable effort to be up-to-date and speculate on the broader implications of advances in glial research." "..this is an excellent book." Robert H. Miller, Trends in Neuroscience, August 1996.

Contents

The Schwann cell lineage: embryonic and early postnatal development, *H.J.S.Stewart et al;* Developmental origins of astrocytes, *J.E.Goldman;* Origins and early development of oligodendrocytes, *W.D.Richardson et al;* Axonal control of oligodendrocyte development, *B.A.Barres & M.C.Raff;* Speculations on myelin sheath evolution, *D.R.Colman et al;* Control of myelin gene expression, *L.D.Hudson et al;* Microglia in the developing and mature central nervous system, *V.H.Perry;* Myelin-specific genes and their mutations in the mouse, *K.-A.Nave;* Axon-Schwann cell interactions during peripheral nerve degeneration and regeneration, *S.S.Scherer & J.L.Salzer;* The role of astrocytes in axon guidance during development and repair, *H.Vickland & J.Silver;* Glial cell transplantation and the repair of demyelinating lesions, *W.F.Blakemore et al;* Transplantation of Schwann cells into the CNS: potential for repair of tract lesions, *G.Raisman;* The role of glia in the development of the insect nervous system, *V.J.Auld.*

Of interest to:

Research and clinical scientists in developmental biology and neurobiology; clinicians and practitioners in neurology and neuropathology.

1872748546; 1995; Hardback; 272 pages

ALSO AVAILABLE FROM BIOS SCIENTIFIC PUBLISHERS LTD

Molecular Biology of the Neuron

W. Davies & B. Morris
Robertson Laboratory of Biotechnology, University of Glasgow, Glasgow, UK

Neurons are central to all neuroscience and the molecular biological investigation of the properties of neurons is now progressing rapidly on many fronts. *Molecular Biology of the Neuron* provides, in a single volume, an up-to-date review of what we know about the molecules that make neurons special. Coverage includes basic molecular processes, molecules involved in special functions of neurons, and molecules involved in development, maintenance, disease or death of neurons.

Contents

(Provisional) Neuronal genes, *W.Davies and G. Stewart*; Molecular genetic of nervous and neuromuscular systems, *H.Jockusch, K.A. Nave, G. Grenningloh and T. Schmitt-John*; Neuron-specific gene expression, *A. Grant and W. Wisden*; The neuronal cytoskeleton, *J. Diaz-Nido and J. Avila*; Voltage-gated ion channels, *A.H.Weston & G.Edwards*; G-protein-coupled receptors, *J.Koenig*; Molecular biology of ligand-gated ion channel receptors, *T. Glencorse and W. Davies*; Cell surface and extracellular matrix glycoproteins implicated in the movements of growth cones and neural cells and in synaptogenesis in the vertebrate nervous system, *H.Volkmer and F.G. Rathgen*; Genetic control of brain development, *N.D. Allen, M.J. Skynner, H. Kato and J. Pratap*; Neuronal plasticity, *B.J. Morris*; Neurotropic factors and the regulation of neuronal survival in the developing peripheral nervous system, *A.M. Davies*; Genetic basis of human neuronal disease, *M.E.S. Bailey, R.T. Moxley III and K.J. Johnson*.

Of interest to:

Researchers; advanced undergraduates in neuroscience.

1859962408; 1997; Hardback; pages

ALSO AVAILABLE FROM BIOS SCIENTIFIC PUBLISHERS LTD

Neurobiology of Alzheimer's Disease

D. Dawbarn & S.J. Allen (Eds)
Department of Medicine, University of Bristol, UK

Several major advances have been made recently in understanding the aetiology of Alzheimer's disease. *Neurobiology of Alzheimer's Disease* contains up-to-date reviews by world authorities on current key issues relating to both the aetiology and treatment of the disease.

"This is a terrific book, covering all the hot areas... highly recommended"
Trends in Neuroscience

Contents

Alzheimer's disease: current controversies - an overview, S.J. Allen & D.Dawbarn; The biology and molecular pathology of β-amyloid protein, D.R. Howlett, S. James, D. Allsop and G.W. Roberts; Molecular genetics of amyloid and apolipoprotein E in Alzheimer's disease, W. Wasco and R.E. Tanzi; Genetics of Alzheimer's disease: age of onset as a key discriminator of etiology, J. Hardy; The tau protein amyloidosis of Alzheimer's disease: its mechanisms, potential trigger factors and consequences, C. Wischik, P.C. Edwards and C.R. Harrington; APP transgenesis: approaches towards the development of animal models for Alzheimer's disease neuropathology, M.J. Savage, D.S. Howland, S.M. Ali, S. Siedlak, G. Perry, R. Siman, R.W. Scott and B.D. Greenberg; The neurochemical pathology of Alzheimer's disease, A.W. Procter, P.T. Francis, C.P.L.-H. Chen, I.P. Chessell, S. Dijk, N.A. Clarke, M.-T. Webster and D.M. Bowen; Cholinergic grafts in primates, H.F. Baker and R.M. Ridley; Neuronal growth factors and Alzheimer's disease, J. Treanor, K. Beck and F. Hefti; Alzheimer's disease and structural imaging: CT and MRI, N. Fox, P. Hartikainen and M. Rossor; Pharmacological approaches to treating Alzheimer's disease, G.K. Wilcock; Index.

Of interest to:

Postgraduates; researchers; medical and neuroscience students; clinicians.

1872748147; 1995; Hardback; 336 pages

ALSO AVAILABLE FROM BIOS SCIENTIFIC PUBLISHERS LTD

Immune Responses in the Nervous System

N.J. Rothwell (Ed.)
University of Manchester, UK

Many molecules and mechanisms classically associated with the peripheral immune system have been found to be active in the brain. The investigation of immune activation is a rapidly expanding field of research, particularly with regard to responses to acute neuronal damage and in degenerative disorders such as Alzheimer's disease and multiple sclerosis.

Contents

The blood-brain barrier and perivascular cells, *B.B.Johansson;* Leukocyte migration into the central nervous system, *L.J.Lawson;* The role of glial cells in immune responses in the brain, *J.Gehrmann & R.B.Banati;* Cytokines and neurodegeneration, *N.Rothwell et al;* Neurotrophins: aspects of their *in vivo* actions and physiology, *M.Meyer;* Immune responses in the central nervous system in inflammatory demyelinating disease, *M.L.Cuzner & T.Smith;* Central nervous system immune reactions in Alzheimer's disease, *P.L.McGeer & E.G.McGeer;* The immunology of brain injury, *M.C.Morganti-Kossmann & T.Kossmann;* Transplantation of neuronal and nonneuronal cells into the brain, *H.Widner.*

Of interest to:

Postgraduates and researchers in neurology and immunology; clinicians.

1872748791; 1995; Hardback; 256 pages